Mineralogy and Mineral Analytical Techniques

Mineralogy and Mineral Analytical Techniques

Edited by **John Wayne**

SYRAWOOD
PUBLISHING HOUSE

New York

Published by Syrawood Publishing House,
750 Third Avenue, 9th Floor,
New York, NY 10017, USA
www.syrawoodpublishinghouse.com

Mineralogy and Mineral Analytical Techniques
Edited by John Wayne

International Standard Book Number: 978-1-68286-141-7 (Hardback)

Printed in the United States of America.

Contents

Preface

The main aim of this book is to educate learners and enhance their research focus by presenting diverse topics covering this vast field. This is an advanced book which compiles significant studies by distinguished experts. This book addresses successive solutions to the challenges arising in the area of application, along with it; the book provides scope for future developments.

Mineralogy focuses on the chemical and physical properties of minerals. This book explores industrial minerals in new light. It delves into mineral metallurgy, advances in analytical techniques, exploration techniques, etc. This book is a vital tool for everyone engaged in this field. Scientists and students of geology, and associated disciplines will find this book full of crucial and unexplored concepts.

It was a great honour to edit this book, though there were challenges, as it involved a lot of communication and networking between me and the editorial team. However, the end result was this all-inclusive book covering diverse themes in the field.

Finally, it is important to acknowledge the efforts of the contributors for their excellent chapters, through which a wide variety of issues have been addressed. I would also like to thank my colleagues for their valuable feedback during the making of this book.

Editor

Long-Term Acid-Generating and Metal Leaching Potential of a Sub-Arctic Oil Shale

Kathryn A. Mumford [1,*], Brendan Pitt [2], Ashley T. Townsend [3], Ian Snape [4] and Damian B. Gore [2]

[1] Department of Chemical and Biomolecular Engineering, The University of Melbourne, Parkville, VIC 3010, Australia

[2] Department of Environment and Geography, Macquarie University, Sydney, NSW 2109, Australia; E-Mails: brendan_pitt@hotmail.com (B.P.); damian.gore@mq.edu.au (D.B.G.)

[3] Central Science Laboratory, University of Tasmania, Hobart, TAS 7001, Australia; E-Mail: ashley.townsend@utas.edu.au

[4] Australian Antarctic Division, Kingston, TAS 7050, Australia; E-Mail: ian.snape@aad.gov.au

* Author to whom correspondence should be addressed; E-Mail: mumfordk@unimelb.edu.au

Abstract: Shales are increasingly being exploited for oil and unconventional gas. Exploitation of sub-arctic oil shales requires the creation of gravel pads to elevate workings above the heaving effects of ground ice. These gravel pads can potentially generate acidic leachate, which can enhance the mobility of metals from the shale. To examine this potential, pyrite-bearing shale originating from sub-Arctic gravel pad sites were subjected to leaching tests for 600 days at initial pH values ranging from 2 to 5, to simulate potential real world conditions. At set times over the 600 day experiment, pH, oxidation reduction potential (ORP), dissolved oxygen and temperature were recorded and small liquid samples withdrawn and analysed for elemental concentrations using total reflection X-ray fluorescence spectrometry (TRXRF). Six of eight shale samples were found to be acid generating, with pH declining and ORP becoming increasingly positive after 100 days. Two of the eight shale samples produced increasingly alkaline leachate conditions with relatively low ORP after 100 days, indicating an inbuilt buffering capacity. By 600 days the buffering capacity of all samples had been consumed and all leachate samples were acidic. TRXRF analyses demonstrated significant potential for the leaching of S, Fe, Ni, Cu, Zn and Mn with greatest concentrations found in reaction vessels with most acidic pH and highest ORP.

Keywords: acid rock drainage; pH; oxidation reduction potential; metal concentration; total reflection X-ray fluorescence spectrometry (TRXRF)

---

1. Introduction

Oil and gas drilling operations in Arctic and sub-Arctic regions often require extensive ground stabilization and construction of gravel pads above the tundra to host drilling equipment, accommodation modules and other buildings. The isolation and transportation costs to these areas effectively restricts "borrow" materials to what is available on or near site, which in sedimentary basins hosting hydrocarbons consists typically of limestone, siltstone and shale. Shale is generally characterised by high concentrations of sulfidic materials such as pyrite (FeS_2), which in the presence of air and sufficient water, are capable of being oxidized to iron sulfates and sulfuric acid [1]. Increased acidity combined with appropriate oxidation-reduction potential (ORP) conditions [2] allows for increased mobility and potential leaching of elements from otherwise stable materials. The elements leached at a particular location are a function of the local geology [3], however elements such as S, Mn, Fe, Cu, Zn, As, Se, Cd, Hg and Pb are common [4,5]. This process is generally referred to as acid rock drainage (ARD). ARD can cause significant environmental damage to many facets of ecosystems including aquatic life [5,6], vegetation [5,7] and groundwater [8]. These elements often bio-accumulate [9]. Many case studies evaluating ARD have been undertaken around the world [10–13], however, to date no studies have been undertaken on gravel construction pads in Arctic or sub-Arctic regions. As shale from gravel pads on oil and gas field sites in Canada is used in this study, Canadian Soil Guidelines will be used to evaluate the risks associated with ARD in this work.

The development of ARD can occur over long time scales and so a number of standard laboratory procedures, with varying degrees of complexity and accuracy, have been developed or adapted to forecast the potential for ARD and possible leachable metals. The most commonly used tests to quantify ARD potential are "static" tests and "kinetic" tests. The most common static test is acid-base accounting (ABA), which is a simple balance of the acid producing and acid neutralizing potential of a rock [14], and is used to determine whether or not a material is acid generating. Importantly, ABA does not account for the various dissolution rates of the constituent materials and hence does not provide a guide to the overall impact on site pH over time.

Full chemical digests of rocks are used to reveal contaminants present and their concentration. Often U.S. Environmental Protection Agency (EPA) Methods are used for this purpose, however as they have not been specifically designed for ARD conditions their applicability is uncertain. For instance, U.S. EPA method 1311 (Toxicity Characteristic Leaching Procedure) was designed to simulate co-disposal in municipal waste while U.S. EPA method 1312 (Synthetic Precipitation Leaching Procedure) was designed to simulate leaching by mildly acidic rain. Conversely, the strong acid leach test (SALT) was designed for ARD conditions [15]. The SALT involves the addition of sulfuric acid to a reaction flask in sufficient quantities to ensure a predetermined final pH irrespective of the buffering capacity of the material. However, similar to the U.S. EPA methods, the SALT contains no provision for the kinetics of ARD reactions as samples are collected and metal concentrations quantified at only one time interval.

Kinetic tests most commonly take the form of laboratory-based columns, humidity cells or field based test pads. These tests can be designed to provide useful information of the relative rates of acid generation and neutralisation as well as the drainage chemistry and resultant loadings downstream [16]. However, these tests do have limitations, specifically matching laboratory set points to site-specific factors such as water flow rates and dissolved oxygen concentrations. Differences in these key set points can result in significant variations between laboratory predictions and site outcomes [17].

This investigation combines features of static and kinetic tests, such that potential for ARD is characterised and the potential for metals to be leached quantified over time. Similar to the SALT method, a specific mass of rock is placed in a reaction vessel with a specified amount of water and pH adjusted to a value between 2 and 5 with sulfuric acid. Flasks are shaken at low speed and samples withdrawn at set intervals for 600 days with temperature, pH, ORP and metal concentrations measured. This procedure enables analysis of reaction kinetics without potential inconsistencies due to poorly selected water flow rates. However, due to the long time scale involved and increased ion concentrations in solution, this procedure does have the risk of being impacted by secondary mineral precipitation processes. The stability of these phases is potentially important in the control of metal distribution [18].

Due to the large number of samples required from each flask to provide adequate kinetic data this work uses a relatively new analytical procedure to quantify metal concentrations leached into solution, total reflection X-ray fluorescence spectrometry (TRXRF). TRXRF can quantify concentrations of elements from Na to U present in a sample down to µg/L levels, much lower than would typically be possible with other forms of XRF spectrometry [19]. Only approximately 1 mL is required per sample, much less than for flame atomic absorption spectroscopy (F-AAS) [20], inductively coupled plasma atomic emission spectroscopy (ICP-AES) [21], or inductively coupled plasma mass spectrometry (ICP-MS) [22]. Sector field ICP-MS was employed as an external analytical comparison method.

2. Materials and Methods

2.1. Sample Collection

Shale samples were collected from an oil and gas lease area in the Canadian Arctic. Sample locations were either in areas of high potential environmental risk, where large volumes of shale were present or collected from creek banks due to their direct contact with local aquatic ecosystems. Samples were collected from a depth of approximately 10 cm, and consisted of gravel (2–50 mm b-axis diameter). The samples were not sterilized, but the mineral fragments were free of soil and plant matter.

2.2. Sample Comminution

Samples of solids were oven dried at 60 °C for 48 h, then crushed without lubricant in a Rocklabs (Onehunga, New Zealand) Enerpac hydraulic crusher with tungsten carbide jaws (Model Number RR-308, 70 MPa pressure) and dry sieved to 0.5–2 mm diameter.

2.3. Material Characterisation

X-ray diffractometry (XRD) was used to identify the mineralogy of every crushed solid sample. X-ray diffractograms were collected with a PANalytical (Almelo, The Netherlands) X'Pert Pro MPD diffractometer, using 45 kV, 40 mA, Cu-Kα radiation and an X'Celerator detector. Scans were collected with 0.05° 2θ steps from 10° to 90° 2θ. Quantitative mineral identification was carried out using PANalytical X'Pert HighScore Plus version 2.2a software with patterns from ICDD PDF-2 and PAN-ICSD version 2008-2 (0.1–0.5 wt % detection limit). A PANalytical Epsilon 5 cartesian geometry energy dispersive XRF spectrometer was used to identify elemental concentrations in crushed shale from every sample. The Epsilon 5 XRF analysed samples in a vacuum, with a dual anode W/Sc tube, and six measurement conditions (Table 1). Measurements were made in triplicate and averaged. Elemental quantification was conducted using "Auto Quantify", PANalytical's automated qualitative spectrum analysis combined with a fundamental parameters model. All major element data are reported as oxides, and the analytical sum was set to 100%. Precipitates were not characterized at any time during the experiment, as the experiment focused solely on the leachate composition which is able to be discharged to the receiving environment. Precipitates, and any other solid phase, were immobile and not of interest for this experiment.

Table 1. Epsilon 5 XRF measurement conditions.

Elements	Secondary Target	Measurement Live Time (s)	Tube (kV)	Tube (mA)
Na-Mg	Al	200	35	17
Al-K	CaF_2	200	40	15
Ca-Mn	Fe	100	75	8
Fe-Ga	Ge	100	75	8
Ge-Y	Zr	100	100	6
Zr-U	Al_2O_3 (Barkla)	100	100	6

2.4. Sample Acidification and Leaching

Concentrated (98%) sulfuric acid (H_2SO_4) was diluted to a 4.9% solution with Type 1 (ASTM D1193-91) deionised water. The diluted acid was added to Type 1 deionised water in 500 mL HDPE square section wide mouth screw top containers in sufficient quantities to produce 350 mL of solution with a pH value of 2, 3, 4, or 5 (monitored with Hanna (Woonsocket, RI, USA) HI9025 microcomputer pH meter with HI1230 probe). 10.00 g of the prepared 0.5–2.0 mm gravel fraction of shale was placed into the prepared solutions, with a duplicate of each.

The prepared reaction vessels and blanks were placed on Ratek platform orbital mixers (Ratek Instruments (Boronia, Australia) Model OM7) operating at 100 rpm. Lids were screwed loosely onto the vessels so oxidising conditions could prevail. At set time intervals (1, 2, 4, 8 h, and 1, 2, 4, 21, 30, 100 and 600 days) 1.5–2 mL of sample was withdrawn from each reaction vessel using a 3 mL syringe, and air and solution temperatures (Hanna HI9025 microcomputer pH meter with temperature probe), pH, dissolved oxygen (DO, Hanna HI9142 Dissolved oxygen meter) and oxidation-reduction potential (Hanna HI9025 microcomputer pH meter with HI3230 ORP probe) was measured and recorded from each reaction vessel. The 1.5–2 mL sample was filtered using a 25 mm diameter 0.45 µm pore size cellulose acetate ("MiniSart") syringe filter into a clean 6 mL HDPE intermediary vessel. 930 µL of sample was then withdrawn from each intermediary vessel and added to a 6 mL sample vial.

50 µL of 20.0 mg/L Gallium and Germanium standard solution (for TRXRF analysis) was added to every sample. 20 µL of concentrated nitric acid (HNO_3; 69%; AnalaR reagent grade) was also added to

each sample vial to acidify samples to pH <2 and keep metals in solution. Every sample vial then contained 1.000 mL of sample spiked with 1.000 mg/L of Ga and Ge. The sample was then placed on a 2000 rpm vortex mixer for 3 s to ensure homogeneity of the solution. 10 µL of solution from every sample vial was pipetted onto a highly polished acrylic disc. This was then placed in a desiccator cabinet (containing silica gel) to allow the liquid to evaporate, leaving a residue on each disc for analysis. Some samples, especially blank and pH 2 samples, did not dry completely in the cabinet, and so were placed on a heating block at 60 °C to dry.

Samples were analysed using a Bruker (Berlin, Germany) S2 Picofox Total Reflection XRF spectrometer. Bruker Spectra software (version 7.0) was used to analyse spectra and derive concentrations of all elements present in samples, using the area of every peak present for all spectra. A multi-element calibration solution was used consisting of certified individual element solutions containing 10.0 mg/L of Ca, 1.00 mg/L of Cr, Fe, Ni, and Cu and 0.5 mg/L of Pb with detection limits <5 mg/kg. A gain correction disc containing 1 µg of arsenic was run every 5 discs to correct for any drift in the detector. At 21 days the first duplicate of each treatment was retired, and the remaining flasks continued to 600 days.

As an independent quality control 26 water samples, selected at random, were analysed for S, Ca, Cr, Mn, Fe, Co, Ni, Cu, Zn, Ga, As and Pb using sector field ICP-MS (ELEMENT 2, Thermo Fisher, Bremen, Germany) at the Central Science Laboratory, University of Tasmania. Three predefined instrumental resolution settings were employed to overcome spectral interferences commonly encountered for isotopes of mass ≤80 amu [23]. Elemental concentrations determined using TXRF and the more sensitive SF-ICP-MS were compared. Analytes with concentrations >50 µg/L had average differences of <5% (Fe, Ni, Cu, Zn), <10% (Mn) and <25% (S, Pb) (further detail may be found in the Supplementary Information).

3. Results

3.1. Material Characterisation

Shale samples were composed mainly of quartz and muscovite, with minor amounts of other minerals (feldspars, carbonates, sulfides and salts; Table 2). Shale B and C contained 11% and 6% anorthite, respectively. Shale H contained 4% orthoclase, and Shale D contained 3% halite. All samples had ~1% pyrite content, except for Shale C, which had <1%. Shale A also had ~1% marcasite, an iron sulfide similar to pyrite, with the same chemical formula (FeS_2), but different chemical structure. Marcasite oxidises more readily than pyrite [24], potentially making Shale A more acid-generating than other samples considered. Shales D and H contained 1% and 3% dolomite respectively, a mineral with pH buffering capacity, known to raise pH levels and reduce acidity [25]. As no other samples contained detectable dolomite, it is likely that Shale D, and particularly Shale H will have greater buffering capacity, potentially producing lesser acidic conditions than the other shales.

Elemental analyses indicate that shale samples were composed mostly of Si, followed by Al, Fe, K, Ca and Ti (Table 3), consistent with the shale mineralogy which is dominated by quartz and other silicates. Of greater interest are trace metals and metalloids of environmental significance. Table 4 presents the concentrations of a selection of these metals and compares them to the Canadian Soil

Quality Guidelines for the Protection of Environmental and Human Health (agricultural) [26]. For all samples, concentrations of Cr, Cu, As, Se, V, Ni and Ba exceeded the agricultural Canadian Soil Guideline limit whilst Shale D and H exceeded the Zn soil guideline limit. All samples contained Pb concentrations below the soil guideline limit. Under acidic conditions, these elements have the potential to become soluble and leach, possibly posing a threat to local soil and aquatic ecosystems.

Table 2. Quantitative mineralogy (wt % abundance) from the shale samples using X-ray diffractometry. Blank cells indicate non-detection.

Shale	Major Minerals (Quartz SiO_2, Muscovite $KAl_2(AlSi_3O_{10})(F,OH)_2$)	Minor Minerals (Anorthite $CaAl_2Si_2O_8$, Halite NaCl, Orthoclase $KAlSi_3O_8$)	Acidity-releasing minerals (Pyrite FeS_2, Marcasite FeS_2)	Neutralizing Minerals (Dolomite $CaMg(CO_3)_2$)
A	Quartz 46%, Muscovite 52%		Pyrite 1%, Marcasite 1%	
B	Quartz 47%, Muscovite 42%	Anorthite 11%	Pyrite 1%	
C	Quartz 45%, Muscovite 49%	Anorthite 11%	Pyrite <1%	
D	Quartz 53%, Muscovite 43%	Halite 3%	Pyrite 1%	Dolomite 1%
E	Quartz 74%, Muscovite 25%		Pyrite 1%	
F	Quartz 56%, Muscovite 44%		Pyrite 1%	
G	Quartz 54%, Muscovite 45%		Pyrite 1%	
H	Quartz 52%, Muscovite 40%	Orthoclase 4%	Pyrite 1%	Dolomite 3%

Table 3. Major element composition (wt % abundance) of the shale samples. Analyses were conducted on crushed solid fragments using X-ray fluorescence spectrometry.

Shale	Na_2O	MgO	Al_2O_3	SiO_2	S	K_2O	CaO	TiO_2	Fe_2O_3
Shale A	<1.0	<1.0	9.8	84.6	<1.0	1.15	0.68	0.30	2.46
Shale B	<1.0	<1.0	10.5	83.7	<1.0	1.29	0.54	0.32	2.66
Shale C	<1.0	<1.0	9.9	85.0	<1.0	1.12	0.69	0.28	2.00
Shale D	<1.0	<1.0	10.2	83.6	<1.0	1.12	1.53	0.28	2.19
Shale E	<1.0	<1.0	<0.1	96.4	<1.0	0.79	0.39	0.23	1.07
Shale F	<1.0	<1.0	9.4	86.0	<1.0	1.02	0.49	0.26	1.83
Shale G	<1.0	<1.0	9.5	85.6	<1.0	1.04	0.65	0.27	1.93
Shale H	<1.0	<1.0	13.1	77.5	<1.0	1.63	3.20	0.40	3.01

Table 4. Composition of selected trace elements (mg/kg). Canadian Soil Quality Guidelines (CSQG) for the Protection of Environmental and Human Health—Agriculture (2007) [26] are given where available. Concentrations above the CSQG are noted in the final row.

Shale	V	Cr	Mn	Ni	Cu	Zn	As	Se	Ba	Pb
Shale A	168	96	93	79	123	133	19	12	1827	38
Shale B	162	103	85	58	126	84	16	12	1890	43
Shale C	305	96	63	66	110	80	24	13	2150	31
Shale D	255	89	115	104	125	251	20	12	2069	30
Shale E	265	66	48	75	96	80	23	13	2051	<5
Shale F	249	82	72	94	113	146	17	13	1881	28
Shale G	198	82	76	102	106	152	20	15	2024	21
Shale H	289	116	205	163	144	244	34	16	2212	32
CSQG (mg/kg)	130	64	N/A	50	63	200	12	1	750	70
Sites failing CSQG	A-H	A-H	N/A	A-H	A-H	D, H	A-H	A-H	A-H	Nil

3.2. Sample Acidification and Leaching

As the experiment was not conducted in a temperature controlled chamber, solution temperatures followed ambient temperatures of 19 to 29 °C. Temperature has been shown to impact on the weathering rates of pyrite [27], however we assume this effect is negligible over the range observed here. Dissolved oxygen (DO) measurements confirmed that conditions remained oxygenated (7.4–9.4 mg/L) for all samples over the sampling period.

pH varied significantly over time and between samples over the 600 day reaction period (Figure 1). Shales A, B, C, E, F, and G experienced an overall decline in solution pH from their initial set point after 100 days. Conversely Shale D and particularly H exhibited significant increases in pH, up to the 100 day sampling time. This variation is likely due to the presence of dolomite in Shales D and H (Table 2). The buffering capacity was overcome by 600 days with all pH values reducing.

ORP results (Figure 2) tended to reflect the pH values, with lower pH solutions generally recording higher ORP values. The ORP values recorded were also consistently within the range considered to be a high oxidation potential often encountered in ARD environments, *i.e.*, >+300 mV [13]. ORP values were relatively stable until between 4 and 21 days (except Shales D and H) and then increased up until 600 days reflecting the drop in pH (data may be found in the Supplementary Information).

3.3. Evolution of Sulfur and Metal Concentrations

TRXRF was used to measure a spectrum of elements, however only a sub-sample of results is reported here for illustrative purposes. Results presented include S, Fe, Ni, Cu, Zn and Mn concentrations and are shown in Figures 3–8, respectively. S in solution correlates to sulfuric acid concentrations and hence acidity. Figure 3 shows that S concentrations generally remained below 200 mg/L for all samples (except Shale A for a short period, decreasing probably due to the formation of sulfur-bearing precipitates) until approximately 10 days after the commencement of the experiment. After this time period, S concentrations increased significantly for all samples except for Shale D which increased after 100 days and H which remained generally unchanged. These concentrations correlate to the pH values observed (Figure 1) as pH values also generally dropped after 10 days. In addition Figure 3 shows there is little change to S concentrations when the initial pH is reduced from 5 to 3, but a large increase in S concentration when the pH is reduced further to 2. As noted below, the increase of S into solution over the experiment is probably due to the dissolution of metal sulfide minerals.

Leachate metal concentrations generally increased for all elements over the 600 days (Figures 4–8), most significantly for Shales A, B, C, E, F and G, reflecting the increased acidity of these samples. Samples with an initial pH of 2 reported significantly higher metal concentrations in the leachate which sequentially reduced with each increase in pH value.

Figure 1. Solution pH recorded over time for all reaction vessels for all samples. Points on lines mark sample intervals from experiment commencement and 1, 2, 3, and 8 h, and 1, 2, 4, 21, 30, 100 and 600 days (logarithmic scale). The legend indicates initial sample pH.

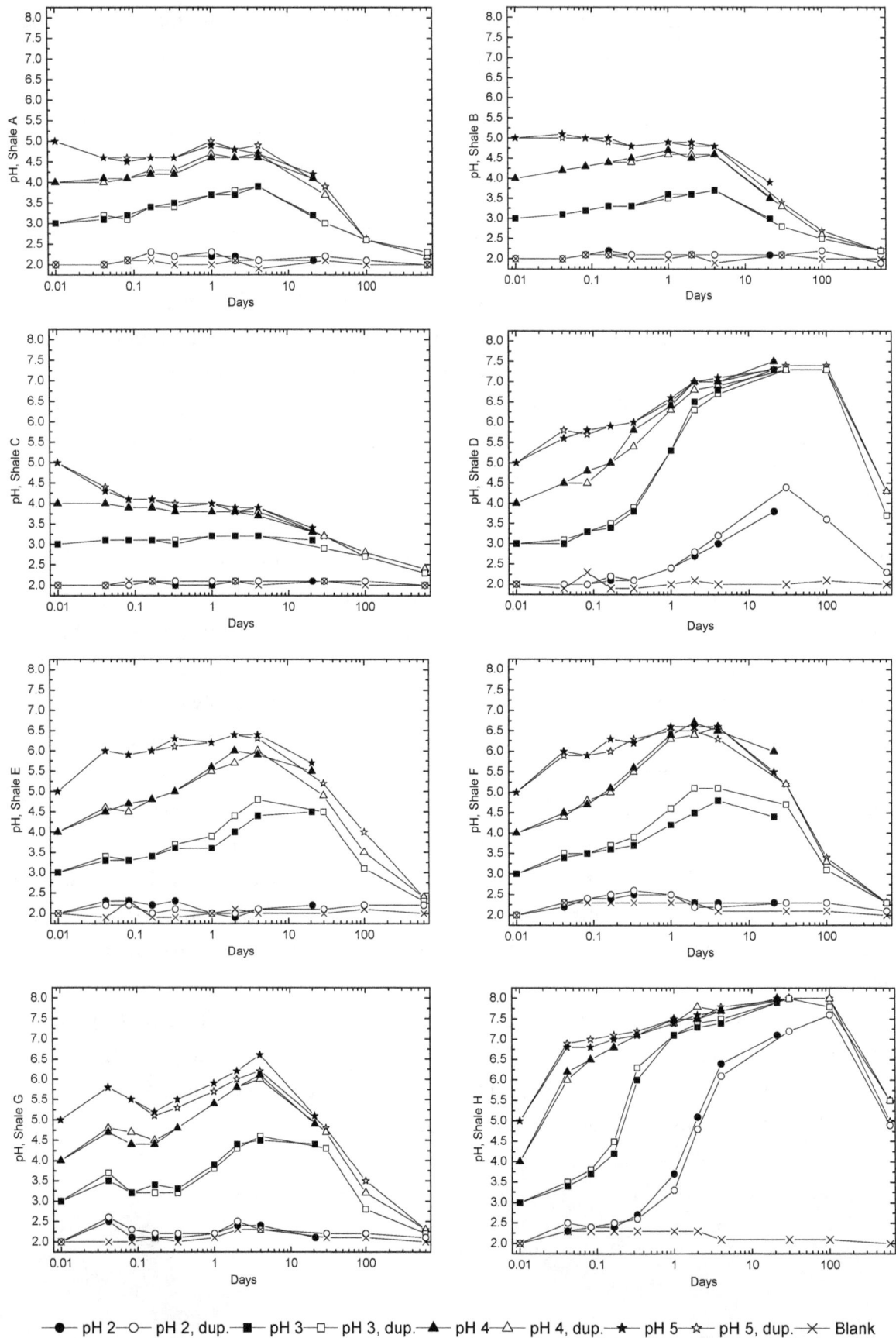

Figure 2. Oxidation reduction potential (ORP) recorded over time for all reaction vessels for all samples. Points on lines mark sample intervals from experiment commencement and 1, 2, 3, and 8 h, and 1, 2, 4, 21, 30, 100 and 600 days (logarithmic scale). The legend indicates initial sample pH.

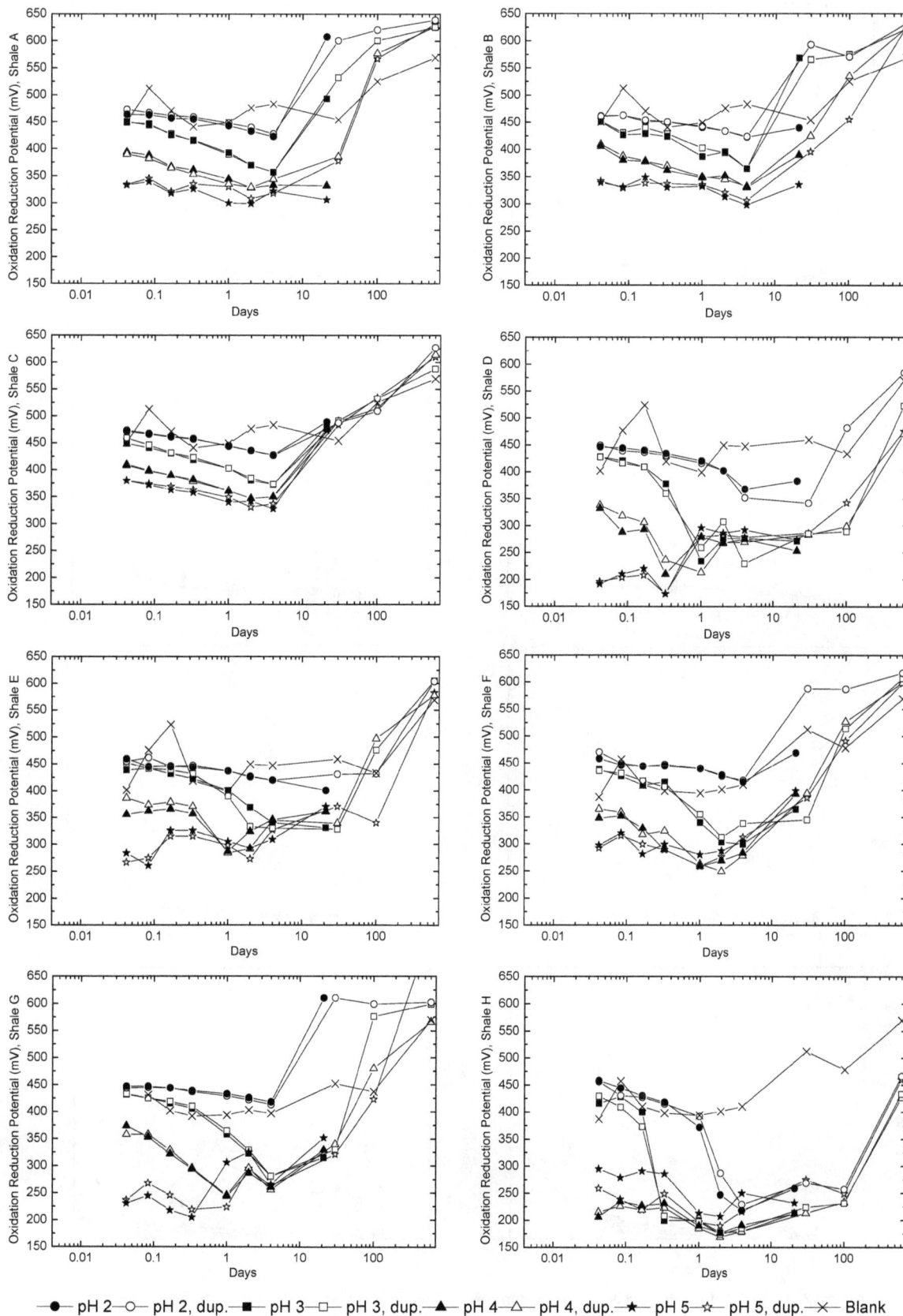

Figure 3. Solution Sulfur concentration recorded over time for all reaction vessels for all samples. Points on lines mark sample intervals from experiment commencement and 1, 2, 3, and 8 h, and 1, 2, 4, 21, 30, 100 and 600 days (logarithmic scale). The legend indicates initial sample pH.

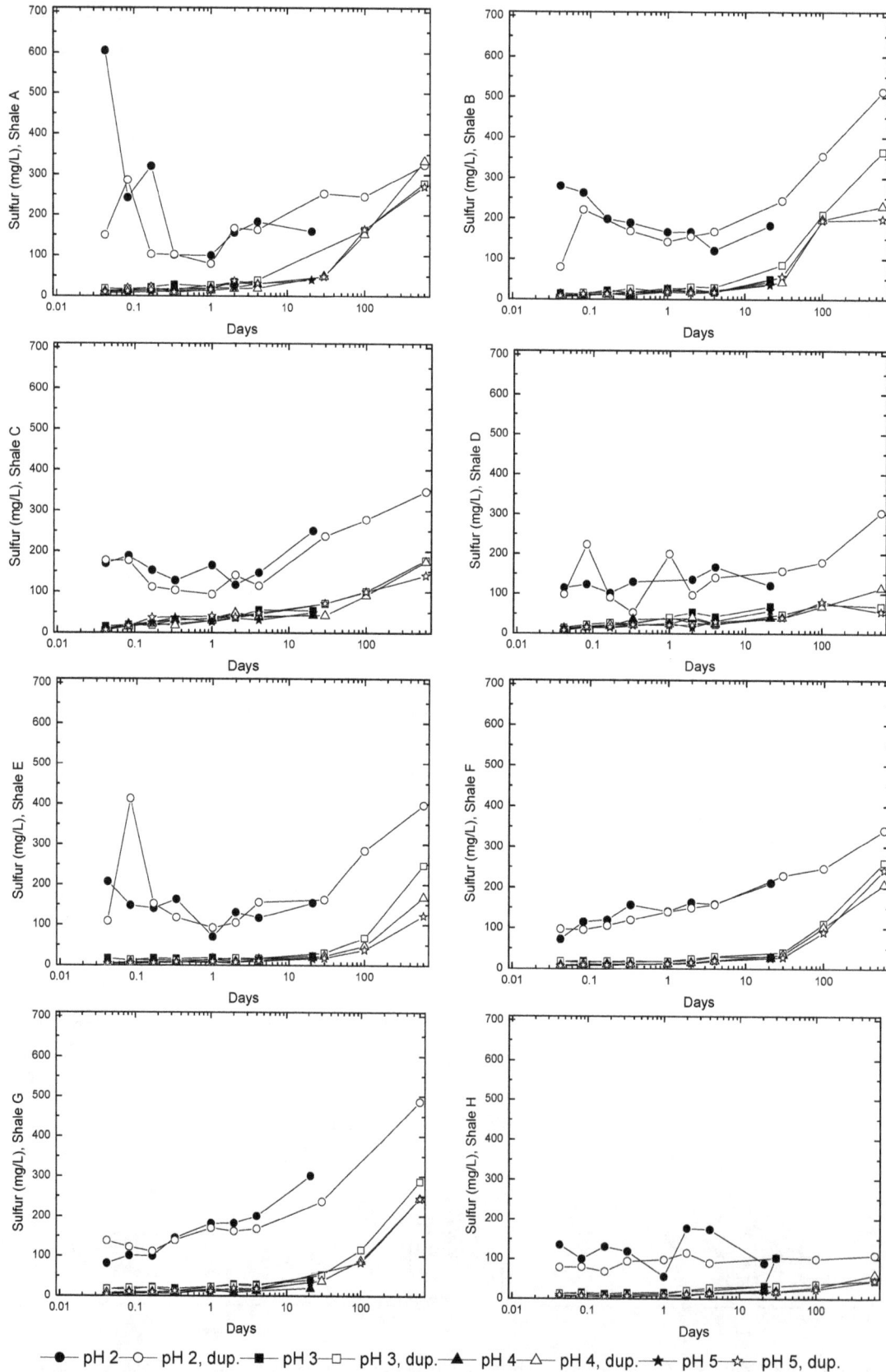

Figure 4. Solution Iron concentration recorded over time for all reaction vessels for all samples. Points on lines mark sample intervals from experiment commencement and 1, 2, 3, and 8 h, and 1, 2, 4, 21, 30, 100 and 600 days (logarithmic scale). The legend indicates initial sample pH.

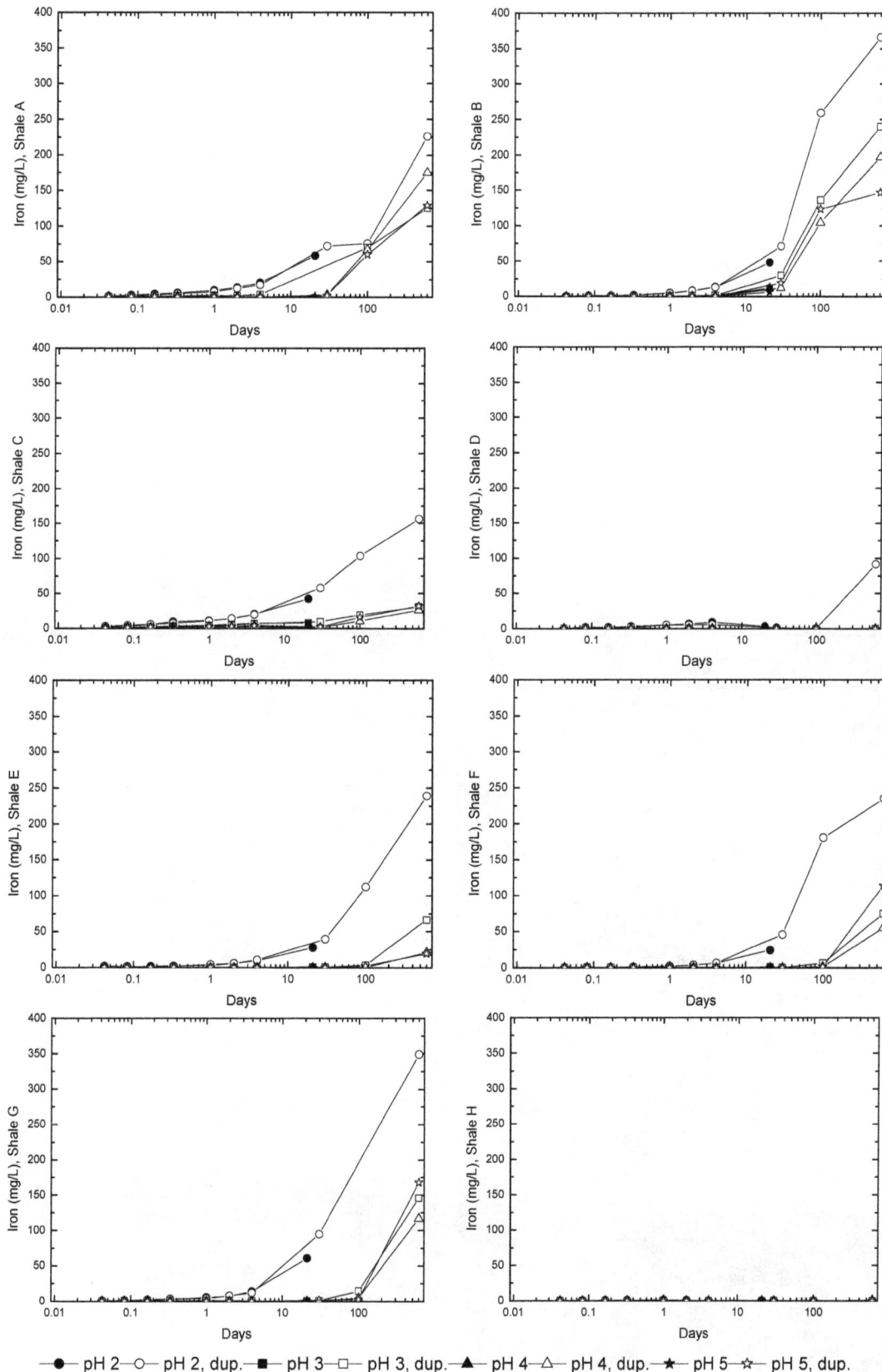

Figure 5. Solution Nickel concentration recorded over time for all reaction vessels for all samples. Points on lines mark sample intervals from experiment commencement and 1, 2, 3, and 8 h, and 1, 2, 4, 21, 30, 100 and 600 days (logarithmic scale). The legend indicates initial sample pH.

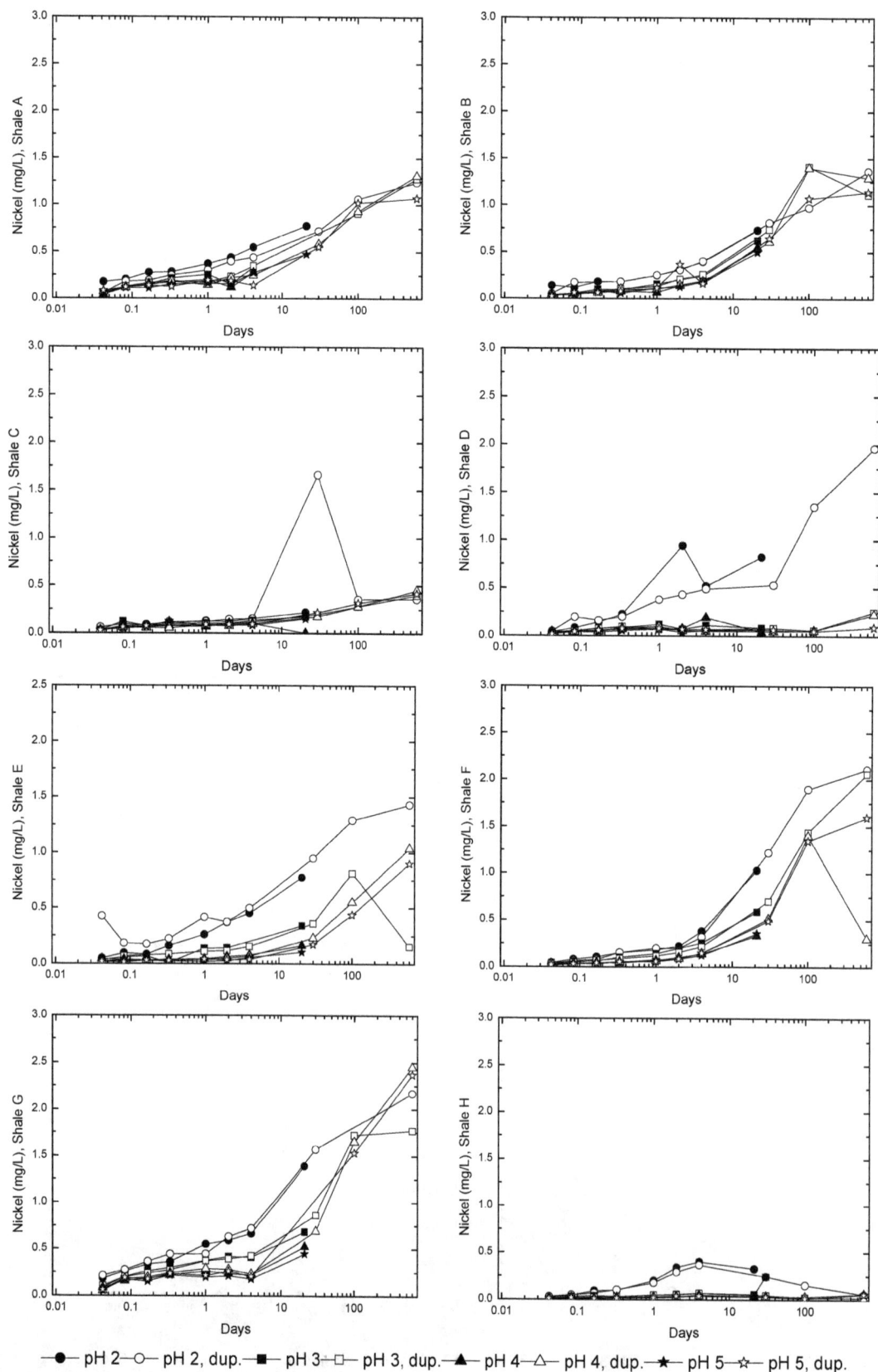

Figure 6. Solution Copper concentration recorded over time for all reaction vessels for all samples. Points on lines mark sample intervals from experiment commencement and 1, 2, 3, and 8 h, and 1, 2, 4, 21, 30, 100 and 600 days (logarithmic scale). The legend indicates initial sample pH.

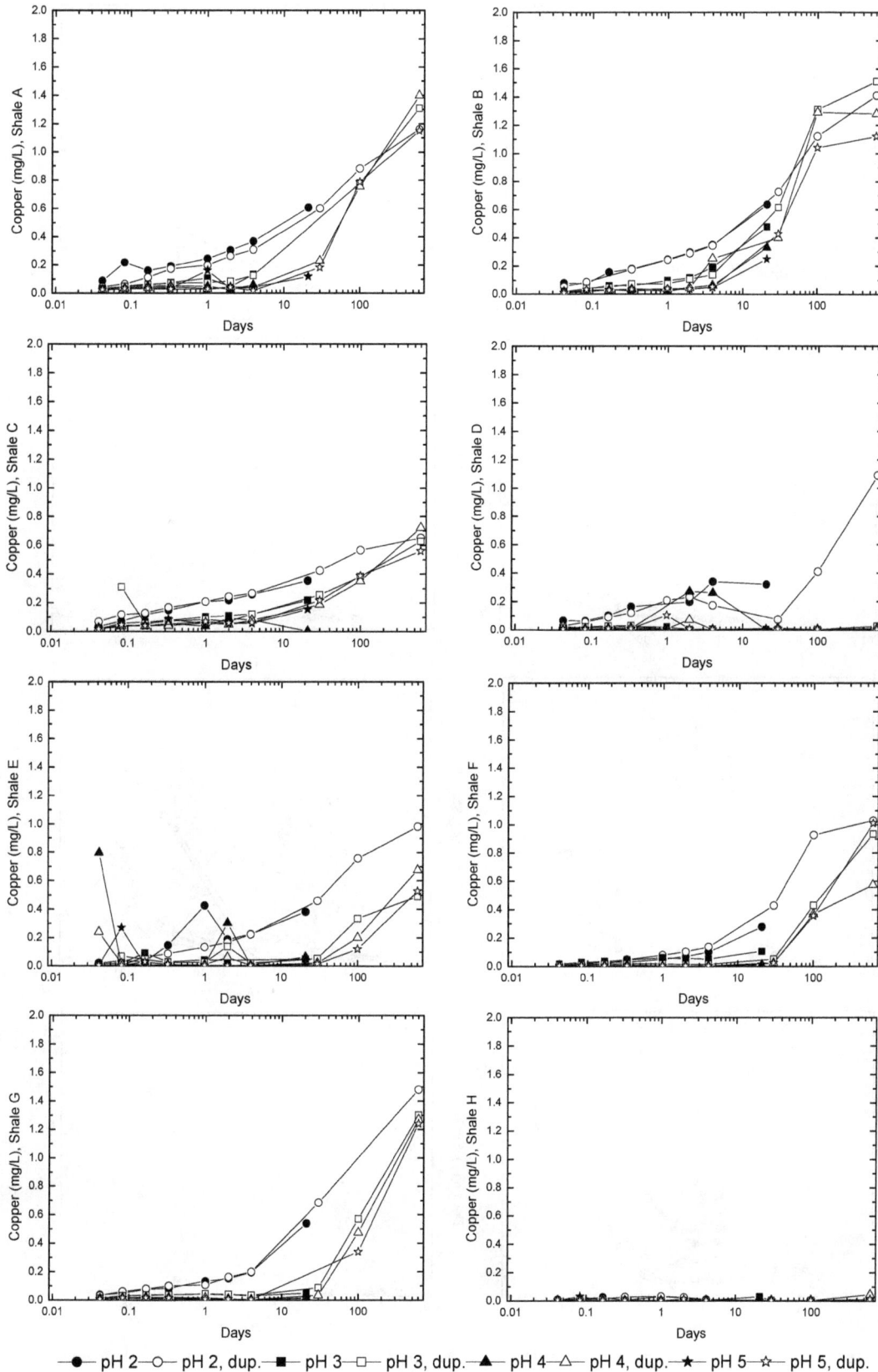

Figure 7. Solution Zinc concentration recorded over time for all reaction vessels for all samples. Points on lines mark sample intervals from experiment commencement and 1, 2, 3, and 8 h, and 1, 2, 4, 21, 30, 100 and 600 days (logarithmic scale). The legend indicates initial sample pH.

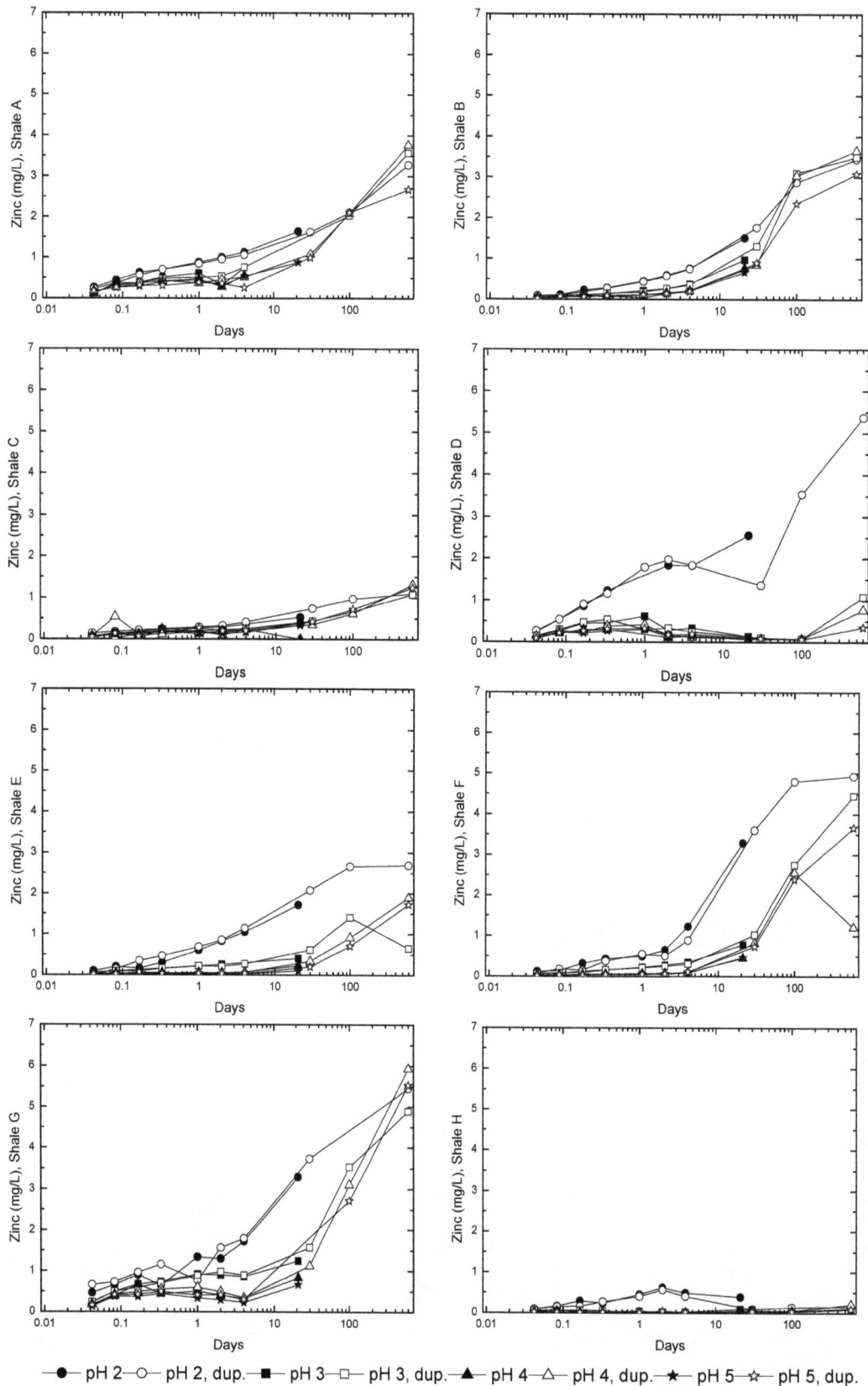

Figure 8. Solution Manganese concentration recorded over time for all reaction vessels for all samples. Points on lines mark sample intervals from experiment commencement and 1, 2, 3 and 8 h, and 1, 2, 4, 21, 30, 100 and 600 days (logarithmic scale). The legend indicates initial sample pH.

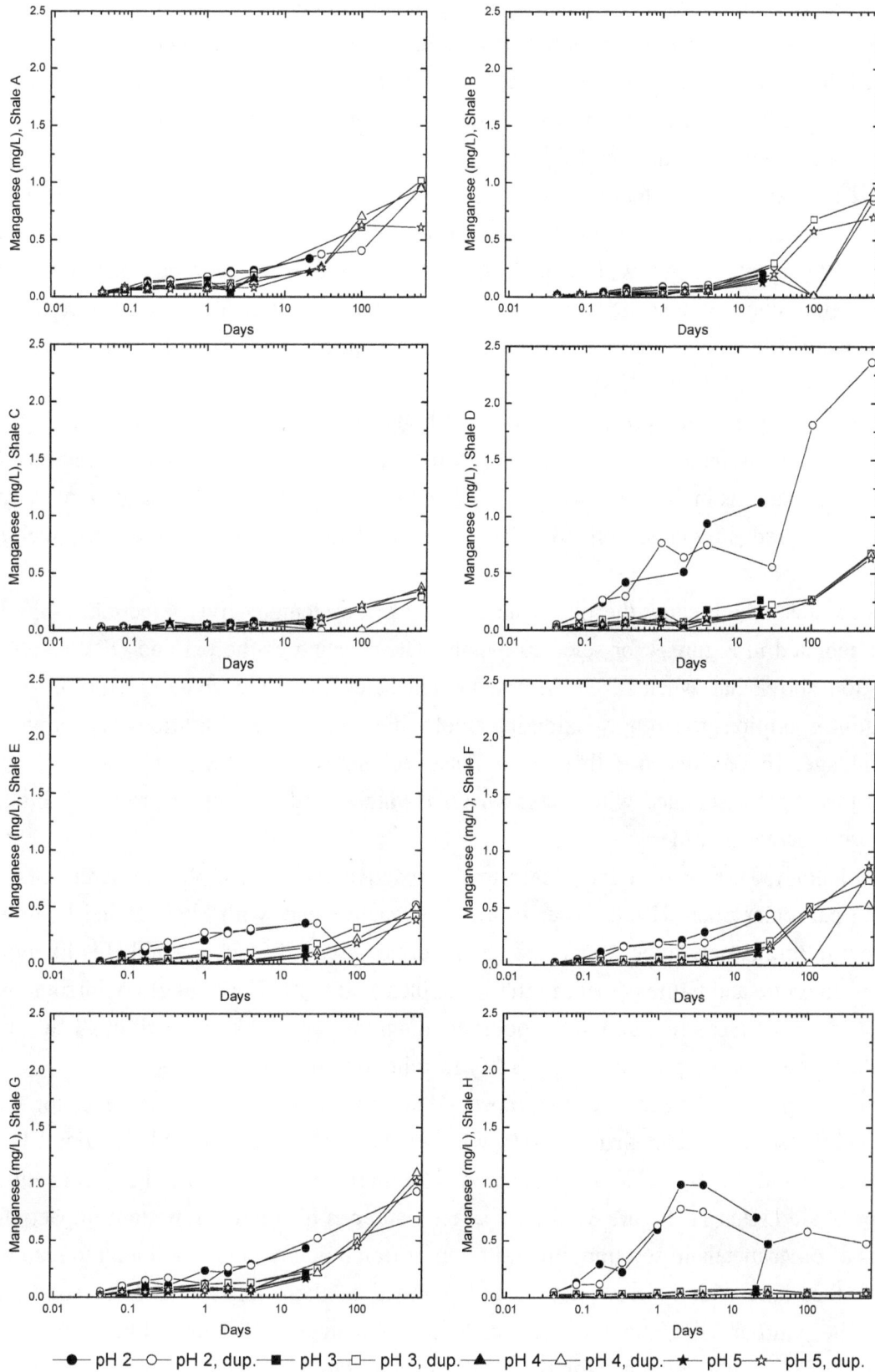

Fe concentrations in solution (Figure 4) behaved similarly to those of S, due to its association with S in the minerals pyrite and marcasite. No other major metal sulfide phases were detected in the X-ray diffractometry of the reactants, which contributed to the reasonably good relationship between S and Fe in Figures 3 and 4. Concentrations of Fe increased most significantly after 10 days for all samples except for Shales D and H, with higher concentrations recorded at lower initial pH values. Ni concentrations increased over the 600 days for all samples (Figure 5) except Shale H even though this sample reported the highest Ni concentration within the raw material (Table 4). Shale C also presented low concentrations within the leachate, however this sample was also characterised by a low nickel concentration in the initial shale sample (Table 4).

Cu followed similar trends to that of sulfur and iron as Shales A, B, C, E, F and G reported increasing copper concentrations throughout the experiment correlating with the overall decline in pH and rise in ORP (Figure 6). As with many other metals Shale H recorded relatively low Cu concentrations, remaining stable at the 0–5 µg/L level. Zn concentrations also mirrored trends in pH, with all shales displaying an increase in Zn concentration except for Shale H which was only found to vary minimally over time (Figure 7) [28].

Compared with other metals reported, Mn showed higher potential to leach from Shale D and H (Figure 8). This is likely to be due to a higher initial Mn concentration in these samples combined with the less oxidising conditions in the vessel (Figure 2). Mn has previously been found to be more mobile with a move to more reducing conditions to a greater extent than other metallic elements such as Fe, Cu and Zn [2].

Pourbaix diagrams, which show the likely stable form of an element given a certain range of pH and ORP, are depicted in Figure 9 for selected metals. The charting of the pH and ORP results from this investigation show that when air is in excess, leachates from the shale can produce mildly oxidizing alkaline conditions to strongly oxidizing acidic (highly reactive) conditions depending on the mineral assemblage. In this instance there is a linear relationship between pH and ORP for the samples, with lower pH associated with increased ORP values, and progression towards this highly reactive state with increasing time.

Generally, Shale A, B, C and G fall closer to the reactive end of sample measurements, while Shale D and in particular Shale H fall closer to the less reactive end, with Shale E and F showing a wider range, but falling largely in between. The less reactive nature of Shales D and H is likely due to the presence of dolomite and halite present in these samples (Table 2). The strongly oxidizing and low pH conditions for all samples presented in the pourbaix diagrams would likely result in all sulfur being present as sulfate, and thus any precipitates probably forming oxy-hydroxysulfates.

The pourbaix diagram for Fe indicates that most of the Fe is present as Fe^{2+}. However for samples at high pH conditions, *i.e.*, greater than 5.8–6.0, which is particularly true for Shale D and H, Fe is likely present as solid Fe_2O_3. This interpretation is reiterated by the low Fe^{2+} concentrations determined for Shale D and H (Figure 3). This characteristic can also have important impacts for the concentrations of other metals in solution. Precipitation of iron oxide minerals will tend to result in the adsorption of other ions with co-precipitation and removal of these other ions from solution as a consequence. The pourbaix diagrams for Mn, Cu, Ni and Zn suggest their most likely ionic form in solution is Mn^{2+}, Cu^{2+}, Ni^{2+} and Zn^{2+} for the range of conditions observed in the reaction vessels. However, the metal concentration results determined and presented in Figures 4 to 8 suggest for

Shale D and H the metal concentrations remain lower than anticipated until the point at which Fe solubilizes, releasing the adsorbed metals into solution. This is in contrast to Shales A, B, C and G which recorded higher metal concentrations in solution throughout the experiment as the lower solution pH and iron solubilization prevented these minerals from forming.

Biofilms were not observed in the reaction vessels, and while some amount of biological activity may have been present, it was not enough to warrant further investigation. It may have been that the prevailing oxidizing conditions, coupled with the low resultant pH and low nutrient content, may have suppressed biological activity. While it possible that microbially-produced H_2S could have left the reaction vessels, this gas would have been particularly noticeable in the laboratory and was not noticed at any time.

Figure 9. Pourbaix diagrams for elements of interest. Phases are based on chemical reaction models which combine the element of interest with H_2O [29].

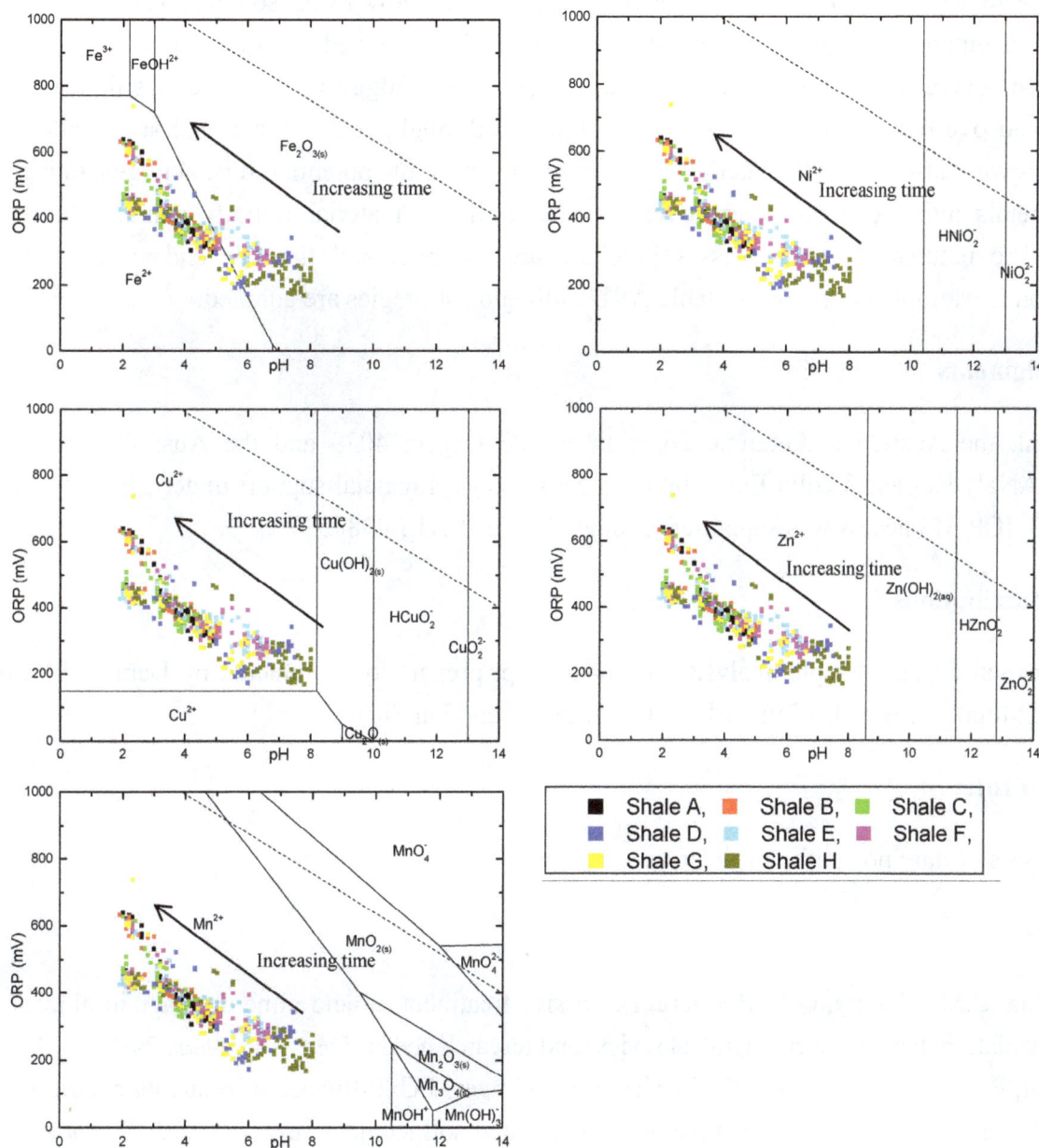

4. Conclusions

This research demonstrates the importance not only of sample buffering capacity, but also dissolution and reaction kinetics on system pH and release of metal leachate from shale materials for ARD processes. pH, ORP and metal concentrations in solution changed significantly throughout the 600 day experiment, which could have significant effects on overall site impacts when occurring on a gravel pad site. Importantly, this research demonstrates the buffering capacity of minerals such as dolomite to halt metal leaching by shales for extended periods of time. However, this capacity does become exhausted over long periods, even years, resulting in a reduction in pH and an increase in metals which eventually leach into solution into groundwater and potentially surface discharge into sensitive receiving waters.

Although this investigation has demonstrated the importance of reaction kinetics on ARD processes there are experimental improvements that may be made. As reaction kinetics are important it follows that water retention times are equally important. Ideally this work would be conducted in flow-through systems on-site whereby the effects of local geology, water flow rates, soil and water chemistry, dilution by uncontaminated groundwater and local could be incorporated into the analysis.

The use of gravel platforms to elevate infrastructure above boggy muskeg areas with seasonally frozen ground overlying permafrost is common practice throughout the Arctic and sub-Arctic. This work has shown that some shale materials from these sites have the potential to be acid generating and to leach metals into the environment over time, even if the material initially contains buffering materials. The natures of the processes involved are complex and time dependant, and careful investigation is warranted to ensure suitable ARD mitigation strategies are adopted.

Acknowledgments

We thank the Australian Antarctic Division ASAC Project 4029 and the Australian Research Council, PANalytical and Veolia Environmental Services for financial support under Linkage Project LP0776373. ICP-MS access was supported through ARC LIEF LE0989539.

Author Contributions

Experimental design, sample analysis and manuscript preparation conducted by Damian B. Gore, Kathryn A. Mumford, Brendan Pitt, Ashley T. Townsend and Ian Snape.

Conflicts of Interest

The authors declare no conflict of interest.

References

1. Neculita, C.M.; Zagury, G.J.; Bussiere, B. Passive treatment of acid mine drainage in bioreactors using sulfate-reducing bacteria: Critical review and research needs. *J. Environ. Qual.* **2007**, *36*, 1–16.

2. Pareuil, P.; Penilla, S.; Ozkan, N.; Bordas, F.; Bollinger, J.C. Influence of reducing conditions on metallic elements released from various contaminated soil samples. *Environ. Sci. Technol.* **2008**, *42*, 7615–7621.

3. Akcil, A.; Koldas, S. Acid Mine Drainage (AMD): Causes, treatment and case studies. *J. Clean Prod.* **2006**, *14*, 1139–1145.

4. Tabak, H.H.; Scharp, R.; Burckle, J.; Kawahara, F.K.; Govind, R. Advances in biotreatment of acid mine drainage and biorecovery of metals: 1. Metal precipitation for recovery and recycle. *Biodegradation* **2003**, *14*, 423–436.

5. Pandey, P.K.; Sharma, R.; Roy, M.; Pandey, M. Toxic mine drainage from Asia's biggest copper mine at Malanjkhand, India. *Environ. Geochem. Health* **2007**, *29*, 237–248.

6. Luis, A.T.; Teixeira, P.; Almeida, S.F.P.; Ector, L.; Matos, J.X.; da Silva, E.A.F. Impact of acid mine drainage (AMD) on water quality, stream sediments and periphytic diatom communities in the surrounding streams of Aljustrel mining area (Portugal). *Water Air Soil Pollut.* **2009**, *200*, 147–167.

7. Askaer, L.; Schmidt, L.B.; Elberling, B.; Asmund, G.; Jonsdottir, I.S. Environmental impact on an arctic soil-plant system resulting from metals released from coal mine waste in Svalbard (78 degrees N). *Water Air Soil Pollut.* **2008**, *195*, 99–114.

8. El Khalil, H.; El Hamiani, O.; Bitton, G.; Ouazzani, N.; Boularbah, A. Heavy metal contamination from mining sites in South Morocco: Monitoring metal content and toxicity of soil runoff and groundwater. *Environ. Monit. Assess.* **2008**, *136*, 147–160.

9. Khan, M.S.; Zaidi, A.; Wani, P.A.; Oves, M. Role of plant growth promoting rhizobacteria in the remediation of metal contaminated soils. *Environ. Chem. Lett.* **2009**, *7*, 1–19.

10. Brake, S.S.; Dannelly, H.K.; Connors, K.A.; Hasiotis, S.T. Influence of water chemistry on the distribution of an acidophilic protozoan in an acid mine drainage system at the abandoned Green Valley coal mine, Indiana, USA. *Appl. Geochem.* **2001**, *16*, 1641–1652.

11. Lee, J.S.; Chon, H.T. Hydrogeochemical characteristics of acid mine drainage in the vicinity of an abandoned mine, Daduk Creek, Korea. *J. Geochem. Explor.* **2006**, *88*, 37–40.

12. Migaszewski, Z.M.; Galuszka, A.; Paslawski, P.; Starnawska, E. An influence of pyrite oxidation on generation of unique acidic pit water: A case study, podwisniowka quarry, Holy Cross Mountains (south-central Poland). *Pol. J. Environ. Stud.* **2007**, *16*, 407–421.

13. Tutu, H.; McCarthy, T.S.; Cukrowska, E. The chemical characteristics of acid mine drainage with particular reference to sources, distribution and remediation: The Witwatersrand Basin, South Africa as a case study. *Appl. Geochem.* **2008**, *23*, 3666–3684.

14. Smith, R.M.; Sobek, A.A.; Arkele, T.; Sencindiver, J.C.; Freeman, J.K. *Extensive Overburden Potentials for Soil and Water Quality*; EPA-600/2-76-185; U.S. Environmental Protection Agency (EPA): Washington, DC, USA, 1976.

15. McDonald, D.M.; Webb, J.A.; Taylor, J. Chemical stability of acid rock drainage treatment sludge and implications for sludge management. *Environ. Sci. Technol.* **2006**, *40*, 1984–1990.

16. Sapsford, D.J.; Bowell, R.J.; Dey, M.; Williams, K.P. Humidity cell tests for the prediction of acid rock drainage. *Miner. Eng.* **2009**, *22*, 25–36.

17. Ardau, C.; Blowes, D.W.; Ptacek, C.J. Comparison of laboratory testing protocols to field observations of the weathering of sulfide-bearing mine tailings. *J. Geochem. Explor.* **2009**, *100*, 182–191.

18. Carbone, C.; Dinelli, E.; Marescotti, P.; Gasparotto, G.; Lucchetti, G. The role of AMD secondary minerals in controlling environmental pollution: Indications from bulk leaching tests. *J. Geochem. Explor.* **2013**, *132*, 188–200.

19. Wobrauschek, P. Total reflection X-ray fluorescence analysis—A review. *X-Ray Spectrom.* **2007**, *36*, 289–300.

20. Zou, H.F.; Xu, S.K.; Fang, Z.L. Determination of chromium in environmental samples by flame AAS with flow injection on-line coprecipitation. *At. Spectrosc.* **1996**, *17*, 112–118.

21. Capota, P.; Baiulescu, G.E.; Constantin, M. The analysis of environmental samples by ICP-AES. *Chem. Anal.* **1996**, *41*, 419–427.

22. Ammann, A.A. Inductively coupled plasma mass spectrometry (ICP-MS): A versatile tool. *J. Mass Spectrom.* **2007**, *42*, 419–427.

23. Townsend, A.T. The accurate determination of the first row transition metals in water, urine, plant, tissue and rock samples by sector field ICP-MS. *J. Anal. At. Spectrom.* **2000**, *15*, 307–314.

24. Wang, H.; Bigham, J.A.; Tuovinen, O.H. Oxidation of marcasite and pyrite by iron-oxidizing bacteria and archaea. *Hydrometallurgy* **2007**, *88*, 127–131.

25. Gurung, S.R.; Stewart, R.B.; Gregg, P.E.H.; Bolan, N.S. An assessment of requirements of neutralising materials of partially oxidised pyritic mine waste. *Aust. J. Soil Res.* **2000**, *38*, 329–344.

26. Canadian Council of Ministers of the Environment (CCME). *Canadian Water Quality Guidelines for the Protection of Aquatic Life: Summary Table, Updated 7.1, December 2007*; Canadian Council of Ministers of the Environment (CCME): Winnipeg, MB, Canada, 2007. Available online: https://www.halifax.ca/environment/documents/CWQG.PAL.summaryTable7.1.Dec2007.pdf (accessed on 20 January 2014).

27. Chandra, A.P.; Gerson, A.R. The mechanisms of pyrite oxidation and leaching: A fundamental perspective. *Surf. Sci. Rep.* **2010**, *65*, 293–315.

28. Chuan, M.C.; Shu, G.Y.; Liu, J.C. Solubility of heavy metals in a contaminated soil: Effects of redox potential and pH. *Water Air Soil Pollut.* **1996**, *90*, 543–556.

29. Takeno, N. *Atlas of Eh-pH Diagrams: Intercomparison of Thermodynamic Databases*; Open File Report No. 419; Geological Survey of Japan: Tsukuba, Japan, 2005. Available online: http://www.gsj.jp/GDB/openfile/files/no0419/openfile419e.pdf (accessed on 20 January 2014).

Evidence for the Multi-Stage Petrogenetic History of the Oka Carbonatite Complex (Québec, Canada) as Recorded by Perovskite and Apatite

Wei Chen * and Antonio Simonetti

Department of Civil & Environmental Engineering & Earth Sciences, University of Notre Dame, 156 Fitzpatrick Hall, Notre Dame, IN 46556, USA; E-Mail: simonetti.3@nd.edu

* Author to whom correspondence should be addressed; E-Mail: wchen2@nd.edu

Abstract: The Oka complex is amongst the youngest carbonatite occurrences in North America and is associated with the Monteregian Igneous Province (MIP; Québec, Canada). The complex consists of both carbonatite and undersaturated silicate rocks (e.g., ijolite, alnöite), and their relative emplacement history is uncertain. The aim of this study is to decipher the petrogenetic history of Oka via the compositional, isotopic and geochronological investigation of accessory minerals, perovskite and apatite, using laser ablation inductively coupled plasma mass spectrometry (LA-ICP-MS). The new compositional data for individual perovskite and apatite grains from both carbonatite and associated alkaline silicate rocks are highly variable and indicative of open system behavior. *In situ* Sr and Nd isotopic compositions for these two minerals are also variable and support the involvement of several mantle sources. U-Pb ages for both perovskite and apatite define a bimodal distribution, and range between 113 and 135 Ma, which overlaps the range of ages reported previously for Oka and the entire MIP. The overall distribution of ages indicates that alnöite was intruded first, followed by okaite and carbonatite, whereas ijolite defines a bimodal emplacement history. The combined chemical, isotopic, and geochronological data is best explained by invoking the periodic generation of small volume, partial melts generated from heterogeneous mantle.

Keywords: carbonatite; geochronology; perovskite; apatite; laser ablation (multi-collector) inductively coupled plasma mass spectrometry

1. Introduction

Previous studies indicate that carbonatites worldwide range in age from the Archean to present day, with the frequency of occurrences increasing with decreasing age (*i.e.*, with those <200 Ma in age being the most abundant) [1]. The oldest known carbonatite on Earth is the Tupertalik complex (3.0 Ga; western Greenland) [2], and the youngest is the active natrocarbonatite volcano, Oldoinyo Lengai, Tanzania [3–7]. Among the 527 carbonatite occurrences identified and compiled by Woolley and Kjarsgaard [1], only 264 have been dated, with most ages determined by the K/Ar method and merely 6% investigated by U-Pb geochronology.

In North America, carbonatite and alkaline magmatism spans ~2.7 Ga [8,9], and Oka (Figure 1) is one of the youngest carbonatite complexes on the basis of available geochronological data for various minerals and/or rock types. Apatite fission track ages reported for Oka vary between 118 ± 4 and 133 ± 11 Ma [10], whereas Shafiquall *et al.* [11] document K-Ar ages that range between 107 and 119 Ma for the intrusive alnöites associated with the complex. Wen *et al.* [12] reported a Rb–Sr biotite-whole rock isochron age of 109 ± 2 Ma obtained by isotope dilution-thermal ionization mass spectrometry (ID-TIMS), whereas Cox and Wilton [13] obtained a U-Pb age of 131 ± 7 Ma for perovskite from carbonatite by laser ablation inductively coupled plasma mass spectrometry (LA-ICP-MS). Cox and Wilton [13] postulated that the variable ages obtained by either Rb-Sr or K-Ar methods for the Oka carbonatite complex most probably result from their lower closure temperatures (relative to the U-Pb isotope system).

In a recent study, Chen and Simonetti [14] conducted a detailed U-Pb geochronological investigation of apatite from carbonatite, okaite and melanite ijolite at Oka. The geochronological results define a bimodal age distribution with peaks at ~117 and ~125 Ma. Chen *et al.* [15] also reported *in-situ* U-Pb ages for niocalite [$Ca_{14}Nb_2(Si_2O_7)_2O_6F_2$] from carbonatite at Oka, a first documented radiometric study for this mineral. The results from the latter study corroborate the variable apatite U-Pb ages [14] since they also define a similar bimodal distribution with peak $^{206}Pb/^{238}U$ weighted mean ages at 133.2 ± 6.1 and 110.1 ± 5.0 Ma. Hence, these detailed investigations reporting *in-situ* U-Pb ages for apatite and niocalite from Oka offer a more comprehensive geochronological view of this complex, and clearly suggest a protracted magmatic history that spanned at least ~10 million years [14]. Of note, the protracted igneous activity outlined for Oka overlaps that for the entire range of ages reported for the remaining Monteregian Igneous Province (MIP)-related intrusions (118–135 Ma; Figure 1) [13–16].

Plutonic igneous bodies, such as the Oka carbonatite complex, may form as a result of sequential and episodic melting and crystallization events. These events may be traced by monitoring the chemical compositions of constituent minerals, whereas the timing of such events can be determined using precise geochronological methods (e.g., U-Pb dating). Rukhlov and Bell [9] emphasized the importance of incorporating data from several isotope systems (and mineral/rock phases) before concluding the definite emplacement ages for carbonatite complexes since these typically result from complicated petrogenetic histories. Moreover, accessory phases such as perovskite, apatite and niocalite are characterized by very high abundances (1000s of ppm) of incompatible elements such as Nd and Sr, which are important isotope tracers for delineating magmatic processes and potential mantle sources. Thus, the combined presence of several U- (and Pb-) bearing accessory minerals, such as apatite, niocalite and perovskite within carbonatite, together with the capability of obtaining precise and accurate, spatially resolved chemical, isotopic, and geochronological data by laser ablation inductively (multi-collector) coupled plasma mass spectrometry (LA-(MC)-ICP-MS) analysis on individual (single) mineral grains renders the investigation of these accessory minerals a powerful tool in deciphering the formational history of carbonatite complexes.

Figure 1. (**A**) Regional map showing distribution of intrusions associated with the Monteregian Igneous Province (MIP) including the Oka carbonatite complex [14,17]. Intrusions identified on the map are: 1-Oka; 2-Royal; 3-Bruno; 4-St. Hilaire; 5-Rougemont; 6-Johnson; 7-Yamaska; 8-Shefford; 9-Brome. (**B**) Geological map of the Oka carbonatite complex [10,15]. The type localities of the Nb deposits within Oka are shown in the map: SLC: St. Lawrence Columbium, NIOCAN: name of mining company with mineral rights, and Bond Zone: NH = New Hampshire. Sample stop numbers correspond to those from the 1986 GAC-MAC-sponsored Oka field excursion [10].

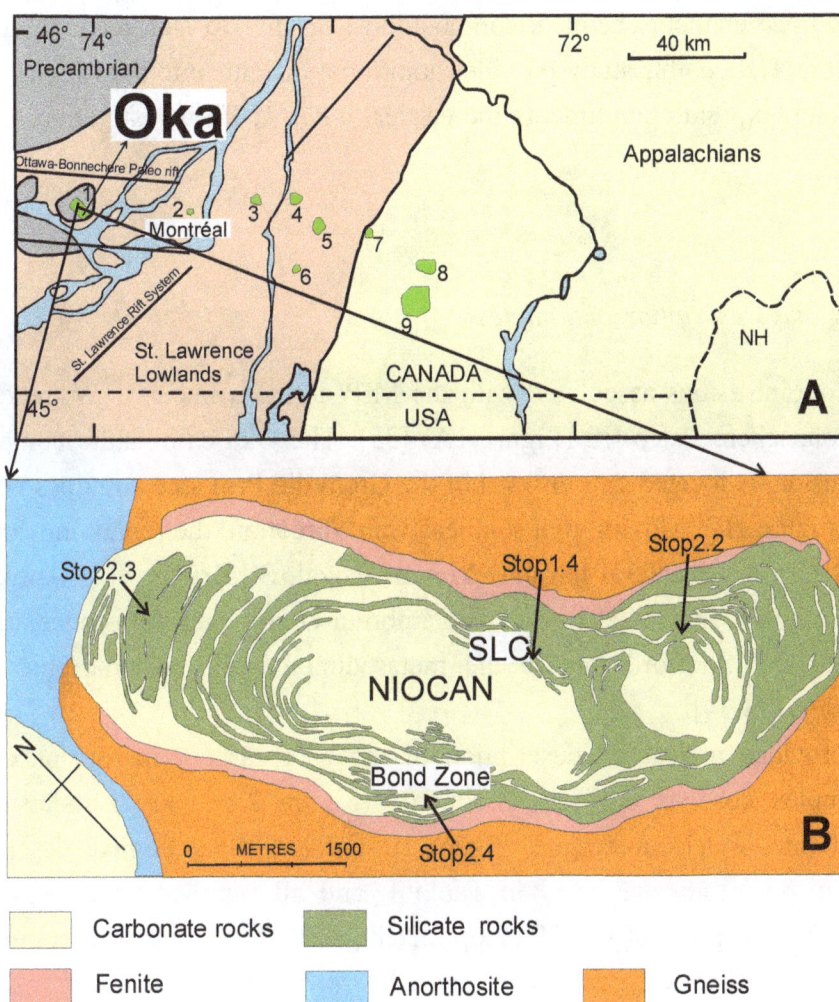

Perovskite, $CaTiO_3$, is a widespread accessory mineral in SiO_2-undersaturated and alkaline magmatic systems, and present within kimberlites, melilitites, foidites, olivinites, clinopyroxenites, ultramafic lamprophyres/carbonatites, and lamproites [18]. It typically forms throughout the melt crystallization sequence and serves as a major host for incompatible trace elements, such as the Rare Earth Elements (REEs) [18,19]. Furthermore, perovskite may preserve the original, magmatic radiogenic isotope signatures that are frequently obscured in whole rock compositions because it is relatively resistant to post-solidification alteration [19]. Recently, perovskite has been the focus of petrogenetic studies involving alkaline magmatic systems because of its capacity to yield combined accurate geochronological, chemical and/or radiogenic isotope information [13,19–22]. Cox and Wilton [13] were the first to report an *in-situ* U-Pb geochronological investigation of perovskite from Oka by LA-ICP-MS; however, their study was not combined with any geochemical data, and hence does not provide any insights into possible melt differentiation processes. For example, the exact petrogenetic relationship between carbonatite and associated alkaline silicate rocks within an individual carbonatite complex remains somewhat elusive; models proposed include liquid immiscibility [23–26], protracted fractional crystallization of a carbonate-rich, Si-undersaturated parental melt [27–29], and small volume partial melts derived from metasomatized mantle [30–32].

This study focuses on a detailed chemical, isotopic and geochronological investigation of perovskite and apatite associated with alnöite and jacupirangite from the Oka carbonatite complex. We report new, *in-situ* major and trace element chemical compositions, Sr and Nd isotopic data, and U-Pb ages for perovskite and apatite. Hence, this study provides additional insights into the formational history of the Oka carbonatite complex, and compliments the earlier *in-situ* U-Pb investigations of apatite [14] and niocalite [15].

2. Background

2.1. Geological Setting and Description of Samples

The series of alkaline intrusions associated with the MIP define a linear E–W trend that roughly follows the Ottawa-Bonnechere Paleo-rift (Figure 1A) [33]. The Oka carbonatite complex, which is the most westerly intrusion, is located entirely within the Grenville Province and does not contain any quartz-bearing rocks (Figure 1B). Moving in a southeastward direction, the remaining complexes have intruded two different tectonic/structural terrains. Mounts Royal, St. Bruno, St. Hilaire, Rougemont, and Johnson are hosted by St. Lawrence Lowlands Cambrian-Ordovician dolostones, carbonates and shales; Mounts Brome and Shefford intrude the metasediments and metamorphic rocks of the Appalachians (Figure 1A) [33,34].

Gold *et al.* [10] provided a detailed description of the Oka carbonatite complex (Figure 1B). In summary, the complex consists of both carbonatite (Figure 2A) and Si-undersaturated rocks (*i.e.*, okaite, ijolite, alnöite, and jacupirangite; Figure 2B–E). The mineralogical descriptions of the samples investigated in this study are listed in Table 1, and all samples were retrieved from the Stop 2.3 locality with the exception of Oka88 (Stop 2.2; Figure 1B). Apatite is a common accessory mineral phase occurring in all rock types, and forms euhedral crystals, which vary between ~1 and ~100 μm in diameter (e.g., Figure 2A,E). Perovskite is present in most of the silicate rocks (*i.e.*, okaite and alnöite) and usually occurs as euhedral crystals characterized by pseudocubo-octahedral habit up to several cm in diameter (*i.e.*, Figure 2F,G). In two okaite samples (Oka229 and Oka137), perovskite exhibits zoning as evidenced in optical microscopy and back-scattered electron (BSE) imagery (e.g., Figure 2G,H). The zoning typically consists of a low average atomic number core and a high average atomic number rim of irregular thickness (Figure 2H).

Table 1. Summary of the mineralogy of the main rock types at Oka.

Rock Type	Carbonatite	Okaite (Melilitite)	Melanite ijolite	Ijolite	Alnoite	Jacupirangite
Calcite	60–95	5–10	2–10	0–5	0–5	-
Apatite	2–10	2–10	2–5	0–5	0–10	2–5
Magnetite	5–10	5–10	0–5	-	5–10	2–10
Perovskite	-	2–10	-	-	-	-
Melilite	-	60–90	-	-	-	-
Biotite	0–5	5–15	-	-	5–10	2–5
Melanite	-	-	20–40	-	-	-
Pyroxene	0–10	-	20–40	40–50	20–40	60–70
Nepheline	-	-	20–40	40–50		15–20
Olivine	-	-	-	-	10–25	-
Niocalite	0–10	-	-	-	-	-
Groundmass	-	-	-	-	30–60	-

Note: Mineral occurrences are reported in volume percent.

Figure 2. Petrographic images illustrating the mineralogy of the different rock types investigated here. (**A**) carbonatite- Oka51; (**B**) okaite- Oka138; (**C**) ijolite- Oka88; (**D**) alnöite- Oka87; (**E**) jacupirangite- Oka78; (**F**) perovskite in okaite Oka208; (**G**) zoned perovskite in okaite Oka229; (**H**) back-scattered electron image of the zoned perovskite in (**G**). Cal: calcite; Ap: apatite; Ox: oxide; Mll: melilite; Cpx: clinopyroxene; Nph: nephline; Prv: perovskite; Ol: olivine.

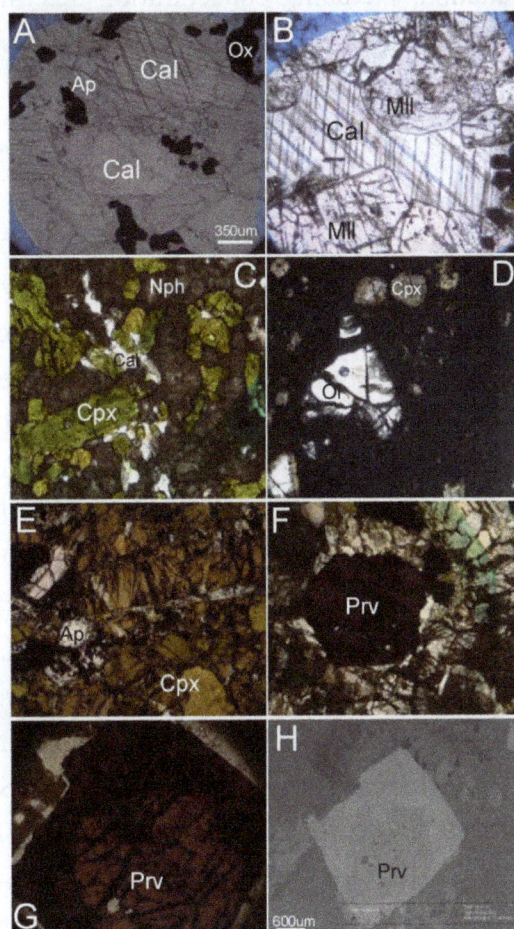

2.2. Analytical Methods

2.2.1. Chemical Analysis

Major and minor element concentrations for apatite and perovskite were determined using a Cameca SX50 electron microprobe (EMP; Cameca, Gennevilliers, France) at the University of Chicago. All thin sections were carbon-coated prior to analysis. The EMP analyses were conducted using a 15 kV accelerating potential and 30–35 nA incident current. The natural and synthetic mineral and glass standards employed for calibration purposes were: natural olivine (for Fe, Mn, Mg, and Si), natural albite (for Na), durango apatite (for P, F, and Ca), synthetic glass of anorthite composition (for Al), strontianite (for Sr), zircon (for Zr), synthetic NB metal (for Nb), synthetic TiO_2 (for Ti), synthetic REE_3 metal (for La, Ce, and Pr), synthetic REE_2 metal (for Nd and Sm), and synthetic TA metal (for Ta). Calculated apatite and perovskite formulae were normalized by stoichiometry.

In-situ trace element analyses of individual apatite and perovskite grains were obtained using a UP213 nm laser ablation system coupled to a Thermo-Finnigan Element2 sector field high-resolution ICP-MS (Thermo Fisher Scientific, Dreieich, Germany) housed within the Midwest Isotope and Trace Element Research Analytical Center (MITERAC) at the University of Notre Dame, and following the protocol by Chen and Simonetti [14] and Chen *et al.* [15]. The NIST SRM 610 international glass standard [35] was used for external calibration and ^{43}Ca ion signal intensities were employed as the internal standard with the CaO content (wt %) obtained by EMP analysis. The sample grains and standard were ablated using a 25 μm spot size, 4–5 Hz repetition rate, and corresponding energy density of ~10–12 J/cm^2. Data reduction, including concentration determinations, method detection limits and individual run uncertainties were obtained with the GLITTER laser ablation software [36].

2.2.2. U-Pb Age Dating by Laser Ablation (Multi-Collector) Inductively Coupled Plasma Mass Spectrometry

The instrumental configuration described above for the trace element determinations was also employed for the *in-situ* U-Pb isotope analyses. The analytical protocol adopted here is similar to that described in Simonetti and Neal [37] and Chen and Simonetti [14]. Data acquisition typically consisted of the first ~30 s for measurement of the background ion signals, followed by 30 s of ablation, and a minimum 15 s of washout time. Single mineral grains were ablated using a 40–55 μm spot size and corresponding fluence of ~3 J/cm^2 and repetition rate of 5 Hz. The following ion signals were acquired: ^{202}Hg, $^{204}(Pb + Hg)$, ^{206}Pb, ^{207}Pb, ^{208}Pb, ^{232}Th, ^{235}U, ^{238}U and $^{232}Th^{16}O$. ^{202}Hg was measured to monitor the ^{204}Hg interference on ^{204}Pb (using a $^{204}Hg/^{202}Hg$ value of 0.229883) [36]. In addition, U-Pb dates for some samples were determined using a NWR193 nm laser ablation system (ESI New Wave Research, Huntingdon, UK) coupled to a Nu Plasma II MC-ICP-MS instrument (Nu Instruments Ltd., Wrexham, UK) within the MITERAC facility at the University of Notre Dame. All masses of interest (^{202}Hg, $^{204}(Pb + Hg)$, ^{206}Pb, ^{207}Pb, and ^{208}Pb) can be simultaneously acquired using a combination of ion counters (Hg and Pb ion signals), and Faraday cups (^{232}Th and ^{238}U). Samples and standards were ablated employing a 55–75 μm spot size with corresponding fluence of ~10 J/cm^2 and 7 Hz repetition rate. Data acquisition consisted of the first ~40 s for background measurement, followed by ~60 s of ablation, and a minimum ~2 min of washout time.

Instrumental drift and laser induced elemental fractionation (LIEF) was monitored using a "standard-sample bracketing" technique. The Madagascar apatite [38] and Ice River perovskite [19] were adopted as the external standards for the U-Pb dating of apatite and perovskite, respectively. Each set of 10–12 unknown analyses was bracketed with five analyses of the pertinent standard both prior and after the unknown analyses. Instrumental drift and Pb-U laser induced fractionation were corrected based on the $^{206}Pb/^{238}U$ and $^{207}Pb/^{235}U$ ratios for the standards, *i.e.*, for the Madagascar apatite, the ratios are 0.0781 and 0.6123 [38], respectively; whereas for the Ice River perovskite, the adopted values are 0.0575 and 0.4270, respectively [19].

Apatite and perovskite are U-bearing accessory minerals that may contain a significant amount of common Pb. The ^{207}Pb-correction method was adopted here, which employs the Tera-Wasserburg Concordia plot and consequently is an approach that does not require knowledge of the accurate abundance of ^{204}Pb. This method was successfully employed in previous studies for a variety of common Pb-bearing accessory minerals, such as titanite [20,39–41], perovskite [13,19,20], apatite [14,41], and niocalite [15]. The ^{207}Pb-correction method does require knowledge, however, of the Pb isotope composition of the common Pb component. In this study, the latter is defined by the Pb isotope composition of the associated and ubiquitously present *U-free* calcite [14]. The $^{207}Pb/^{206}Pb$ ratio of 0.792 ± 0.06 [14,42] obtained for the latter is then applied to correct the measured $^{206}Pb/^{238}U$ ratios using well established common lead–radiogenic lead mixing equations [13,39].

Fragments of Emerald Lake and Durango apatites were used as secondary apatite standards and both are well characterized with ages of 90.5 ± 3.1 and 30.6 ± 2.3 Ma, respectively [43]. Repeated analyses of these two standards obtained during the course of this study yielded weighted mean $^{206}Pb/^{238}U$ ages of 92.6 ± 1.6 Ma ($n = 17$) and 31.9 ± 1.3 Ma ($n = 10$), respectively, and both are identical (given their associated uncertainties) to the ages reported by Chew *et al.* (Figure 3A,B) [41]. Of importance, U-Pb ages were obtained by both laser ablation multi-collector inductively coupled plasma mass spectrometry (LA-MC-ICP-MS) and LA-ICP-MS methods for the same apatite grains in one carbonatite sample (Oka147), and these yield identical dates (Figure 3C,D). These corroborative results in turn serve to further validate the analytical methods employed here.

Uncertainties associated with individual analyses, which include propagation of errors from individual measurements (based on counting statistics) and the relative standard deviation associated with repeated analyses of the Madagascar apatite and Ice River perovskite standards, were determined using the quadratic equation [38,44,45]. Isoplot v3.0 program was employed for constructing Tera-Wasserburg diagrams and determination of Concordia lower intercept ages, and $^{206}Pb/^{238}U$ weighted mean age calculations [46].

Figure 3. U-Pb isotopic ages for apatite secondary standards and sample. The U-Pb age of the secondary standard—Durango apatite is shown (**A**), and it is identical to the reported value (**B**) by Chew *et al.* [41]. Apatite from carbonatite Oka147 was dated by the laser ablation multi-collector inductively coupled plasma mass spectrometry (LA-MC-ICP-MS) (**C**) and LA-ICP-MS (**D**).

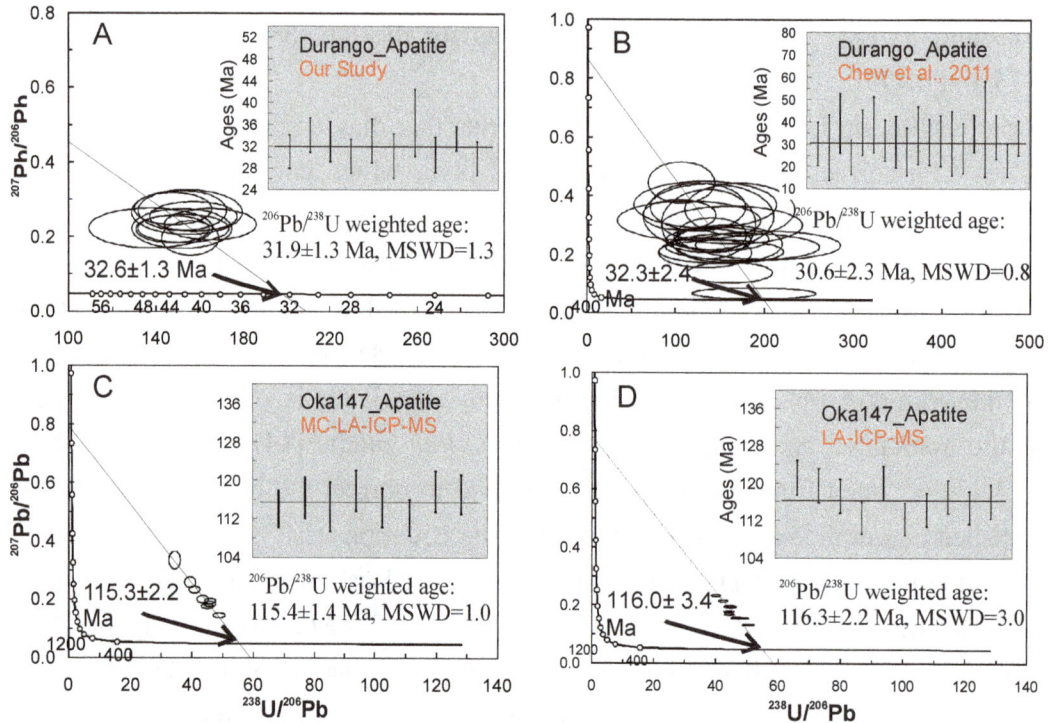

2.2.3. Sr and Nd Analysis by LA-(MC)-ICP-MS

In-situ Sr and Nd isotope ratios for perovskite and apatite were determined with a NWR193 laser ablation system coupled to a Nu Plasma II MC-ICP-MS instrument (Nu Instruments Ltd., Wrexham, UK). *In-situ* Sr isotope measurements involve correction of critical spectral interferences that include Kr, Rb, and doubly charged REEs [47,48]. These detailed corrections are adopted in this study and are identical to those reported in Chen *et al.* [15]. A modern-day coral (Indian Ocean) served as an external, in-house standard, which is well characterized for its $^{87}Sr/^{86}Sr$ isotopic composition by ID-TIMS [49]. The coral standard and perovskite grains were analyzed using a 75~100 μm spot size, 7 Hz repetition rate, and an energy density ~11 J/cm^2. The average $^{87}Sr/^{86}Sr$ ratio obtained for the coral standard is 0.70915 ± 0.00003 based on five measurements, and is indistinguishable (given uncertainties) compared to the corresponding TIMS value of 0.70910 ± 0.00002 [49].

In order to obtain accurate measurement of Nd isotope ratios, it is important to identify and correct isobaric interferences and monitor for instrumental mass discrimination. The isobaric interferences for *in-situ* Nd isotope determinations are principally related to Sm, Ce and Ba, and the correction of the ^{144}Sm ion signal on ^{144}Nd is critical. The $^{146}Nd/^{144}Nd$ ratio is traditionally selected to correct for instrumental mass discrimination with the $^{146}Nd/^{144}Nd_{ref} = 0.7219$ [50]. As described in Yang *et al.* [51], the mass bias for both Sm and Nd were set to be identical (βSm = βNd). The mass bias correction for Sm is based on the $^{149}Sm/^{144}Sm$, and consequently the ^{144}Nd ion signal can be calculated according to the following equation:

$$I_{144Nd} = I_{144(Nd+Sm)} - I_{149Sm} \times \left(\frac{^{144}Sm}{^{149}Sm} \right) \times \left(\frac{m_{149Sm}}{m_{144Sm}} \right)^{\beta Sm} \qquad (1)$$

The Durango apatite was adopted as the external standard with the accepted $^{143}Nd/^{144}Nd = 0.512483$ [52]. The "standard-sample" bracketing method was used and both standard and samples were analyzed using a 75–100 μm spot size, 7 Hz repetition rate, and corresponding energy density of ~9 J/cm^2.

3. Results

3.1. Major and Trace Element Data

The major and trace element data for perovskite investigated in this study are listed in Tables 2 and 3. In total, >30 chemical analyses of perovskite from okaite (Oka229, Oka137, and Oka208), alnöite (Oka73), and jacupirangite (Oka70) were obtained. Based on their major element compositions, the data for perovskite plot into two groups, in particular relative to their Nb_2O_5 wt % abundances (Figure 4A). One group (Nb-E: Nb-enriched) contains high Nb_2O_5 contents (between 7.25 and 10.80 wt %), whereas the other (Nb-D: Nb-depleted) is characterized by lower Nb_2O_5 abundances (between 1.56 and 4.92 wt %). Table 2 shows that perovskite from both alnöite and jacupirangite belongs to the Nb-D group, whereas compositions for those from okaite are variable. Of importance, both Nb-E and Nb-D types of perovskite are present within individual samples (*i.e.*, Oka137 and Oka229) and even within singular zoned grains (e.g., Figure 5; in total two well-zoned grains have been identified). For example, the zoned perovskite grain shown in Figure 5 has a rim that consists of a Nb-E composition, whereas the central area is characterized by the Nb-D component. In general, Nb-E perovskite is characterized by high contents of Na, Al, Fe, Sr, Ta and REEs (Figure 4B,C; Tables 2 and 3). The composition for the Nb-E perovskite is not that of the ideal end-member, but can be described by involving components of lueshite ($NaNbO_3$), latrappite ($CaNb_{0.5}Fe_{0.5}O_3$), and (LREE)FeO_3 (light REEs), which form by elemental substitutions into the structure at both Ca and Ti sites [53,54]. Molar percentages of different perovskite endmembers are also listed in Table 2. The most significant substituents in the Ca site are REEs and Na (Figure 4B), which comprise up to 8.70 mol % of the (LREE)FeO_3 component in some Nb-E perovskite (Table 2). Another example is the coupled substitution between Ti and Nb + Fe^{3+} or Nb + Al, which accounts for up to 10.96 mol % of $CaNb_{0.5}Fe_{0.5}O_3$ (latrappite) or $CaNb_{0.5}Al_{0.5}O_3$ (Figure 4C; Table 2).

The total REE budget for perovskite is dominated by the light REEs (LREEs; *i.e.*, La, Ce, Pr, Nd) abundances with (La/Yb)$_{CN}$ ratios that vary between 364 and 1652 (Table 3), and as illustrated by the pronounced, negatively-sloped chondrite-normalized REE patterns (Figure 6). Pb and Th abundances for perovskite both define negative correlations with Ca contents and suggest their substitution within the same Ca site (Figure 7A,B); in contrast, U abundances do not show any covariance with Ca contents (not shown). Of note, some elements are reported for both EMP and LA-ICP-MS analysis (e.g., LREEs). For example, the relative difference in the measured abundances of Pr obtained by these two methods is ~10%.

Table 2. Major element compositions of perovskite from the Oka complex.

Sample	Oka70 (Jacupirangite)				Oka73 (Alnoite)				
	PV1	PV2	PV3	PV4	PV3	PV4	PV7	PV8	PV9
Nb_2O_5	2.15	2.12	2.26	1.67	2.90	1.66	1.56	3.99	1.26
Ta_2O_5	0.01	0.00	0.18	0.05	0.10	0.43	0.15	0.14	0.07
SiO_2	0.01	0.00	0.01	0.01	0.00	0.00	0.00	0.00	0.00
TiO_2	52.45	52.72	52.27	52.09	49.75	48.81	53.07	45.98	53.82
ZrO_2	0.07	0.08	0.06	0.05	0.25	0.06	0.02	0.28	0.14
Al_2O_3	0.40	0.40	0.43	0.48	0.48	0.54	0.37	0.52	0.21
Fe_2O_3	2.66	2.48	2.72	2.68	3.34	4.03	2.16	4.94	1.82
La_2O_3	0.85	0.90	0.93	0.88	1.72	2.08	1.33	2.32	0.93
Ce_2O_3	1.49	1.46	2.04	1.82	3.19	4.48	2.70	5.24	1.52
Pr_2O_3	0.12	0.13	0.22	0.23	0.39	0.50	0.22	0.54	0.14
Nd_2O_3	0.42	0.50	0.65	0.61	0.98	1.72	0.93	1.88	0.45
Sm_2O_3	0.09	0.04	0.08	0.06	0.16	0.09	0.00	0.19	0.05
CaO	37.98	38.22	37.58	38.01	36.36	34.58	36.94	33.13	38.75
SrO	0.52	0.54	0.48	0.49	0.38	0.25	0.43	0.39	0.40
Na_2O	0.31	0.31	0.35	0.28	0.38	0.38	0.38	0.70	0.23
Total	99.01	99.43	99.74	98.87	99.74	98.83	99.84	99.29	99.43
Structural formulae									
Nb^{5+}	0.023	0.022	0.024	0.018	0.031	0.018	0.017	0.042	0.013
Ta^{5+}	0.000	0.000	0.001	0.000	0.001	0.003	0.001	0.001	0.000
Si^{4+}	0.000	0.000	0.000	0.000	0.000	0.000	0.000	0.000	0.000
Ti^{4+}	0.925	0.929	0.922	0.918	0.877	0.861	0.936	0.811	0.949
Zr^{4+}	0.000	0.000	0.000	0.000	0.001	0.000	0.000	0.002	0.001
Al^{3+}	0.011	0.011	0.012	0.013	0.013	0.015	0.010	0.014	0.006
Fe^{3+}	0.047	0.044	0.048	0.047	0.059	0.071	0.038	0.087	0.032
La^{3+}	0.007	0.008	0.008	0.008	0.015	0.018	0.011	0.020	0.008
Ce^{3+}	0.013	0.013	0.018	0.016	0.027	0.038	0.023	0.045	0.013
Pr^{3+}	0.001	0.001	0.002	0.002	0.003	0.004	0.002	0.005	0.001
Nd^{3+}	0.004	0.004	0.005	0.005	0.008	0.014	0.008	0.016	0.004
Sm^{3+}	0.001	0.000	0.001	0.000	0.001	0.001	0.000	0.002	0.000
Ca^{2+}	0.954	0.960	0.944	0.954	0.913	0.868	0.927	0.832	0.973
Sr^{2+}	0.007	0.007	0.007	0.007	0.005	0.003	0.006	0.005	0.005
Na^{+}	0.014	0.014	0.016	0.013	0.017	0.017	0.017	0.032	0.011
mol % of the endmembers									
$CaTiO_3$	93.68	93.63	93.00	93.25	89.46	89.62	93.79	85.43	94.84
$CaNb_{0.5}(Fe,Al)_{0.5}O_3$	2.89	3.00	2.18	3.61	3.51	0.75	0.00	2.06	2.33
$NaNbO_3$	0.86	0.76	1.44	0.01	1.44	1.73	1.73	3.36	0.21
$(LREE)FeO_3$	2.57	2.61	3.37	3.12	5.59	7.89	4.48	9.15	2.62

Table 2. *Cont.*

Sample	Oka229 (Okaite)									
	PV1	PV1_2	PV2_1	PV2_2	PV2_3	PV2_4	PV2_5	PV2_6	PV2_7	PV2_8
			Rim	Core	Core	Core	Rim	Rim		
Nb_2O_5	3.71	3.60	10.43	4.92	4.76	4.74	9.77	10.14	7.25	7.25
Ta_2O_5	0.31	0.17	0.80	0.17	0.20	0.21	0.92	0.90	0.38	0.43
SiO_2	0.00	0.00	0.17	0.00	0.02	0.05	0.12	0.04	0.02	0.05
TiO_2	47.01	48.15	38.86	46.26	46.18	46.72	39.62	39.44	42.95	43.26
ZrO_2	1.35	0.98	0.35	0.90	0.81	0.83	0.34	0.21	0.34	0.39
Al_2O_3	0.74	0.75	0.87	0.73	0.70	0.74	0.85	0.74	0.86	0.77
Fe_2O_3	4.30	4.04	6.18	4.33	4.09	4.35	6.04	6.02	5.26	5.51
La_2O_3	1.02	1.03	2.33	1.23	1.18	1.07	2.35	2.41	1.94	1.67
Ce_2O_3	2.71	2.65	4.64	2.72	2.55	2.65	4.75	4.94	4.27	4.17
Pr_2O_3	0.29	0.27	0.46	0.34	0.22	0.24	0.44	0.45	0.46	0.46
Nd_2O_3	1.11	1.12	1.38	0.97	0.99	0.95	1.44	1.46	1.41	1.42
Sm_2O_3	0.15	0.16	0.08	0.15	0.07	0.04	0.16	0.11	0.14	0.14
CaO	37.15	37.24	31.39	37.02	37.15	37.20	31.45	31.07	33.48	33.76
SrO	0.22	0.22	0.52	0.22	0.20	0.21	0.52	0.57	0.44	0.40
Na_2O	0.08	0.09	1.15	0.11	0.12	0.12	1.18	1.34	0.72	0.65
Total	99.32	99.71	98.44	99.24	98.45	99.29	98.81	98.68	98.93	99.29
Structural formulae										
Nb^{5+}	0.039	0.038	0.110	0.052	0.050	0.050	0.104	0.107	0.077	0.077
Ta^{5+}	0.002	0.001	0.005	0.001	0.001	0.001	0.006	0.006	0.002	0.003
Si^{4+}	0.000	0.000	0.004	0.000	0.000	0.001	0.003	0.001	0.000	0.001
Ti^{4+}	0.829	0.849	0.685	0.816	0.814	0.824	0.699	0.695	0.757	0.763
Zr^{4+}	0.008	0.006	0.002	0.005	0.005	0.005	0.002	0.001	0.002	0.002
Al^{3+}	0.020	0.021	0.024	0.020	0.019	0.020	0.023	0.020	0.024	0.021
Fe^{3+}	0.076	0.071	0.109	0.076	0.072	0.077	0.107	0.106	0.093	0.097
La^{3+}	0.009	0.009	0.020	0.011	0.010	0.009	0.020	0.021	0.017	0.014
Ce^{3+}	0.023	0.023	0.040	0.023	0.022	0.023	0.041	0.042	0.037	0.036
Pr^{3+}	0.002	0.002	0.004	0.003	0.002	0.002	0.004	0.004	0.004	0.004
Nd^{3+}	0.009	0.009	0.012	0.008	0.008	0.008	0.012	0.012	0.012	0.012
Sm^{3+}	0.001	0.001	0.001	0.001	0.001	0.000	0.001	0.001	0.001	0.001
Ca^{2+}	0.933	0.935	0.788	0.930	0.933	0.934	0.790	0.780	0.841	0.848
Sr^{2+}	0.003	0.003	0.007	0.003	0.003	0.003	0.007	0.008	0.006	0.005
Na^+	0.004	0.004	0.052	0.005	0.006	0.005	0.054	0.061	0.033	0.030
mol % of the endmembers										
$CaTiO_3$	86.76	87.41	75.41	84.58	85.14	85.08	76.31	75.71	80.48	81.12
$CaNb_{0.5}(Fe,Al)_{0.5}O_3$	8.57	8.03	10.59	10.96	10.74	10.58	9.36	8.99	8.60	8.64
$NaNbO_3$	0.00	0.00	5.70	0.00	0.00	0.00	5.85	6.61	3.47	3.14
$(LREE)FeO_3$	4.67	4.56	8.30	4.46	4.13	4.34	8.48	8.70	7.45	7.11

Table 2. *Cont.*

Sample	Oka137 (Okaite)					Oka209 (Okaite)				
	PV1_1	PV1_2	PV3	PV4	PV5	PV2	PV3_1	PV3_2	PV4	PV5
	Core	Rim								
Nb_2O_5	3.53	10.80	10.62	9.61	9.93	1.63	1.78	1.60	1.73	1.66
Ta_2O_5	0.28	1.02	0.87	0.76	0.67	0.06	0.14	0.02	0.04	0.04
SiO_2	0.01	0.00	0.00	0.01	0.00	0.00	0.01	0.00	0.00	0.00
TiO_2	47.82	39.23	39.48	40.01	40.12	52.46	52.53	52.94	52.39	51.84
ZrO_2	0.99	0.19	0.40	0.40	0.42	0.07	0.06	0.03	0.08	0.07
Al_2O_3	0.75	0.75	0.74	0.84	0.73	0.42	0.44	0.41	0.46	0.48
Fe_2O_3	4.09	6.11	5.65	5.80	5.90	2.40	2.54	2.36	2.37	2.69
La_2O_3	0.92	2.46	2.03	1.93	2.14	1.18	1.15	1.16	1.20	0.91
Ce_2O_3	2.66	4.74	4.21	4.23	4.44	2.11	2.10	2.06	2.10	1.95
Pr_2O_3	0.32	0.43	0.45	0.40	0.43	0.17	0.20	0.12	0.15	0.21
Nd_2O_3	1.05	1.42	1.30	1.32	1.34	0.72	0.64	0.65	0.67	0.56
Sm_2O_3	0.08	0.09	0.11	0.02	0.08	0.07	0.06	0.09	0.01	0.03
CaO	37.12	31.24	31.62	32.82	32.35	37.79	37.75	38.05	37.62	37.86
SrO	0.20	0.56	0.51	0.48	0.51	0.44	0.52	0.47	0.47	0.47
Na_2O	0.09	1.38	1.31	1.00	1.19	0.29	0.30	0.28	0.29	0.29
Total	99.14	99.24	98.23	98.52	99.12	99.35	99.71	99.79	99.15	98.55
Structural formulae										
Nb^{5+}	0.037	0.114	0.113	0.102	0.105	0.017	0.019	0.017	0.018	0.018
Ta^{5+}	0.002	0.006	0.006	0.005	0.004	0.000	0.001	0.000	0.000	0.000
Si^{4+}	0.000	0.000	0.000	0.000	0.000	0.000	0.000	0.000	0.000	0.000
Ti^{4+}	0.843	0.692	0.696	0.705	0.707	0.925	0.926	0.933	0.924	0.914
Zr^{4+}	0.006	0.001	0.002	0.002	0.002	0.000	0.000	0.000	0.000	0.000
Al^{3+}	0.021	0.021	0.020	0.023	0.020	0.012	0.012	0.011	0.013	0.013
Fe^{3+}	0.072	0.108	0.100	0.102	0.104	0.042	0.045	0.042	0.042	0.047
La^{3+}	0.008	0.021	0.018	0.017	0.018	0.010	0.010	0.010	0.010	0.008
Ce^{3+}	0.023	0.041	0.036	0.036	0.038	0.018	0.018	0.018	0.018	0.017
Pr^{3+}	0.003	0.004	0.004	0.003	0.004	0.001	0.002	0.001	0.001	0.002
Nd^{3+}	0.009	0.012	0.011	0.011	0.011	0.006	0.005	0.005	0.006	0.005
Sm^{3+}	0.001	0.001	0.001	0.000	0.001	0.001	0.000	0.001	0.000	0.000
Ca^{2+}	0.932	0.784	0.794	0.824	0.812	0.949	0.948	0.955	0.945	0.951
Sr^{2+}	0.003	0.008	0.007	0.007	0.007	0.006	0.007	0.006	0.006	0.006
Na^+	0.004	0.063	0.060	0.046	0.054	0.013	0.014	0.013	0.013	0.013
mol % of the endmembers										
$CaTiO_3$	87.50	74.85	75.69	75.54	75.59	93.37	93.37	93.68	93.50	93.17
$CaNb_{0.5}(Fe,Al)_{0.5}O_3$	8.07	9.90	10.35	12.40	10.93	2.36	2.13	2.19	2.07	3.63
$NaNbO_3$	0.00	6.81	6.47	4.87	5.80	0.60	0.92	0.62	0.85	0.00
$(LREE)FeO_3$	4.42	8.44	7.50	7.19	7.68	3.67	3.58	3.50	3.59	3.19

Table 2. *Cont.*

Sample	Oka208 (Okaite)					
	PV1	PV2	PV3	PV5	PV6	PV7
Nb_2O_5	2.55	3.13	2.86	3.30	3.14	2.70
Ta_2O_5	0.18	0.16	0.13	0.20	0.09	0.20
SiO_2	0.02	0.01	0.00	0.00	0.00	0.01
TiO_2	49.63	48.17	49.81	49.52	49.99	49.88
ZrO_2	0.04	0.07	0.06	0.07	0.05	0.07
Al_2O_3	0.55	0.53	0.61	0.51	0.52	0.58
Fe_2O_3	3.10	3.05	3.18	3.23	3.02	3.04
La_2O_3	1.27	1.32	1.37	1.35	1.38	1.35
Ce_2O_3	2.83	2.97	2.92	2.99	2.86	2.89
Pr_2O_3	0.27	0.33	0.35	0.28	0.26	0.30
Nd_2O_3	0.96	1.04	0.95	1.02	1.00	0.95
Sm_2O_3	0.08	0.09	0.17	0.04	0.12	0.10
CaO	35.60	36.32	36.45	35.75	36.01	36.21
SrO	0.47	0.49	0.52	0.51	0.53	0.51
Na_2O	0.54	0.42	0.39	0.45	0.45	0.39
Total	97.50	97.53	99.15	98.61	98.84	98.61
Structural formulae						
Nb^{5+}	0.027	0.033	0.030	0.035	0.033	0.029
Ta^{5+}	0.001	0.001	0.001	0.001	0.001	0.001
Si^{4+}	0.000	0.000	0.000	0.000	0.000	0.000
Ti^{4+}	0.875	0.849	0.878	0.873	0.881	0.879
Zr^{4+}	0.000	0.000	0.000	0.000	0.000	0.000
Al^{3+}	0.015	0.015	0.017	0.014	0.014	0.016
Fe^{3+}	0.055	0.054	0.056	0.057	0.053	0.054
La^{3+}	0.011	0.011	0.012	0.012	0.012	0.012
Ce^{3+}	0.024	0.026	0.025	0.026	0.025	0.025
Pr^{3+}	0.002	0.003	0.003	0.002	0.002	0.003
Nd^{3+}	0.008	0.009	0.008	0.009	0.008	0.008
Sm^{3+}	0.001	0.001	0.001	0.000	0.001	0.001
Ca^{2+}	0.894	0.912	0.915	0.898	0.904	0.909
Sr^{2+}	0.006	0.007	0.007	0.007	0.007	0.007
Na^+	0.025	0.019	0.018	0.020	0.020	0.018
mol % of the endmembers						
$CaTiO_3$	91.29	89.23	89.90	90.38	90.63	90.51
$CaNb_{0.5}(Fe,Al)_{0.5}O_3$	1.89	6.52	3.77	2.49	2.33	2.99
$NaNbO_3$	1.99	0.32	1.30	2.11	2.10	1.58
$(LREE)FeO_3$	4.83	3.93	5.03	5.02	4.93	4.92

Notes: Major element compositions are presented in wt %. Structural formulae are calculated based on 3 atoms of oxygen. The mol % of the endmembers are calculated following the sequence: (1) Ti^{4+} is assigned for $CaTiO_3$; (2) latrappite is calculated based on the availability of Ca^{2+}, Nb^{5+}, Fe^{3+} and Al^{3+}; (3) depends on the available Na^+ and Nb^{5+}, mol % of $NaNbO_3$ is calculated; and (4) the rest of the Fe^{3+} is combined with the $LREE^{3+}$ to $(LREE)FeO_3$.

Table 3. Trace element abundances of perovskite.

Sample	Oka70 (Jacupirangite)				Oka73 (Alnoite)				
	PV1	PV2	PV3	PV4	PV3	PV4	PV7	PV8	PV9
Mn	490	475	493	435	452	440	327	637	506
Mn	490	475	493	435	452	440	327	637	506
Ga	12	11	13	11	18	20	14	30	24
Sr	3,599	3,412	3,616	3,444	2,951	1,862	3,073	2,828	3,709
Y	232	240	307	205	156	407	144	268	347
Zr	288	243	247	214	433	2,488	68	1,494	600
Ba	55	50	64	51	91	104	70	143	111
La	6,020	8,086	8,177	6,054	10,489	14,881	8,296	14,658	10,781
Ce	10,488	13,001	1,7787	12,401	19,381	31,900	16,769	32,891	17,526
Pr	914	1,287	1,641	1,260	1,858	3,382	1,801	3,341	1,521
Nd	2,891	4,298	5,504	4,057	5,784	11,918	5,641	11,492	5,114
Sm	387	513	673	468	571	1,323	559	1,187	628
Eu	116	133	173	129	138	284	136	244	183
Gd	198	250	321	217	222	545	215	457	310
Tb	26	29	39	26	25	59	24	48	37
Dy	98	104	135	91	75	190	71	135	138
Ho	14	15	18	12	10	25	9	17	20
Er	28	30	40	26	24	58	22	42	42
Tm	2.4	2.4	3.0	2.0	1.5	4.8	1.3	2.7	3.9
Yb	11.2	11.2	14.9	9.8	8.7	24.7	6.3	13.9	18.5
Lu	1.1	0.9	1.3	0.8	0.8	2.6	0.4	1.5	1.7
Hf	7.5	6.9	6.4	5.6	10.0	48.9	1.8	44.1	15.8
Ta	243	395	376	353	460	5,412	622	427	177
Pb	5	8	10	9	21	36	24	45	14
Th	219	478	688	426	761	1403	1,282	2,417	229
U	144	110	111	112	124	124	113	101	206
(La/Yb)N	364	490	372	421	819	409	895	715	396

Sample	Oka229 (Okaite)									
	PV1	PV1_2	PV2_1	PV2_2	PV2_3	PV2_4	PV2_5	PV2_6	PV2_7	PV2_8
			Rim	Core	Core	Core	Rim	Rim		
Mn	306	536	802	338	525	332	853	750	681	791
Ga	13	15	22	16	16	13	23	22	19	21
Sr	1,689	1,962	3,465	1,623	1,710	1,770	3,568	3,668	2,813	3,072
Y	312	255	145	299	293	262	147	115	214	199
Zr	5,862	3,661	1,347	4,039	3,975	3,967	1,350	622	1,660	1,616
Ba	71	83	114	83	81	70	117	116	100	111
La	7,549	7,565	15,129	10,849	9,278	7,877	15,016	13,962	13,081	11,088
Ce	19,887	19,407	29,922	23,870	19,961	19,381	30,154	28,404	28,579	27,510
Pr	2,299	2,455	2,946	2,524	2,257	2,097	3,046	2,862	2,944	2,951
Nd	7,905	7,954	8,672	8,267	7,527	6,781	8,893	8,182	9,185	9,116
Sm	922	850	769	915	862	751	790	684	914	881
Eu	229	209	165	231	224	200	173	147	209	200
Gd	381	336	274	375	368	318	286	236	341	330

Table 3. *Cont.*

Sample	Oka229 (Okaite)									

Sample	PV1	PV1_2	PV2_1	PV2_2	PV2_3	PV2_4	PV2_5	PV2_6	PV2_7	PV2_8
			Rim	Core	Core	Core	Rim	Rim		
Tb	45	39	30	43	42	37	31	25	38	36
Dy	151	127	82	142	143	123	83	65	111	106
Ho	20	16	10	19	18	16	11	8	14	13
Er	42	37	27	42	41	36	27	23	35	33
Tm	3.1	2.8	1.7	3.2	3.0	2.8	1.4	1.1	2.3	2.2
Yb	17.6	14.3	7.4	17.1	16.8	15.0	8.4	5.7	13.1	10.3
Lu	1.4	1.1	0.7	1.3	1.5	1.2	0.6	0.4	1.1	0.9
Hf	50.4	20.1	29.9	21.1	25.3	21.8	27.9	16.5	19.7	23.0
Ta	1,412	1,264	3,945	1,055	1,165	1,141	3,833	3,485	1,934	2,412
Pb	5	9	37	6	5	5	43	29	31	35
Th	194	238	1,746	255	97	95	2,224	1,162	1,558	1,889
U	191	186	71	209	188	210	70	58	183	152
(La/Yb)$_N$	291	360	1,391	431	375	356	1,219	1,652	680	728

Sample	Oka137 (Okaite)					Oka209 (Okaite)				
	PV1_1	PV1_2	PV3	PV4	PV5	PV2	PV3_1	PV3_2	PV4	PV5
	core	rim								
Mn	289	719	865	807	683	348	353	416	319	374
Ga	16	23	25	23	21	8	9	9	13	8
Sr	1,656	3,473	3,711	3,611	3,492	4,025	3,155	3,352	3,047	3,095
Y	367	182	165	189	144	190	214	258	204	185
Zr	5,427	1,508	1,481	1,817	1,535	240	279	276	250	213
Ba	73	106	119	111	87	49	49	54	84	52
La	7,515	16,177	15,177	13,939	8,859	6,870	8,551	8,771	8,335	7,076
Ce	21,468	31,025	31,244	30,380	18,276	12,187	15,534	15,508	14,481	15,036
Pr	2,299	2,926	3,093	2,974	1,735	944	815	794	829	893
Nd	8,239	9,030	9,391	9,275	5,398	4,066	4,643	4,747	4,503	4,210
Sm	969	817	849	847	562	467	532	551	511	492
Eu	232	171	178	177	135	119	137	147	129	133
Gd	459	337	321	329	211	227	292	301	250	245
Tb	53	38	35	36	24	26	33	35	29	29
Dy	170	97	92	98	72	91	114	121	92	94
Ho	23	12	11	13	9	12	15	16	12	12
Er	53	35	32	33	23	28	36	36	30	27
Tm	3.6	1.8	1.7	1.9	1.6	2.0	2.1	2.8	1.9	1.8
Yb	20.5	9.4	7.9	9.6	8.1	9.0	10.7	12.0	8.8	9.4
Lu	1.7	0.8	0.7	0.9	0.7	0.7	0.9	1.0	0.9	0.7
Hf	46.9	35.0	32.3	41.3	15.8	7.1	9.5	7.6	7.5	6.3
Ta	1,527	4,131	4,305	3,761	1,249	353	409	438	394	341
Pb	5	40	46	40	19	8	9	9	9	9
Th	192	2,504	2,630	2,327	845	409	494	517	487	413
U	164	79	63	86	103	100	102	127	113	110
(La/Yb)N	249	1,165	1,313	991	745	516	545	496	641	510

Table 3. *Cont.*

Sample	Oka208 (Okaite)					
	PV1	PV2	PV3	PV5	PV6	PV7
Mn	346	318	342	358	352	415
Ga	13	17	14	15	13	17
Sr	3,219	3,592	3,532	3,338	3,548	3,488
Y	201	169	186	171	162	174
Zr	236	222	242	233	215	279
Ba	82	94	78	81	77	97
La	8,624	7,369	9,318	8,351	8,341	10,879
Ce	19,084	16,432	19,665	18,405	17,216	23,169
Pr	1,872	1,734	1,899	1,890	1,852	2,394
Nd	6,320	5,601	6,255	6,102	5,887	7,449
Sm	662	590	630	619	581	705
Eu	154	139	150	151	143	167
Gd	273	230	257	245	231	270
Tb	30	26	29	27	26	31
Dy	94	79	90	84	77	88
Ho	13	10	12	11	10	11
Er	28	23	27	24	24	26
Tm	2.0	1.4	1.8	1.6	1.4	1.7
Yb	9.4	7.1	8.4	7.8	7.3	8.9
Lu	0.8	0.6	0.7	0.6	0.6	0.6
Hf	5.5	5.5	5.6	5.4	4.8	6.1
Ta	776	724	801	866	722	864
Pb	17	15	16	22	17	19
Th	1,179	1,002	1,125	1,337	1,016	1,149
U	79	76	81	91	88	97
(La/Yb)N	624	708	754	724	778	833

Note: Trace element concentrations are listed in ppm.

Figure 4. Compositional variation diagrams for perovskite from Oka. (**A**) TiO_2 *vs.* Nb_2O_5, with Nb-E and Nb-D groups identified; (**B**) ($Na^+ + LREE^{3+}$) *vs.* Ca^{2+}; (**C**) ($Al^{3+} + Fe^{3+} + Nb^{5+}$) *vs.* Ti^{4+}.

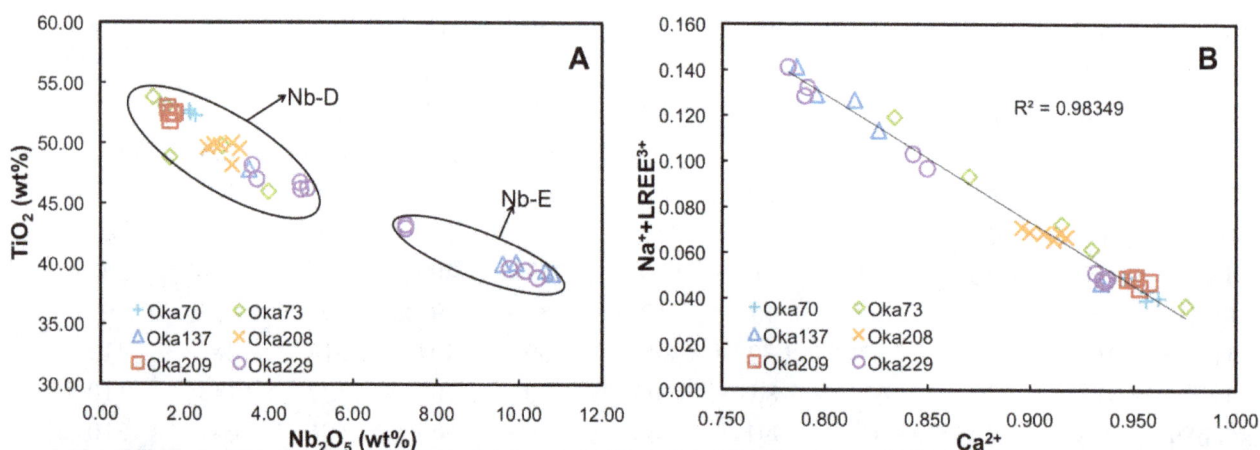

Figure 4. *Cont.*

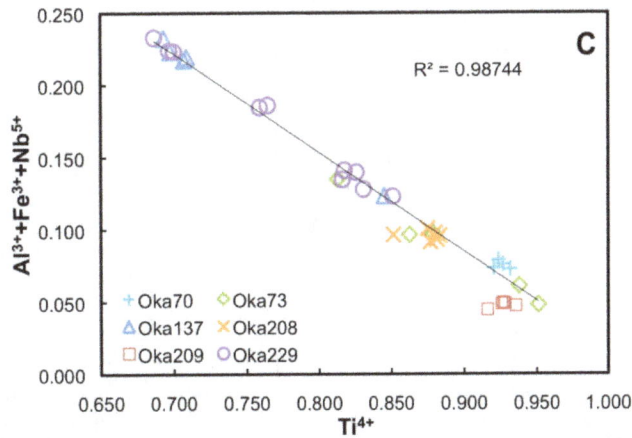

Figure 5. Petrographic image showing the zonation of perovskite. (**A**) U (ppm) concentrations obtained by LA-ICPMS with a 25 μm spot size; (**B**) Nb_2O_5 (wt %) abundances analyzed by EMP with a 5 μm beam; and (**C**) $^{206}Pb/^{238}U$ weighted mean ages (determined by LA-MC-ICPMS using a spot size of 75 μm) across a zoned perovskite grain from sample Oka229 (okaite).

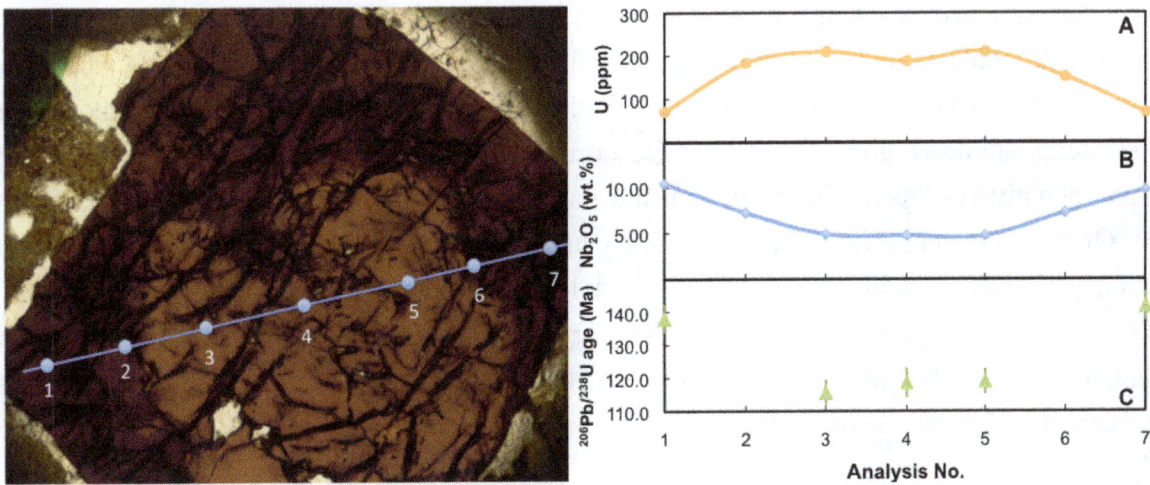

Figure 6. Chondrite normalized Rare Earth Element (REE) patterns for perovskite from alkaline silicate samples at Oka. Chondrite values are from McDonough and Sun [55].

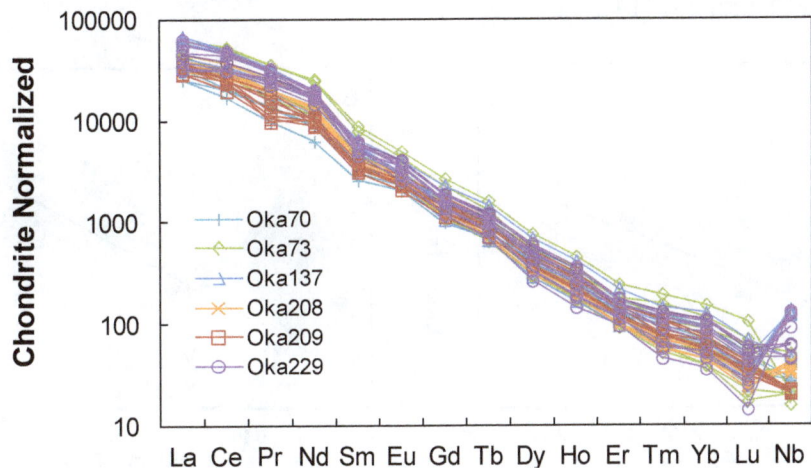

Figure 7. Plots exhibiting the correlations between Pb (**A**) and Th (**B**) abundances *vs.* Ca^{2+} (a.p.f.u.) for perovskite from Oka.

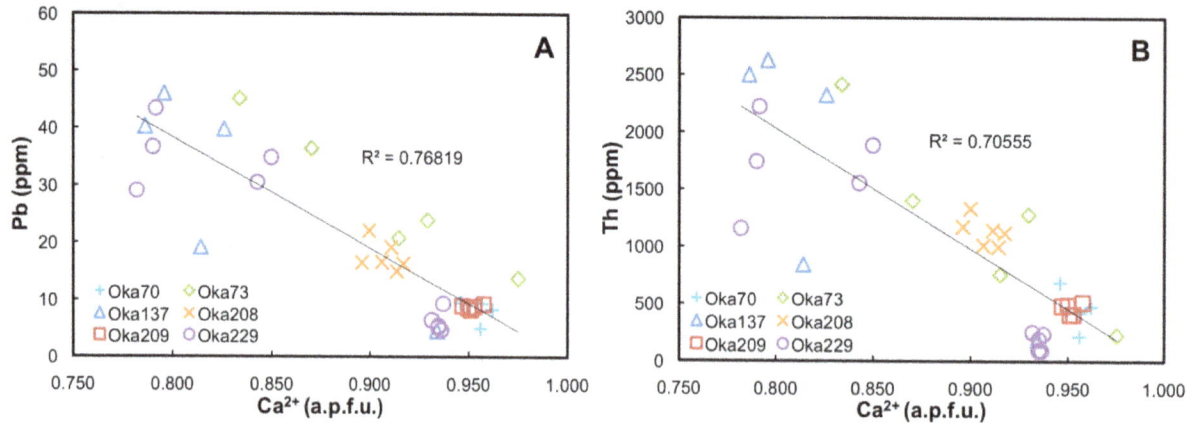

Newly obtained chemical compositions for apatite from alnöite, ijolite and jacupirangite are listed in Tables 4 and 5. Figure 8 plots the major element and LREE compositions for apatite from all rock types investigated here, along with those from carbonatite and okaite (Chen and Simonetti [14]). The compositions of fluorapatite from alnöite and jacupirangite are chemically distinct relative to the remaining rock types at Oka (Figure 8), *i.e.*, they contain a higher Ca content for a given P abundance (Figure 8A). As explained by Chen and Simonetti [14], REE abundances for apatite exhibit a positive correlation with Si contents due to their co-substitution within Ca and P structural sites. Once again, the same substitution scheme is evidenced here for all the apatites with the exception of those from alnöite and jacupirangite (Figure 8B). Chondrite normalized REE patterns for apatite investigated here are also negatively-sloped (Figure 9), but are more variable compared to those for perovskite (Figure 6). Of note, the chondrite normalized REE patterns for apatite from alnöite and jacupirangite exhibit less negative slopes among all rock types, with lower LREE abundances and comparable heavy REE (HREE) contents (Figure 9). $(La/Yb)_N$ ratios vary from 45 to 161 for apatite from alnöite and jacupirangite (Table 5), whereas ratios range between 106 and 695 for the remaining apatite [14]; these ratios for apatite are generally lower compared to those for perovskite (Table 3).

Figure 8. Chemical variation diagrams for apatite from different rock types (carbonatite (Carb.), alnöite, ijolite, okaite, and jacupirangite (Jac.)). (**A**) Ca^{2+} *vs.* P^{5+}; (**B**) $LREE^{3+}$ *vs.* Si^{4+}. Additional chemical compositions for apatite from carbonatite, okaite and melanite ijolite are from Chen and Simonetti [14].

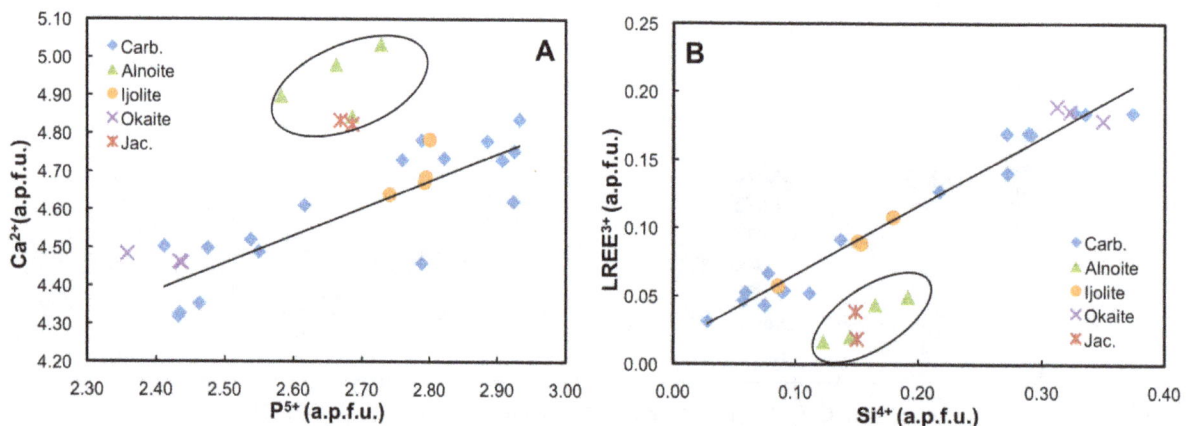

Figure 9. Chondrite normalized REE patterns for apatite from different rock types at Oka. As with Figure 8, additional REE abundances for apatite from carbonatite, okaite and melanite ijolite are from Chen and Simonetti [14]. The grey shaded area outlines the normalized patterns for apatite from alnöite and jacupirangite. Chondrite values are from McDonough and Sun [55].

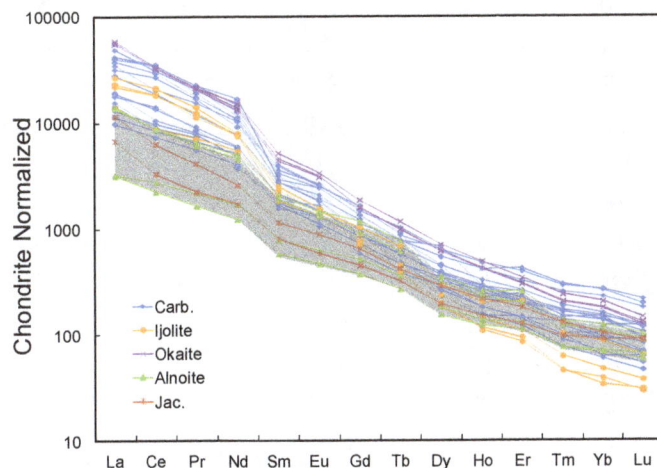

Table 4. Major element abundances of apatite from ijolite, alnöite and jacupirangite.

Sample	Oka132 alnöite n = 5	Oka134 alnöite n = 10	Oka73 alnöite n = 4	Oka75 alnöite n = 17	Oka88 ijolite n = 9	Oka70 jacupirangite n = 5	Oka78 jacupirangite n = 2
P_2O_5	37.93	36.47	37.60	38.53	39.56	37.93	37.69
SiO_2	1.73	2.29	1.98	1.47	1.02	1.79	1.80
La_2O_3	0.19	0.48	0.40	0.17	0.54	0.43	0.20
Ce_2O_3	0.27	0.73	0.62	0.20	0.90	0.61	0.29
Pr_2O_3	0.02	0.06	0.14	0.05	0.07	0.03	0.05
Nd_2O_3	0.19	0.38	0.29	0.13	0.38	0.19	0.07
MgO	0.04	0.03	0.03	0.03	0.01	0.02	0.04
CaO	54.07	54.66	55.58	56.19	53.39	53.83	53.95
MnO	0.02	0.01	0.02	0.01	0.05	0.03	0.04
FeO	0.04	0.04	0.03	0.04	0.04	0.13	0.03
SrO	0.48	0.46	0.46	0.47	0.85	0.83	0.52
F	1.77	2.24	2.10	2.05	4.00	2.31	1.93
Total	96.75	97.84	99.24	99.33	100.82	98.15	96.60
P^{5+}	2.686	2.582	2.662	2.728	2.801	2.686	2.668
Ca^{2+}	4.845	4.898	4.980	5.035	4.784	4.824	4.834
Sr^{2+}	0.023	0.022	0.022	0.023	0.041	0.040	0.025
Si^{4+}	0.145	0.192	0.165	0.123	0.086	0.149	0.150
La^{3+}	0.006	0.015	0.012	0.005	0.017	0.013	0.006
Ce^{3+}	0.008	0.022	0.019	0.006	0.028	0.019	0.009
Pr^{3+}	0.001	0.002	0.004	0.002	0.002	0.001	0.002
Nd^{3+}	0.006	0.011	0.009	0.004	0.011	0.006	0.002
$LREE^{3+}$	0.021	0.050	0.044	0.017	0.058	0.039	0.019

Notes: Major element compositions are presented in wt %; Structural formulae of apatite are calculated based on 12 atoms of oxygen.

Table 5. Trace element abundances of apatite from ijolite, alnöite and jacupirangite.

Sample	Oka132 alnöite n = 5	Oka134 alnöite n = 10	Oka73 alnöite n = 4	Oka75 alnöite n = 17	Oka88 ijolite n = 9	Oka70 jacupirangite n = 5	Oka78 jacupirangite n = 2
Mn	120	163	167	119	358	194	164
Rb	0.26	0.20	b.d.	1.37	b.d.	b.d.	b.d.
Sr	3798	3716	3393	3842	5774	8938	4236
Y	195	358	327	187	256	325	243
Ba	19.80	27.21	28.71	10.19	92.66	32.36	11.83
La	780	3423	3346	763	3361	2772	1624
Ce	1728	5496	5515	1411	5513	3961	2060
Pr	208	623	584	157	677	390	212
Nd	788	2320	2149	572	2540	1201	801
Sm	120	314	285	87	312	173	122
Eu	35.38	82.65	76.00	26.35	82.89	51.03	33.12
Gd	104	238	186	75.15	209	130	90.48
Tb	13.10	28.42	22.19	9.79	24.69	15.16	11.67
Dy	45.59	87.78	77.74	37.85	74.42	71.85	48.47
Ho	7.50	14.02	12.57	6.70	11.23	11.86	8.22
Er	21.14	41.28	31.95	18.09	33.37	29.24	20.35
Tm	1.88	3.37	3.04	1.89	2.54	3.21	2.39
Yb	11.12	20.04	17.54	11.54	14.15	16.73	14.58
Lu	1.46	2.54	2.24	1.63	1.60	2.19	2.23
Pb	2.93	4.69	4.61	3.20	9.79	2.96	2.50
Th	105	240	161	140	600	52.29	59.41
U	20.86	36.87	25.71	22.52	6.73	21.45	11.29
$(La/Yb)_N$	48	116	130	45	161	113	76

Notes: Trace element concentrations are listed in ppm; b.d. = below detection limit.

3.2. Geochronological Data

New, *in-situ* U-Pb ages for apatite are reported here from three alnöites, one ijolite, two okaites, and one jacupirangite (Table 6). As with the *in-situ* U-Pb dating results documented previously for apatite from Oka by Chen and Simonetti [14], the newly obtained ages for several alkaline silicate rock samples also indicate a bimodal distribution (e.g., Oka132 and Oka229; Figure 10B,C). For example, apatite from alnöite sample Oka75 yields bimodal $^{206}Pb/^{238}U$ weighted mean ages of 111.7 ± 3.4 and 131.6 ± 2.4 Ma. In general, samples with only one age peak (*i.e.*, Oka209, Figure 10A) yield a relatively young age.

In total, ~40 U-Pb analyses for perovskite from four okaites and one alnöite obtained here are listed in Table 7. Of interest, U-Pb ages for perovskite from okaite sample Oka229 also yields a bimodal distribution (Figure 10D). Moreover, individual ages correlate with their corresponding chemical compositions, *i.e.*, older perovskites that define a $^{206}Pb/^{238}U$ weighted mean age of 139.4 ± 2.5 Ma are characterized by high Nb_2O_5 contents (Nb-E group), whereas younger perovskites with an age of 115.7 ± 3.9 Ma belong to the Nb-D group (Figure 5). Of note, the young ages for perovskite obtained in this study (Table 4 and Figure 10E) are younger than the previously reported (single) U-Pb age of

131 ± 7 Ma for the same mineral from carbonatite at Oka [13]. Thus, as with the recently published apatite and niocalite ages for Oka [14,15], the U-Pb perovskite ages obtained here also suggest a rather protracted crystallization history for Oka. Of importance, the Th/U ratios for perovskite investigated in this study are all >1 with the highest value up to 31. Chew *et al.* [41] pointed out that using the [208]Pb-correction method in conjunction with determining [208]Pb-[232]Th ages only yields reliable geochronological results when [232]Th/[238]U ratios are <0.5. Consequently, we do not report the [232]Th-[208]Pb ages for perovskite investigated here.

Figure 10. U-Pb isotopic ages for apatite and perovskite from the Oka carbonatite complex. Examples are illustrated for samples with a single age for apatite (**A**) and bimodal age distributions (**B,C**). Diagrams (**D**) and (**E**) illustrate examples of perovskite age results, (**D**) gives a bimodal distribution age and (**E**) yields a single young age. All reported uncertainties are at 2σ level as determined by Isoplot [46]. The Mean Square Weighted Deviation (MSWD) is used as a statistical validity of the regression line according to the criteria defined by Wendt and Carl [56].

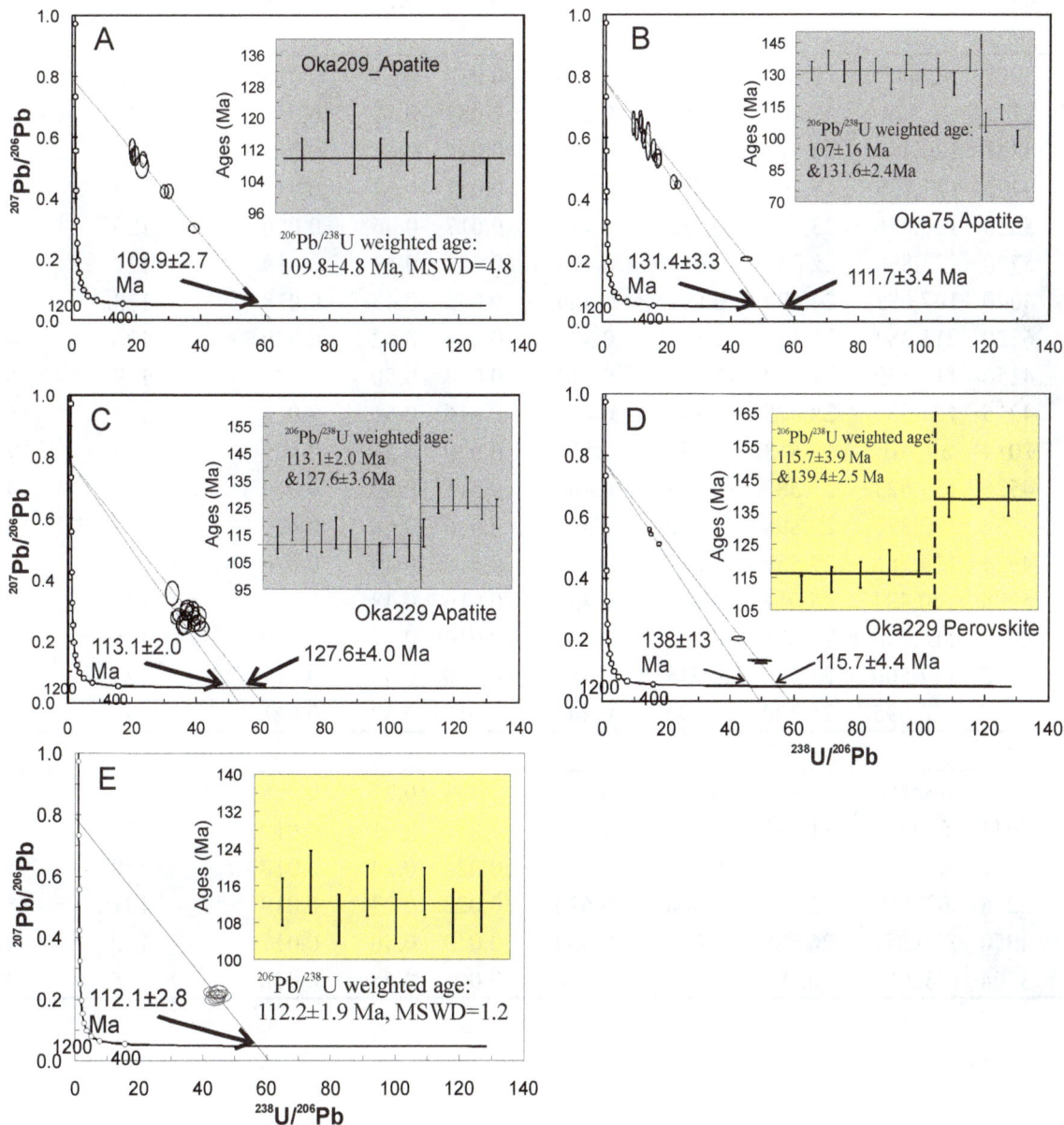

Table 6. *In-situ* U-Pb geochronological results for apatite by LA-ICP-MS. Rad.: Radiogenic.

Analyses	^{206}Pb (cps)	^{238}U (cps)	$^{238}U/^{206}Pb$	2σ	$^{207}Pb/^{206}Pb$	2σ	F206	Rad. $^{206}Pb/^{238}U$	Rad. $^{206}Pb/^{238}U$ Age (Ma)	2σ
Oka75										
ap2	3860	73,757	17.297	0.581	0.528	0.016	0.35	0.020	130	4
ap5	3846	67,692	16.153	0.531	0.536	0.013	0.34	0.021	135	4
ap6	2005	27,238	12.429	0.474	0.603	0.018	0.25	0.020	129	5
ap7	2118	39,197	17.029	0.874	0.532	0.024	0.35	0.020	130	7
ap8	2793	19,763	9.551	0.395	0.647	0.033	0.19	0.020	129	5
ap12	2241	32,451	12.690	0.448	0.595	0.025	0.26	0.021	132	5
ap13	3514	93,575	23.239	0.786	0.447	0.011	0.46	0.020	127	4
ap16	2105	51,591	21.907	0.857	0.455	0.019	0.45	0.021	131	5
ap17	1034	16,038	13.968	0.601	0.589	0.051	0.27	0.019	124	5
ap18	1542	17,890	10.492	0.402	0.627	0.028	0.22	0.021	134	5
ap10	2247	31,047	11.825	0.485	0.646	0.025	0.19	0.016	105	4
ap11	8678	436,411	44.920	1.414	0.205	0.006	0.79	0.018	112	4
Oka132										
ap1	2987	42,448	12.778	0.613	0.618	0.037	0.23	0.018	116	6
ap2	2542	26,579	9.461	0.486	0.660	0.045	0.18	0.019	118	6
ap3	4431	120,443	24.373	0.858	0.452	0.017	0.46	0.019	119	4
ap4	4405	127,135	26.202	1.067	0.414	0.017	0.51	0.019	124	5
ap6	5230	136,876	23.515	0.833	0.448	0.017	0.46	0.020	125	4
ap7	5770	123,850	22.334	1.045	0.473	0.023	0.43	0.019	122	6
ap8	4040	107,693	24.499	0.891	0.469	0.035	0.43	0.018	113	4
ap9	6320	213,332	31.331	1.108	0.331	0.014	0.62	0.020	126	4
ap10	4256	111,559	24.952	1.074	0.417	0.024	0.50	0.020	129	6
ap12	4215	127,533	28.325	1.113	0.411	0.016	0.51	0.018	115	5
ap13	2014	39,761	18.813	1.039	0.513	0.029	0.37	0.020	127	7
ap15	4526	134,623	27.585	0.916	0.402	0.015	0.52	0.019	121	4
ap16	5532	182,371	29.898	1.000	0.362	0.016	0.58	0.019	123	4
ap17	4954	134,766	24.557	0.859	0.426	0.019	0.49	0.020	128	4
ap18	5098	140,484	25.011	1.026	0.431	0.016	0.48	0.019	124	5
ap19	4482	118,700	24.078	0.896	0.440	0.019	0.47	0.020	125	5
ap20	4578	117,506	23.534	0.935	0.455	0.020	0.45	0.019	123	5
ap21	4537	126,595	25.390	0.899	0.402	0.017	0.52	0.021	131	5
Oka134										
ap3	2633	46,091	11.623	0.512	0.632	0.033	0.21	0.018	117	5
ap4	2047	57,963	31.557	1.069	0.337	0.011	0.61	0.019	124	4
ap6_2	1881	48,510	12.961	0.513	0.613	0.024	0.24	0.018	118	5
ap7	2226	63,035	12.111	0.460	0.620	0.028	0.23	0.019	121	5
ap8	4950	220,877	26.580	1.025	0.450	0.019	0.46	0.017	110	4
ap10	3204	96,354	36.193	1.188	0.306	0.008	0.65	0.018	115	4

Table 6. *Cont.*

Analyses	^{206}Pb (cps)	^{238}U (cps)	^{238}U/^{206}Pb	2σ	^{207}Pb/^{206}Pb	2σ	F206	Rad. ^{206}Pb/^{238}U	Rad. ^{206}Pb/^{238}U Age (Ma)	2σ
Oka78										
ap1	2515	44,472	15.505	0.655	0.576	0.031	0.29	0.019	119	5
ap2	2419	36,383	13.013	0.545	0.636	0.033	0.21	0.016	102	4
ap2_2	2682	49,979	15.883	0.719	0.590	0.018	0.27	0.017	109	5
ap5	3147	36,786	10.981	0.475	0.616	0.041	0.23	0.021	136	6
ap6	2235	26,384	9.809	0.432	0.650	0.038	0.19	0.019	123	5
ap7	2127	25,446	10.862	0.638	0.629	0.050	0.22	0.020	127	7
ap8	3479	86,154	20.688	0.770	0.472	0.016	0.43	0.021	132	5
Oka88										
ap1	1928	24,364	10.634	0.571	0.630	0.034	0.22	0.020	130	7
ap5	3191	52,289	14.240	0.526	0.599	0.023	0.26	0.018	116	4
ap9	2266	22,820	8.590	0.379	0.671	0.029	0.16	0.019	119	5
ap12	2087	28,603	11.739	0.460	0.639	0.034	0.20	0.017	111	4
Oka209										
ap1	3806	136,013	28.715	1.044	0.423	0.017	0.50	0.017	110	4
ap2	2669	61,596	18.711	0.627	0.537	0.022	0.34	0.018	116	4
ap3	2792	82,285	21.569	1.675	0.505	0.031	0.38	0.018	114	9
ap5	5696	262,191	37.865	1.239	0.302	0.013	0.66	0.017	111	4
ap6	2535	61,132	19.526	0.857	0.540	0.027	0.34	0.017	110	5
ap7	2545	55,895	18.549	0.728	0.564	0.026	0.30	0.016	105	4
ap8	2418	63,624	21.972	0.892	0.527	0.020	0.35	0.016	103	4
ap9	2622	93,867	30.042	1.094	0.423	0.019	0.49	0.016	105	4

Table 7. *In-situ* U-Pb geochronological results for perovskite by LA-(MC)-ICP-MS.

Analyses	^{206}Pb (V)	^{238}U (V)	^{238}U/^{206}Pb	2σ	^{207}Pb/^{206}Pb	2σ	F206	Rad. ^{206}Pb/^{238}U	Rad. ^{206}Pb/^{238}U Age (Ma)	2σ
Oka229										
PV1	0.0014	0.075	51.008	1.791	0.130	0.001	0.89	0.017	111	4
PV1_2	0.0012	0.065	49.588	1.661	0.132	0.001	0.89	0.018	114	4
PV2_1	0.0016	0.029	17.390	0.569	0.511	0.005	0.38	0.022	138	5
PV2_2	0.0012	0.064	49.735	1.689	0.122	0.001	0.90	0.018	116	4
PV2_3	0.0011	0.049	42.489	1.631	0.205	0.006	0.79	0.019	119	5
PV2_4	0.0010	0.052	47.546	1.537	0.133	0.001	0.89	0.019	119	4
PV2_5	0.0014	0.022	14.934	0.470	0.544	0.006	0.33	0.022	142	4
PV2_6	0.0013	0.020	14.402	0.470	0.558	0.006	0.31	0.022	138	5
Oka208										
PV1	2,006	90,727	44.557	2.004	0.209	0.016	0.78	0.018	112	5
PV2	2,127	92,231	43.726	2.517	0.198	0.014	0.80	0.018	117	7
PV3	1,928	89,696	45.028	2.215	0.223	0.018	0.76	0.017	108	5
PV4	2,163	92,836	42.614	2.005	0.223	0.015	0.76	0.018	115	5
PV5	2,329	104,011	46.008	2.261	0.211	0.016	0.78	0.017	108	5
PV6	2,079	91,906	43.869	1.961	0.207	0.018	0.79	0.018	115	5
PV7	2,130	94,597	44.558	2.351	0.225	0.017	0.76	0.017	109	6
PV7_2	2,137	96,161	44.041	2.579	0.216	0.011	0.77	0.018	112	7

Table 7. *Cont.*

Analyses	^{206}Pb (V)	^{238}U (V)	^{238}U/^{206}Pb	2σ	^{207}Pb/^{206}Pb	2σ	F206	Rad. ^{206}Pb/^{238}U	Rad. ^{206}Pb/^{238}U Age (Ma)	2σ
Oka209										
PV2	0.0004	0.017	38.887	1.274	0.244	0.003	0.74	0.019	121	4
PV2_2	0.0004	0.019	48.124	1.593	0.159	0.002	0.85	0.018	113	4
PV3	0.0006	0.029	48.942	1.685	0.144	0.002	0.87	0.018	114	4
PV3_2	0.0006	0.033	50.112	1.714	0.140	0.002	0.88	0.017	112	4
PV4	0.0006	0.028	47.879	1.605	0.147	0.002	0.87	0.018	116	4
PV4_2	0.0006	0.030	43.947	1.397	0.189	0.003	0.81	0.018	118	4
PV5	0.0006	0.028	44.530	1.509	0.162	0.002	0.85	0.019	121	4
PV5_2	0.0006	0.028	44.981	1.481	0.158	0.002	0.85	0.019	121	4
Oka137										
PV1	0.0009	0.042	45.790	1.615	0.165	0.002	0.84	0.018	118	4
PV1_2	0.0012	0.018	13.855	0.500	0.560	0.006	0.31	0.022	143	5
PV3	0.0012	0.019	14.283	0.478	0.552	0.006	0.32	0.022	143	5
PV3_2	0.0011	0.017	14.344	0.464	0.559	0.006	0.31	0.022	138	4
PV4	0.0010	0.020	18.015	0.587	0.504	0.005	0.39	0.021	137	4
PV4_2	0.0011	0.020	17.352	0.636	0.517	0.005	0.37	0.021	135	5
PV5	0.0011	0.028	23.512	0.936	0.427	0.005	0.49	0.021	133	5
PV5_2	0.0009	0.041	42.112	1.336	0.228	0.004	0.76	0.018	115	4
Oka73										
PV3	0.0009	0.039	39.781	1.357	0.235	0.003	0.75	0.019	120	4
PV4	0.0014	0.027	17.431	0.686	0.515	0.005	0.37	0.021	136	5
PV7	0.0006	0.028	42.803	1.349	0.148	0.002	0.87	0.020	129	4
PV7_2	0.0007	0.033	45.200	1.532	0.160	0.002	0.85	0.019	120	4
PV8	0.0021	0.059	26.176	1.013	0.395	0.007	0.53	0.020	130	5
PV9	0.0018	0.093	50.231	1.697	0.123	0.002	0.90	0.018	115	4

Notes: F206 is the proportion of common ^{206}Pb; Sample Oka208 is determined by LA-ICP-MS, and all other samples are analyzed by LA-MC-ICP-MS. Rad.: Radiogenic.

A recent geochronological study by Chen *et al.* [15] focused on the Nb-disilicate mineral, niocalite, for which Oka is the type locality. Niocalite from one of the carbonatite samples investigated by Chen *et al.* [15] also indicates a bimodal age distribution with weighted mean ^{206}Pb/^{238}U ages of 110.1 ± 5.0 and 133.2 ± 6.1 Ma, and overlaps that of co-existing apatite for the same sample [15]. Niocalite from two other carbonatite samples yield younger ages of 110.6 ± 1.2 and 115.0 ± 1.9 Ma [15].

In summary, a total of 293 *in-situ* U-Pb apatite ages yield a bimodal distribution pattern using the Kernel Density Estimation (KDE) diagram (Figure 11A; KDE is a standard statistical technique used for estimating the density distribution in geochronlogical studies) [57], with two peaks at ~126 and ~115 Ma. The variable perovskite ages indicate an additional older age peak at 135.4 ± 3.2 Ma (Figure 11D), which is similar (given the associated uncertainties) to the age of 131 ± 7 Ma for perovskite obtained by Cox and Wilton [13]. In contrast, the niocalite U-Pb dating results tend to converge toward the younger age, with a peak at 112.6 ± 1.2 Ma (Figure 11C) [15]. The majority of the combined *in situ* U-Pb dating results for apatite, perovskite, and niocalite from Oka clearly support a protracted history of magmatic activity in the order of ~10–15 million years (Figure 11A).

Figure 11. Kernel Density Estimation (KDE) plots for the weighted mean $^{206}Pb/^{238}U$ ages for the different rock/mineral groups. (**A**) The entire geochronological data for Oka ($n = 363$); (**B**) apatite ($n = 293$); (**C**) niocalite ($n = 38$); (**D**) perovskite ($n = 32$); (**E**) carbonatite ($n = 160$); (**F**) okaite ($n = 101$); (**G**) ijolite ($n = 44$); (**H**) alnöite ($n = 41$).

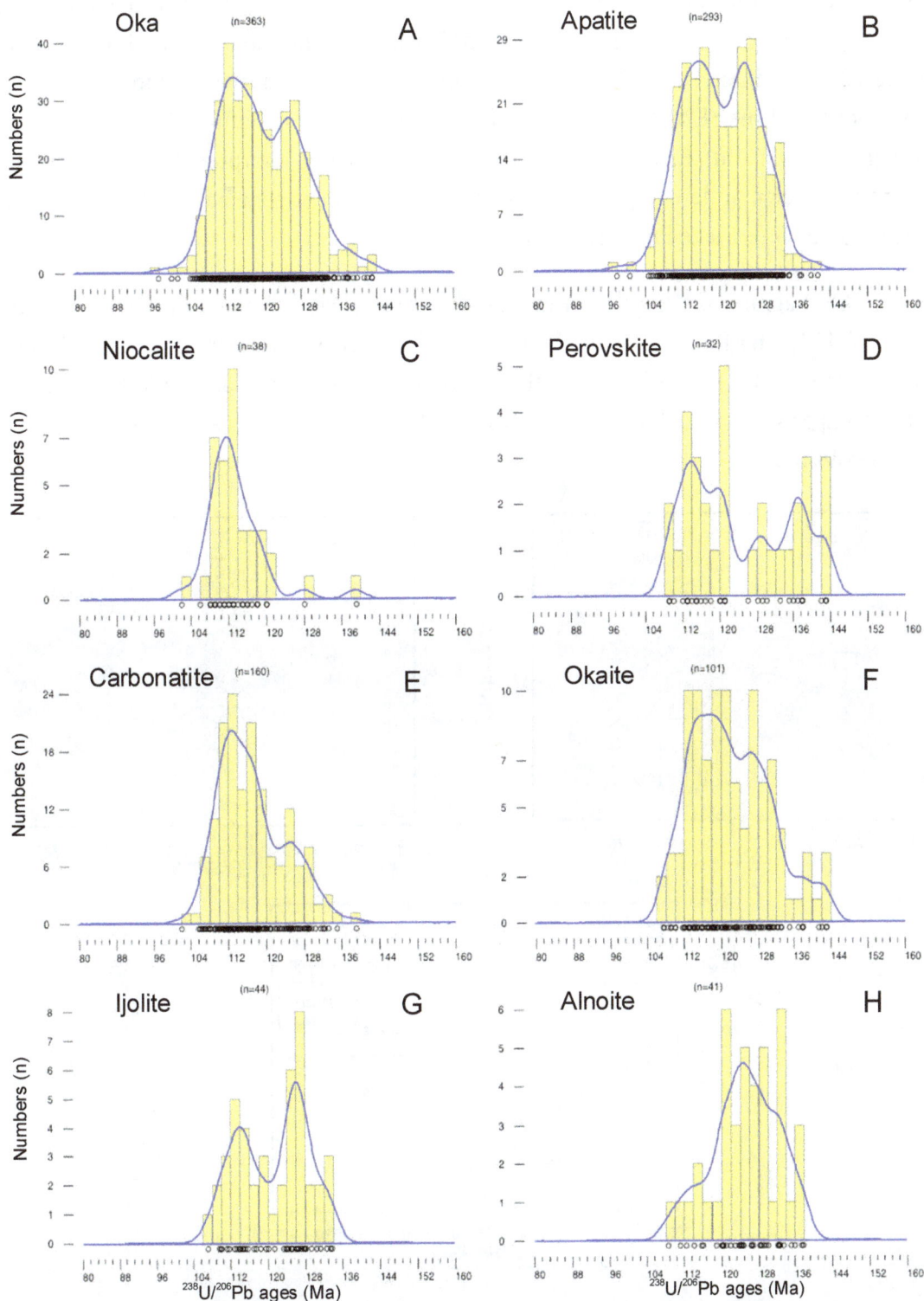

3.3. Radiogenic Isotope Data

The Sr and Nd isotope results for perovskite and apatite obtained here are listed in Table 8 and shown in Figure 12. Overall, Rb concentrations are below (or close to) the detection limit, and consequently the calculated Rb/Sr ratios are extremely low so that the age correction of the measured $^{87}Sr/^{86}Sr$ ratio is negligible. For the Sm-Nd data, a correction for radiogenic ^{143}Nd was applied, and ages used for the correction were based on the U/Pb dating results obtained here. The *in-situ* Sr and Nd isotope data for both perovskite and apatite overlap the entire range defined by previously reported whole rock data for carbonatite from Oka [12], but definitely indicate a larger variation ($^{87}Sr/^{86}Sr$: 0.70312–0.70367; $^{143}Nd/^{144}Nd$: 0.51270–0.51286), and is not consistent with closed-system behavior (Figure 12A). Of interest, the Nd and Sr isotope data from Oka overlap the upper end of the East African Carbonatite Line (EACL; Figure 12B) [58]. The EACL is defined by the Nd-Sr isotope values for young (<40 Ma old) East African carbonatites, and was interpreted to represent mixing between two end-member mantle components: HIMU (mantle component with time integrated, high $^{238}U/^{204}Pb$ ratio)- and EMI (enriched mantle 1)-like.

Figure 12. (A) Diagram of $^{143}Nd/^{144}Nd$ *vs.* $^{87}Sr/^{86}Sr$ shows data obtained in this study and by Wen *et al.* [12]. **(B)** Plot of $^{143}Nd/^{144}Nd$ *vs.* $^{87}Sr/^{86}Sr$. Also shown are the East African Carbonatite Line (EACL) from Bell and Blenkinsop [58], and CHUR and Bulk Earth (BE) values for comparison. **(C)** Diagram of $^{143}Nd/^{144}Nd$ *vs.* $^{87}Sr/^{86}Sr$ values for the different groups of perovskite.

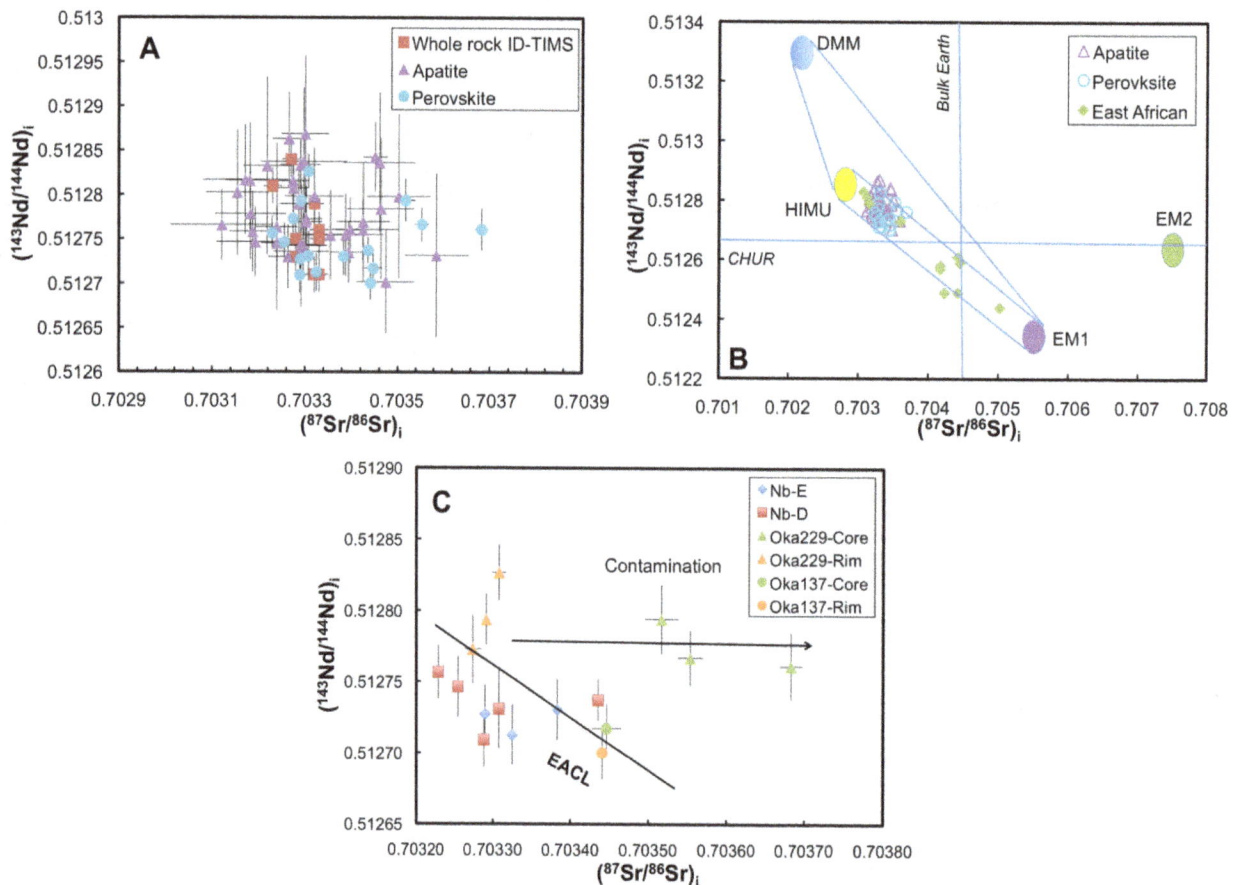

Table 8. *In-situ* Sr and Nd isotopic compositions for perovskite and apatite by LA-MC-ICP-MS.

Sample	Analysis	$(^{87}Sr/^{86}Sr)_i$	2σ	$^{143}Nd/^{144}Nd$	$(^{143}Nd/^{144}Nd)_i$	2σ
Oka4b	AP1	0.70327	0.00004	0.51283	0.51279	0.00004
	AP7	0.70326	0.00004	0.51277	0.51273	0.00004
	AP12	0.70330	0.00003	0.51281	0.51277	0.00004
Oka51	AP1	0.70329	0.00006	0.51279	0.51274	0.00006
	AP5	0.70330	0.00005	0.51292	0.51287	0.00009
	AP8	0.70343	0.00006	0.51282	0.51277	0.00007
	AP14	0.70319	0.00005	0.51281	0.51276	0.00005
Oka72	AP2	0.70349	0.00003	0.51275	0.51270	0.00018
	AP17	0.70346	0.00007	0.51283	0.51278	0.00008
Oka153	AP22	0.70345	0.00003	0.51277	0.51272	0.00003
Oka200a	AP6	0.70322	0.00005	0.51288	0.51283	0.00010
	AP12	0.70350	0.00008	0.51284	0.51275	0.00013
Oka206	AP1	0.70327	0.00003	0.51285	0.51281	0.00003
	AP13_1	0.70329	0.00004	0.51287	0.51283	0.00003
Oka21	AP10	0.70332	0.00006	0.51284	0.51280	0.00005
	AP11	0.70315	0.00007	0.51285	0.51280	0.00007
	AP14	0.70330	0.00009	0.51288	0.51284	0.00008
Oka31	AP1	0.70329	0.00004	0.51285	0.51281	0.00004
	AP5	0.70324	0.00003	0.51281	0.51276	0.00004
	AP7	0.70330	0.00003	0.51283	0.51279	0.00004
	AP11	0.70327	0.00003	0.51279	0.51275	0.00004
	AP12	0.70329	0.00004	0.51281	0.51277	0.00003
Oka89	AP2	0.70359	0.00007	0.51277	0.51273	0.00009
	AP3	0.70347	0.00006	0.51274	0.51270	0.00006
	AP9	0.70346	0.00008	0.51288	0.51284	0.00008
Oka138	AP4	0.70340	0.00003	0.51281	0.51276	0.00004
	AP6	0.70330	0.00005	0.51279	0.51274	0.00004
	AP7	0.70339	0.00004	0.51278	0.51274	0.00003
	AP13	0.70339	0.00008	0.51280	0.51275	0.00005
	AP14	0.70342	0.00008	0.51281	0.51276	0.00003
	AP19	0.70335	0.00006	0.51280	0.51275	0.00005
Oka229	AP7	0.70319	0.00008	0.51280	0.51275	0.00004
	AP18	0.70318	0.00008	0.51283	0.51278	0.00005
	AP19	0.70312	0.00011	0.51282	0.51277	0.00004
	AP20	0.70317	0.00009	0.51286	0.51282	0.00006
Oka75	AP8	0.70336	0.00004	0.51279	0.51272	0.00016
	AP12	0.70349	0.00005	0.51278	0.51271	0.00021
	AP13	0.70342	0.00005	0.51290	0.51283	0.00007
Oka132	AP8	0.70353	0.00007	0.51287	0.51280	0.00013
	AP9	0.70350	0.00007	0.51286	0.51280	0.00009
	AP10	0.70364	0.00008	0.51299	0.51293	0.00016
	AP18	0.70367	0.00006	0.51288	0.51282	0.00013

Table 8. *Cont.*

Sample	Analysis	$(^{87}Sr/^{86}Sr)_i$	2σ	$^{143}Nd/^{144}Nd$	$(^{143}Nd/^{144}Nd)_i$	2σ
Oka137	PV1_1	0.70345	0.00002	0.51277	0.51272	0.00002
	PV1_2	0.70344	0.00001	0.51275	0.51270	0.00002
	PV3_2	0.70338	0.00001	0.51277	0.51273	0.00002
	PV4_2	0.70332	0.00000	0.51276	0.51271	0.00002
	PV5	0.70329	0.00000	0.51277	0.51273	0.00002
	PV5_2	0.70344	0.00001	0.51278	0.51274	0.00001
Oka209	PV2_1	0.70323	0.00000	0.51281	0.51276	0.00002
	PV2_2	0.70326	0.00001	0.51280	0.51275	0.00002
	PV4	0.70329	0.00000	0.51276	0.51271	0.00002
	PV5	0.70331	0.00001	0.51279	0.51273	0.00003
Oka229	PV2_1	0.70329	0.00000	0.51284	0.51279	0.00002
	PV2_2	0.70352	0.00002	0.51285	0.51279	0.00002
	PV2_3	0.70355	0.00002	0.51282	0.51277	0.00002
	PV2_4	0.70368	0.00001	0.51281	0.51276	0.00002
	PV2_5	0.70327	0.00001	0.51282	0.51277	0.00002
	PV2_6	0.70331	0.00001	0.51287	0.51283	0.00002

Notes: AP = apatite; PV = perovskite.

4. Discussions

4.1. Timing of Magmatism at Oka

The timing of magmatism associated with the Oka carbonatite complex was previously investigated by apatite fission track, *in-situ* U-Pb dating for apatite, niocalite, and a single perovskite age determination [13–15]. All the data from these previous geochronological studies are indicative of a protracted petrogenetic history with a duration in the order of ~10–15 million years (given the associated uncertainties). However, this study is the first to report a thorough *in-situ* U-Pb geochronological investigation for apatite and perovskite from the associated silica-undersaturated rocks at Oka, *i.e.*, ijolite, alnöite, and jacupirangite.

Based on the geochronological data shown in Figure 11, it is clear that all the rock types display a protracted crystallization history (Figure 11A); however, their respective age distribution patterns vary slightly. U-Pb dating results for alnöite are shifted slightly towards the older ages with the majority falling between ~124 and ~135 Ma, and a minor peak at ~115 Ma (Figure 11H). In contrast, a majority of the U-Pb ages for carbonatite yield a younger age signature with the main peak at ~114 Ma, and a minor older peak at ~126 Ma (Figure 11E). The ages for okaite are more evenly distributed and vary between ~114 and ~127 Ma (Figure 11F). The U-Pb dating results for ijolite define the most distinctive bimodality with two peaks at ~114 and ~127 Ma (Figure 11G). Thus, the older age peak for perovskite at 135.4 ± 3.2 Ma and the younger niocalite age at 112.6 ± 1.2 Ma may define the absolute "maximum" duration of magmatism at Oka.

Overall, the magmatic history for Oka may be summarized as follows: (1) An early igneous event occurred at ~135 Ma, which corresponds to the main period of formation for the alkaline silicate rocks, in particular the alnöite and ijolite; (2) This was followed by the main period of emplacement for okaite between 120 and 127 Ma; and (3) Lastly, at ~114 Ma, emplacement of the vast majority of the carbonatite, along with okaite, ijolite, and a minor amount of alnöite occurred.

Oka is not the sole alkaline complex that is characterized by an extended formational history spanning millions of years. Several previous studies of carbonatite and kimberlite alkaline complexes also define a protracted history of magmatic activity up to 40 million years [36,59–67]. Of note, based on U-Pb ages for ~30 kimberlite complexes in North America, Heaman and Kjarsgaard [59] stated that discrete kimberlite emplacement events within individual fields can occur over time intervals of up to 20 Myrs. For example, the majority of the kimberlite complexes located within the region of Timiskaming, which is located ~1000 km northwest from the MIP, were emplaced between 155 and 134 Ma (*i.e.*, over ~20 Myrs period).

The protracted emplacement history (and ensuing melt differentiation) that occurred at Oka may be explained by invoking either one of two models: (1) Melt generation occurred at ~135 Ma, followed by magma differentiation in a closed-system over a period of ~10–15 million years; or (2) There was periodic generation of small volume, partial melts from a metasomatized, CO_2-bearing mantle source over a period ~10–15 million years, with each melt fraction undergoing an independent crystallization/differentiation path. Given the extremely large variations in trace element abundances recorded by apatite (Table 5) [14], and those depicted by perovskite investigated here (Figures 4–7; Tables 2 and 3), it is difficult if not impossible to attribute these variations to closed-system melt differentiation involving solely one parental melt, regardless of whether this melt was carbonatitic, or a carbonate-rich, alkaline, silica-undersaturated parental melt [14]. Chen and Simonetti [14] advocated for open-system behavior, possibly involving magma mixing, which is an interpretation also put forward by Zurevinski and Mitchell [68] to explain the chemical variations documented by pyrochlore from Oka. In this study, Figures 8 and 9 (and Tables 4 and 5) clearly indicate that the chemical compositions for apatite from alnöite and jacupirangite are distinct relative to those from other rock types. Their major and trace element contents and REE chondrite normalized patterns suggest derivation from a different mantle source. Of interest, Nb-E perovskites are only present in okaite and some represent the rim of zoned perovskite grains (Figure 5). The latter texture has been described as reverse zoning (*i.e.*, an increase of REE and Th contents from core to rim) [53], which is uncommon and possibly results from re-equilibration of perovskite with magma modified by assimilation or contamination processes, or later surrounded by a melt of different composition [53]. Thus, based on the combined chemical and geochronological data obtained for all rock types at Oka, we believe that the second hypothesis involving periodic generation of small volume melts and subsequent magma mixing best explains the petrogenetic history of the complex.

4.2. Relationship between Oka, Monteregian Igneous Province (MIP)-Related Intrusions, and Mantle Plumes?

There exist two competing hypotheses for the formation of Oka and the associated MIP-related intrusions in southeastern Québec (Figure 1A). One model proposes that they formed as the result of intraplate melting in an extensional setting associated with opening of the Atlantic Ocean [69,70]. The alternative view is that the MIP results from the passage of the North American plate over the Great Meteor hotspot [59,71–74]. The main criticism with the latter is the lack of a precise correlation between the radiometric ages of the MIP-related intrusions and lithospheric plate migration (*i.e.*, geographic position). However, the majority of the geochronological data for the MIP-related intrusions were obtained either by apatite fission-track or K-Ar methods, and only a small number of analyses were conducted for each intrusion. Thus, a more thorough and robust geochronological evaluation is required for each of the MIP intrusions before the plume hypothesis is completely ruled out. Moreover, the results from this study and Chen and Simonetti [14] both report ages for Oka that overlap the entire MIP age range, which further complicate matters in relation to evaluating a temporal relationship for the MIP intrusions relative to a plume hypothesis.

Carbonatites can provide valuable information for deciphering the geochemical nature of the upper mantle as their isotopic ratios inherited from their source region are buffered against crustal contamination due to their extremely high concentrations of incompatible elements (e.g., Sr and Nd). For example, in their study of the carbonatites and associated Si-undersaturated rocks from the Chilwa Island carbonatite complex, Simonetti and Bell [75] clearly indicate that an unrealistic amount of crustal assimilation is needed in order to explain the variable Nd and Sr isotope ratios. Hence, they advocated for melt derivation from a chemically and isotopically heterogeneous (metasomatized) mantle source region. In this study, the Nd and Sr isotope data for both apatite and perovskite overlap those previously obtained for whole rock samples from Oka (Figure 12) [12], but the former are clearly much more variable. This simply reflects the fact that whole rock analyses represent a weighted average of the Sr and Nd isotope composition of the (Sr- and Nd-bearing) constituent minerals (e.g., apatite, calcite, perovskite, and niocalite), and mask subtle differences between phases; however, the latter provide important details for deciphering the petrogenetic history of a complex. Evidence for "open-system" behavior at Oka was already evident from the TIMS generated whole rock data as these define a range of Sr and Nd isotope values that are well outside the typical in-run analytical precision (Figure 12). In Figure 12B, the Nd and Sr isotope compositions for apatite and perovskite from Oka are compared to those for well-established mantle components (*i.e.*, HIMU, EMI, EMII (enriched mantle 2), and DMM (depleted mid-ocean ridge basaltic mantle)) [76] and East African carbonatite complexes [58]. The Nd and Sr isotope data from Oka plot proximal to the field for the HIMU mantle component and most lie along a HIMU-EMI mixing array. Both HIMU and EMI are prevalent mantle components that underlie most of East Africa and also characterize the isotope compositions of ocean island basalts (OIBs) worldwide. Several previous investigations have advocated for the involvement of HIMU, EMI, and FOZO (Focus Zone) mantle components in the generation of most young (<200 Ma) carbonatites on a global scale [77–81]. On the basis of a compilation of both radiogenic and stable isotopic data from carbonatites worldwide, Bell and Simonetti [82] made the argument that parental carbonatitic magmas are derived from a sub-lithospheric source that is associated with either asthenospheric "upwellings" or more deep-seated, plume-related activity. Amongst the important evidences that support the generation of carbonated melts from sub-lithospheric mantle are: the petrogenetic and temporal association of carbonatites with large igneous provinces (LIPs; e.g., Deccan, Parana), carbonatites with primitive noble gas isotopic signatures, and their radiogenic isotope ratios similar to OIBs.

Numerous previous studies have advocated for a direct link between carbonatite melt generation and mantle plumes [74,78,82,83]. As pointed out by Rukhlov and Bell [9], the presence of carbonatites may mark the initiation of mantle-generated magmatism because of the very fluid nature of carbonatitic melts, and the fact that they are produced by low degrees of partial melting (*i.e.*, precursors to basaltic activity, and perhaps are associated with changes in mantle dynamics). In relation to the southeastern region of Québec and location of the MIP intrusions, tomographic data clearly indicates the presence of a low-velocity anomaly in the upper mantle region beneath the Ottawa-Bonnechere Rift [84]. This anomaly is further interpreted to extend over a broad region at a depth of ~200 km beneath the Great Lakes, where lithosphere was partially breached by the Great Meteor plume [85]. Hence, in relation to the MIP-related magmatism, we propose that carbonatite-like melts and volatile-bearing fluids first metasomatized the upper mantle at ~135 Ma, which gave rise to the older alkaline silicate rocks at Oka (e.g., alnöite). Subsequently, based on the limited geochronological data for the remaining MIP-related intrusions (Table 9) [16,86], the slightly undersaturated to critically saturated complexes of Mounts Royal, Bruno, Rougemont, Yamaska, Shefford, and Brome were emplaced between ~135 Ma and ~128 Ma. The last magmatism to occur involved generation of the moderately-to-strongly undersaturated silicate melts at Royal, Johnson, Yamaska, Shefford, and Brome that occurred ~117 Ma.

Table 9. Ages for Monteregian Intrusions.

Intrusion	Rock type/phase	Mineral dated	Method	Age	2σ
Royal	Nepheline diorite	Sphene	Fission track	117	3
	Leucogabbro	Apatite	Fission track	138	6
	Pyroxenite	Apatite	Fission track	134	9
Bruno	Gabbro	Apatite	Fission track	135	11
	Pyroxenite	Apatite	Fission track	135	11
Johnson	Essexite	Apatite	Fission track	117	9
	Pulaskite	Apatite	Fission track	120	8
Rougemont	Gabbro	Apatite	Fission track	136	10
	Pyroxenite	Apatite	Fission track	138	11
Yamaska	Essexite	Apatite	Fission track	119	8
	Nepheline syenite	Sphene	Fission track	120	10
	Younger pyroxenite	Apatite	Fission track	132	10
	Older pyroxenite	Apatite	Fission track	140	9
	Gabbro	Apatite	Fission track	140	11
	Gabbro	Apatite	Fission track	141	9
Shefford	Nordmarkite	Whole-rock isochron	Rb-Sr	120.3	1
	Pulaskite	Whole-rock isochron	Rb-Sr	128.5	3
	Nepheline diorite	Apatite	Fission track	119	8
	Diorite	Apatite	Fission track	131	8
Brome	Nepheline diorite	Whole-rock isochron	Rb-Sr	118.4	2.2
	Pulaskite	Whole-rock isochron	Rb-Sr	136.2	1.7
	Nepheline diorite	Apatite	Fission track	117	9
	Gabbro	Apatite	Fission track	139	13

4.3. Chemical Zoning of Perovskite

Two perovskite grains (from a total of 25) investigated here display reverse zoning, with Nb and REE abundances that are enriched in the rim relative to their respective central regions of the crystals; the latter may have undergone Pb loss since these are characterized by younger ages resulting from higher U abundances relative to the rim (Figure 5). Minerals yielding younger ages within any given sample should always be carefully examined for Pb loss or U addition, especially since the Nb-D and Nb-E groups of perovskite have different trace element compositions (e.g., Pb, Th; Figure 7 and Tables 2 and 3). Figure 13A plots U-Pb ages against their respective U abundances for perovskite and it is possible that the young ages corresponding to the higher U contents in samples Oka229, Oka73, and Oka137 can be attributed to Pb loss. The reason being that perovskite with higher U abundances (~200 ppm in this case) will undergo a higher amount of alpha decay from the radioactive disintegration of U, which consequently damages the crystal structure; this then enhances the possibility of losing loosely bound radiogenic Pb. In contrast, perovskite from okaite samples Oka208 and Oka209 yield relatively young ages of 112.2 ± 1.9 Ma (Figure 10E) and 116.4 ± 2.9 Ma, respectively, with U contents comparable to those for the older perovskite (Figure 13A). A previous study of perovskite attributed the higher abundances of incompatible elements at grain boundaries to secondary processes; e.g., alteration in intergranular regions as a result of interaction with a melt, an aqueous fluid, or a gas phase [53]. As stated above, the core areas of two large perovskite grains may have undergone a Pb loss event, and these are characterized by more radiogenic Sr (and comparable Nd) isotopic ratios compared to the remaining perovskite (Figure 12C). In addition, the core regions are marked by distinct chemical compositions (i.e., lower Nb/Zr; Figure 14A,B). Hence, the distinct, more radiogenic $^{87}Sr/^{86}Sr$ isotope compositions for the cores of these two perovskite grains may be attributed to either crystallization from a melt derived from a distinct mantle source, or perturbation by a contamination/alteration process (Figures 12C and 14). Thus, a possible formational history for the reversely zoned perovskite grains is as follows: (1) the cores formed from a first batch of magma; (2) this was followed by the influx of a new (distinct) batch of magma with lower $^{87}Sr/^{86}Sr$ ratio (relative to the cores), which resulted in the crystallization of the rims; and (3) the latter was associated with an "autometasomatic" event in which fluids scavenged the Nb (and certain other trace elements) from the core towards the rim. Obviously, this process was not widespread since only two of the perovskite grains investigated here exhibit this anomalous, reversely zoned texture. It is possible that this "autometasomatic" activity produced the vast majority of the late-stage pyrochlore and/or niocalite resulting in the Nb ore deposits at Oka. Chen et al. [15] discussed the issue of late-stage replacement of niocalite by pyrochlore (or vice versa) at Oka. In addition, Samson et al. [87] also advocated for the occurrence of late-stage hydrothermal activity at Oka as recorded by fluid inclusions within constituent minerals.

The highly variable chemical compositions, Nd and Sr isotope ratios, and ages documented for perovskite, pyrochlore, niocalite, and apatite from the different rock types associated with the Oka carbonatite complex indicate that these formed as a result of episodic, small volume partial melting and subsequent magma mixing [14,15,68]. However, Figure 13B shows that there is a positive correlation between the total REE contents and U-Pb ages for the perovskite grains investigated here (excluding the two reversely zoned grains). Thus, a possible interpretation is that the Nb-E perovskite formed first within a melt produced at ~135 Ma, and their enriched geochemical nature is the result of low-degree partial melting of a metasomatized (carbonated) mantle source region. The Nb-D perovskite formed later at ~114 Ma from a less-enriched magma and possibly reflects derivation from a more depleted (less metasomatized) mantle source region.

Figure 13. (A) Plots illustrating the chemical and geochronological data for perovskite: **(A)** U abundances *vs.* $^{206}Pb/^{238}U$ ages; and **(B)** REE contents *vs.* $^{206}Pb/^{238}U$ ages.

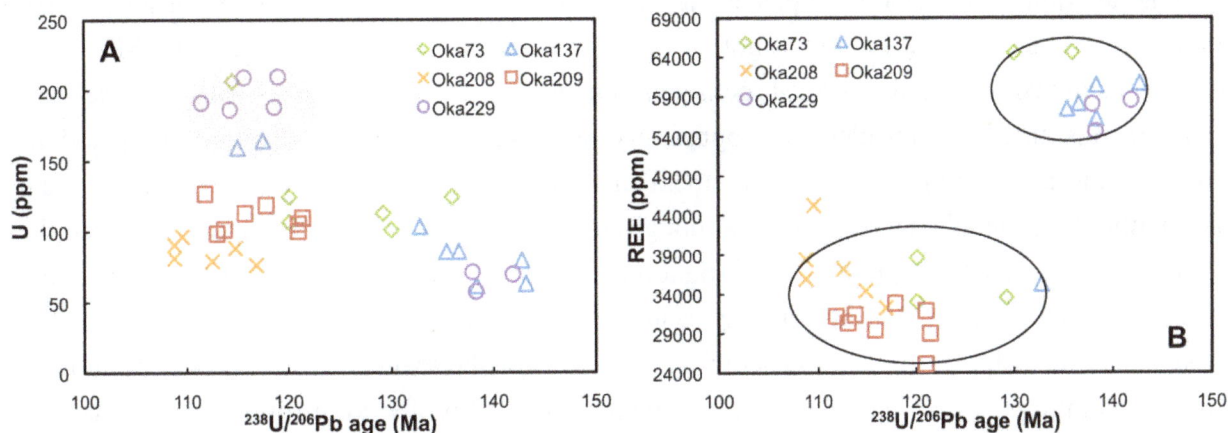

Figure 14. (A) Diagram of $^{87}Sr/^{86}Sr$ *vs.* Nb/Zr illustrating the different groups of perovskite; and **(B)** Plot of $^{143}Nd/^{144}Nd$ *vs.* Nb/Zr for the different perovskite groups.

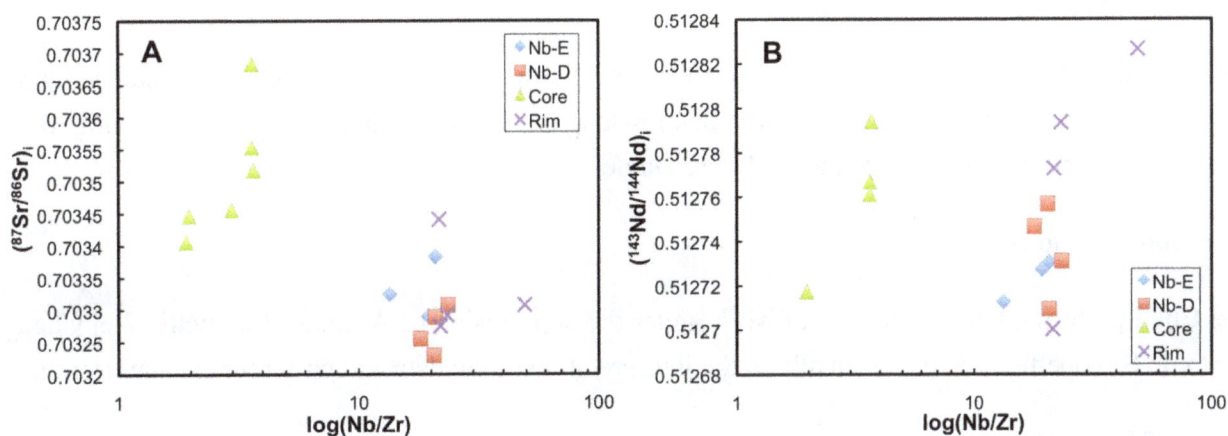

5. Conclusions

This study reports combined geochemical, isotopic, and geochronological data for both perovskite and apatite from the Oka carbonatite complex, and clearly demonstrates that a more detailed petrogenetic history can be deciphered for complexly zoned igneous centers. Of importance, the U-Pb results from this study indicate the need for conducting a thorough geochronological investigation rather than defining the age of any one alkaline intrusive complex solely on the basis of a single or small number of radiometric age determinations.

The combined chemical, isotopic and geochronological data for apatite suggest that its crystallization occurred during the entire magmatic history of the complex. Moreover, the lack of any significant correlations between geochemical and geochronological results for apatite and niocalite indicate a complicated petrogenetic history involving magma mixing [14,15]. On the basis of correlations between chemical compositions and U-Pb ages for the perovskite investigated in this study, these formed during two main episodes of melt generation.

The geochronological results from this study offer valuable insights into the emplacement relationships between carbonatite, okaite, ijolite and alnöite at Oka. It is proposed that carbonatite-like melts/fluids were the first to emanate from an enriched, volatile-bearing mantle plume, and these interacted with the overlying lithosphere; ensuing alkaline silicate melts were formed and generated the first alnöite, ijolite and okaite emplaced at ~135 Ma. Periodic, small volume partial melting subsequently continued, with a second major pulse of magmatism that occurred at ~114 Ma. Later generation melts mixed with earlier-formed rocks and minerals so as to yield samples with multiple-aged accessory minerals; these are considered as cognate crystals [14]. The Sr and Nd isotopic compositions for perovskite and apatite indicate the involvement of at least two mantle endmembers, HIMU- and EMI-like, within their mantle source region, although dominated by the former component. Given the results reported here and from previous investigations on Oka [14,15], it is difficult to assign either component to a mantle region; *i.e.*, lithosphere *vs.* asthenosphere (or plume). Alternatively, the mantle plume itself may be isotopically heterogeneous as proposed for the magmatic/tectonic regime for the East African alkaline province [58]. Regardless of which model is preferred, infiltration/refertilization of the lithosphere by enriched, volatile-bearing melts/fluids from a plume component will "swamp" the geochemical and isotopic composition of the overlying lithosphere [82].

Acknowledgments

We thank Ian Steele, University of Chicago Electron Microprobe Laboratory, for his assistance with EMP data collection. Wei Chen gratefully acknowledges receiving financial support during her doctoral dissertation from the University of Notre Dame.

Author Contributions

Wei Chen performed all the analytical work under the supervision of Antonio Simonetti. Wei Chen and Antonio Simonetti participated equally in the interpretation of the data and co-wrote the manuscript.

Conflicts of Interest

The authors declare no conflict of interest.

References

1.   Woolley, A.R.; Kjarsgaard, B.A. *Carbonatite Occurrences of the World: Map and Database*; Open File 5796; Natural Resources Canada: Sherbrooke, QC, Canada, 2008.

2.   Bizzarro, M.; Simonetti, A.; Stevenson, R.K.; David, J. Hf isotope evidence for a hidden mantle reservoir. *Geology* **2002**, *30*, 771–774.

3.   De Moor, J.M.; Fischer, T.P.; King, P.L.; Botcharnikov, R.E.; Hervig, R.L.; Hilton, D.R.; Barry, P.H.; Mangasini, F.; Ramirez, C. Volatile-rich silicate melts from Oldoinyo Lengai volcano (Tanzania): Implications for carbonatite genesis and eruptive behavior. *Earth Planet. Sci. Lett.* **2013**, *361*, 379–390.

4.   Bell, K.; Dawson, J.B. Nd and Sr isotope systematics of the active carbonatite volcano, Oldoinyo Lengai. In *Carbonatite Volcanism: Oldoinyo Lengai and the Petrogenesis of Natrocarbonatites*; Bell, K., Keller, J., Eds.; Springer: Berlin, Germany, 1995; pp. 100–112.

5. Dawson, J.B.; Pinkerton, H.; Norton, G.E.; Pyle, D.M.; Browning, P.; Jackson, D.; Fallick, A.E. Petrology and geochemistry of Oldoinyo Lengai lavas extruded in November 1988: Magma source, ascent and crystallization. In *Carbonatite Volcanism: Oldoinyo Lengai and the Petrogenesis of Natrocarbonatites*; Bell, K., Keller, J., Eds.; Springer: Berlin, Germany, 1995; pp. 47–69.

6. Peterson, T.D.; Kjarsgaard, B. What are the parental magmas at Oldoinyo Lengai? In *Carbonatite Volcanism: Oldoinyo Lengai and the Petrogenesis of Natrocarbonatites*; Bell, K., Keller, J., Eds.; Springer: Berlin, Germany, 1995; pp. 148–162.

7. Keller, J.; Spettel, B. The trace element composition and petrogenesis of natrocarbonatites. In *Carbonatite Volcanism: Oldoinyo Lengai and the Petrogenesis of Natrocarbonatites*; Bell, K., Keller, J., Eds.; Springer: Berlin, Germany, 1995; pp. 70–86.

8. Bell, K.; Blenkinsop, J.; Cole, T.J.S.; Menagh, D.P. Evidence from Sr isotopes for long-lived heterogeneities in the upper mantle. *Nature* **1982**, *298*, 251–253.

9. Rukhlov, A.S.; Bell, K. Geochronology of carbonatites from the Canadian and Baltic Shields, and the Canadian Cordillera: Clues to mantle evolution. *Mineral. Petrol.* **2010**, *98*, 11–54.

10. Gold, D.P.; Eby, G.N.; Bell, K.; Vallée, M. Carbonatites, diatremes and ultra-alkaline rocks in the Oka area, Quebec. In *Geological Association of Canada Guidebook*; Geological Association of Canada: St. John's, NL, Canada, 1986.

11. Shafiqullah, M.; Tupper, W.; Cole, T. K-Ar age of the carbonatite complex, Oka, Quebec. *Can. Mineral.* **1970**, *10*, 541–552.

12. Wen, J.; Bell, K.; Blenkinsop, J. Nd and Sr isotope systematics of the Oka complex, Québec, and their bearing on the evolution of the sub-continental upper mantle. *Contrib. Mineral. Petrol.* **1987**, *97*, 433–437.

13. Cox, R.A.; Wilton, D.H.C. U-Pb dating of perovskite by LA-ICP-MS: An example from the Oka carbonatite, Quebec, Canada. *Chem. Geol.* **2006**, *235*, 21–32.

14. Chen, W.; Simonetti, A. *In-situ* determination of major and trace elements in calcite and apatite, and U-Pb ages of apatite from the Oka carbonatite complex: Insights into a complex crystallization history. *Chem. Geol.* **2013**, *353*, 151–172.

15. Chen, W.; Simonetti, A.; Burns, P.C. A combined geochemical and geochronological investigation of niocalite from the Oka carbonatite complex, Canada. *Can. Mineral.* **2013**, *51*, 785–800.

16. Eby, G. Geochronology of the Monteregian Hills alkaline igneous province, Quebec. *Geology* **1984**, *12*, 468–470.

17. Foland, K.A.; Gilbbert, L.A.; Sebring, C.A.; Chen, J.-F. $^{40}Ar/^{39}Ar$ ages for plutons of the Monteregian Hills, Quebec: Evidence for a single episode of Cretaceous magmatism. *Geol. Soc. Am. Bull.* **1986**, *97*, 966–974.

18. Mitchell, R.H.; Chakhmouradian, A.R. Instability of peroskite in a CO_2-rich environment: Examples from carbonatite and kimberlite. *Can. Mineral.* **1998**, *36*, 939–951.

19. Tappe, S.; Simonetti, A. Combined U-Pb geochronology and Sr-Nd isotope analysis of the Ice River perovskite standard, with implications for kimberlite and alkaline rock petrogenesis. *Chem. Geol.* **2012**, *304–305*, 10–17.

20. Simonetti, A.; Heaman, L.M.; Chacko, T. Use of discrete-dynode secondary electron multipliers with Faradays—A "reduced volume" approach for *in-situ* U-Pb dating of accessory minerals within petrographic thin section by LA-MC-ICP-MS. *Mineral. Assoc. Can. Short Course Ser.* **2008**, *40*, 241–264.

21. Wu, F.; Yang, Y.; Mitchell, R.H.; Li, Q.; Yang, J.; Zhang, Y. *In situ* U-Pb age determination and Nd isotopic analysis of perovkites from kimberlites in southern Africa and Somerset Island, Canada. *Lithos* **2010**, *115*, 205–222.

22. Zhang, D.; Zhang, Z.; Santosh, M.; Cheng, Z.; Huang, H.; Kang, J. Perovskite and baddeleyite from kimberlitic intrusions in the Tarim large igneous province signal the onset of an end-Carboniferous mantle plume. *Earth Planet. Sci. Lett.* **2013**, *361*, 238–248.

23. Koster van Groos, A.F.; Wyllie, P.J. Liquid immiscibility in the join $CaAl_2Si_2O_8$-$NaAlSi_3O_8$-Na_2CO_3-H_2O. *Am. J. Sci.* **1973**, *273*, 465–487.

24. Kjarsgaard, B.A.; Hamilton, D.L. Liquid immiscibility and the origin of alkali-poor carbonatites. *Mineral. Mag.* **1988**, *52*, 43–55.

25. Halama, R.; Vennemann, T.; Siebel, W.; Markl, G. The Grønnedal-Ika carbonatite-syenite complex, South Greenland: Carbonatite formation by liquid immiscibility. *J. Petrol.* **2005**, *46*, 191–217.

26. Brooker, R.A. The effect of CO_2 saturation on immiscibility between silicate and carbonate liquids: An experimental study. *J. Petrol.* **1998**, *39*, 1905–1915.

27. Lee, W.; Wyllie, P.J. Experimental data bearing on liquid immiscibility, crystal fractionation, and the origin of calciocarbonatite and natrocarbonatites. *Int. Geol. Rev.* **1994**, *36*, 797–819.

28. Korobeinikov, A.N.; Mitrofanov, F.P.; Gehör, S.; Laajoki, K.; Pavlov, V.P.; Mamontov, V.P. Geology and copper sulphide mineralization of the Salmagorskii ring igneous province, Kola Peninsula, NW Russia. *J. Petrol.* **1998**, *39*, 2033–2041.

29. Veksler, I.V.; Nielsen, T.F.D.; Sokolov, S.V. Mineralogy of crystallized melt inclusions from Gardiner and Kovdor ultramafic alkaline complexes: Implications for carbonatite genesis. *J. Petrol.* **1998**, *39*, 2015–2031.

30. Harmer, R.E.; Gittins, J. The Case for primary, mantle-derived carbonatite magma. *J. Petrol.* **1998**, *39*, 1895–1903.

31. Dalton, J.A.; Presnall, D.C. The continuum of primary carbonatitic-kimberlitic melt compositions in equilibrium with lherzolite: Data from the system CaO-MgO-Al_2O_3-SiO_2-CO_2 at 6 GPa. *J. Petrol.* **1998**, *39*, 1953–1964.

32. Bell, K.; Kjarsgaard, B.; Simonetti, A. Carbonatites-into the twenty-first century. *J. Petrol.* **1998**, *39*, 1839–1845.

33. Roulleau, E.; Pinti, D.L.; Stevenson, R.K.; Takahata, N.; Sano, Y.; Pitre, F. N, Ar and Pb isotopic co-variations in magmatic minerals: Discriminating fractionation processes from magmatic sources in Monteregian Hills, Québec, Canada. *Chem. Geol.* **2012**, *326–327*, 123–131.

34. Eby, G. Monteregian Hills I. Petrology, major and trace element geochemistry, and strontium isotopic chemistry of the western intrusions: Mounts Royal, St. Bruno, and Johnson. *J. Petrol.* **1984**, *25*, 421–452.

35. Pearce, N.J.G.; Perkins, W.T.; Westgate, J.A.; Gorton, M.P.; Jackson, S.E.; Neal, C.R.; Chenery, S.P. A compilation of new and published major and trace element data for NIST SRM 610 and NIST SRM 612 glass reference materials. *Geostand. Newsl.* **1997**, *21*, 115–144.

36. Van Acherbergh, E.; Ryan, C.G.; Jackson, S.E.; Griffin, W. Data reduction software for LA-ICP-MS. In *Laser Ablation-ICPMS in the Earth Science*; Sylvester, P., Ed.; Mineralogical Association of Canada: Quebec, QC, Canada, 2001; Volume 29, pp. 239–243.

37. Simonetti, A.; Neal, C.R. *In-situ* chemical, U-Pb dating, and Hf isotope investigation of megacrystic zircons, Malaita (Solomon Islands): Evidence for multi-stage alkaline magmatic activity beneath the Ontong Java Plateau. *Earth Planet. Sci. Lett.* **2010**, *295*, 251–261.

38. Thomson, S.N.; Gehrels, G.E.; Cecil, R.; Ruiz, J. Exploring routine laser ablation multicollector ICP-MS U-Pb dating of apatite. In Proceedings of the American Geophysical Union (AGU) Fall Meeting, San Francisco, CA, USA, 14–18 December 2009.

39. Storey, C.D.; Jeffries, T.E.; Smith, M. Common lead-corrected laser ablation ICP-MS U-Pb systematics and geochronology of titanite. *Chem. Geol.* **2006**, *227*, 37–52.

40. Simonetti, A.; Heaman, L.M.; Chacko, T.; Banerjee, N. *In-situ* petrographic thin section U-Pb dating of zircon, monazite, and titanite using laser ablation-MC-ICP-MS. *Int. J. Mass Spectrom.* **2006**, *253*, 87–97.

41. Banerjee, N.R.; Simonetti, A.; Furnes, H.; Muehlenbachs, K.; Staudigel, H.; Heaman, L.; van Kranendonk, M.J. Direct dating of Archean microbial ichnofossils. *Geology* **2007**, *35*, 487–490.

42. Grünenfelder, M.H.; Tilton, G.R.; Bell, K.; Blenkinsop, J. Lead and strontium isotope relationships in the Oka carbonatite complex, Quebec. *Geochim. Cosmochim. Acta* **1986**, *50*, 461–468.

43. Chew, D.M.; Sylvester, P.J.; Tubrett, M.N. U-Pb and Th-Pb dating of apatite by LA-ICPMS. *Chem. Geol.* **2011**, *280*, 200–216.

44. Horstwood, M.S.A.; Foster, G.L.; Parrish, R.R.; Noble, S.R.; Nowell, G.M. Common-Pb corrected *in situ* U-Pb accessory mineral geochronology by LA-MC-ICP-MS. *J. Anal. At. Spectrom.* **2003**, *18*, 837–846.

45. Simonetti, A.; Heaman, L. U-Pb zircon dating by laser ablation-MC-ICP-MS using a new multiple ion counting Faraday collector array. *J. Anal. At. Spectrom.* **2005**, *20*, 677–686.

46. Ludwig, K.R. *User's Manual for Isoplot 3.00: A Geochronological Toolkit for Microsoft Excel*; Berkeley Geochronology Center: Berkeley, CA, USA, 2003.

47. Paton, C.; Woodhead, J.D.; Hergt, J.M.; Phillips, D.; Shee, S. Strontium isotope analysis of groundmass perovskite via LA-MC-ICP-MS. *Geostand. Geoanal. Res.* **2007**, *31*, 321–330.

48. Ramos, F.C.; Wolf, J.A.; Tollstrup, D.L. Measuring $^{87}Sr/^{86}Sr$ variations in minerals and groundmass from basalts using LA-MC-ICP-MS. *Chem. Geol.* **2004**, *211*, 135–158.

49. Bizzarro, M.; Simonetti, A.; Stevenson, R.K.; Kurszlaukis, S. *In situ* $^{87}Sr/^{86}Sr$ investigation of igneous apatites and carbonates using laser ablation MC-ICP-MS. *Geochim. Cosmochim. Acta* **2003**, *67*, 289–302.

50. Wasserburg, G.J.; Jacobsen, S.B.; DePaolo, D.J.; McCulloch, M.T.; Wen, T. Precise determination of Sm/Nd ratios, Sm and Nd isotopic abundances in standard solutions. *Geochim. Cosmochim. Acta* **1981**, *45*, 2311–2323.

51. Yang, Y.H.; Zhang, H.F.; Xie, L.W.; Wu, F. Accurate measurement of neodymium isotopic composition using Neptune multiple collector inductively coupled plasma mass spectrometry. *Chin. J. Anal. Chem.* **2007**, *35*, 71–74.

52. McFarlane, C.; McCulloch, M. Sm-Nd and Sr isotope systematics in LREE-rich accessory minerals using LA-MC-ICP-MS. *V.M. Goldschmidt Laser Ablation Short Course Vol.* **2008**, *40*, 117–134.

53. Chakhmouradian, A.R.; Mitchell, R.H. Occurrence, alteration patterns and compositional variation of perovskite in kimberlites. *Can. Mineral.* **2000**, *38*, 975–994.

54. Chakhmouradian, A.R.; Mitchell, R.H. Three compositional varieties of perovskite from kimberlites of the Lac de Gras field (Northwest Territories, Canada). *Mineral. Mag.* **2001**, *65*, 133–148.

55. McDonough, X.F.; Sun, S. The composition of the Earth. *Chem. Geol.* **1995**, *120*, 223–253.

56. Wendt, I.; Carl, C. The statistical distribution of the mean squared weighted deviation. *Chem. Geol.* **1991**, *35*, 696–698.

57. Vermeesch, P. On the visualisation of detrital age distributions. *Chem. Geol.* **2012**, *312–313*, 190–194.

58. Bell, K.; Blenkinsop, J. Nd and Sr isotopic compositions of East African carbonatites: Implications for mantle heterogeneity. *Geology* **1987**, *15*, 99–102.

59. Heaman, L.M.; Kjarsgaard, B.A. Timing of eastern North American kimberlite magmatism: Continental extension of the Great Meteor hotspot track? *Earth Planet. Sci. Lett.* **2000**, *178*, 253–268.

60. Secher, K.; Heaman, L.M.; Nielsen, T.F.D.; Jensen, S.M.; Schjøth, F.; Creaser, R.A. Timing of kimberlite, carbonatite, and ultramafic lamprophyre emplacement in the alkaline province located 64°–67° N in southern West Greenland. *Lithos* **2009**, *112*, 400–406.

61. Tappe, S. Genesis of Ultramafic Lamprophyres and Carbonatites at Aillik Bay, Labrador: A Consequence of Incipient Lithospheric Thinning beneath the North Atlantic Craton. *J. Petrol.* **2006**, *47*, 1261–1315.

62. Tappe, S.; Pearson, D.G.; Nowell, G.; Nielsen, T.; Milstead, P.; Muehlenbachs, K. A fresh isotopic look at Greenland kimberlites: Cratonic mantle lithosphere imprint on deep source signal. *Earth Planet. Sci. Lett.* **2011**, *305*, 235–248.

63. Tappe, S.; Steenfelt, A.; Heaman, L.M.; Simonetti, A. The newly discovered Jurassic Tikiusaaq carbonatite-aillikite occurrence, West Greenland, and some remarks on carbonatite–kimberlite relationships. *Lithos* **2009**, *112*, 385–399.

64. Smith, C. Pb, Sr and Nd isotope evidence for sources of southern African Cretaceous kimberlites. *Nature* **1983**, *304*, 51–54.

65. Zartman, R.E.; Richardson, S.H. Evidence from kimberlitic zircon for a decreasing mantle Th/U since the Ardean. *Chem. Geol.* **2005**, *220*, 263–283.

66. Zurevinski, S.E.; Heaman, L.M.; Creaser, R.A. The origin of Triassic/Jurassic kimberlite magmatism, Canada: Two mantle sources revealed from the Sr-Nd isotopic composition of groundmass perovskite. *Geochem. Geophys. Geosyst.* **2011**, *12*, doi:10.1029/2011GC003659.

67. Zurevinski, S.E.; Heaman, L.M.; Creaser, R.A.; Strand, P. The Churchill kimberlite field, NU, Canada: Petrography, mineral chemistry and geochronology. *Can. J. Earth Sci.* **2008**, *45*, 1039–1059.

68. Zurevinski, S.E.; Mitchell, R.H. Extreme compositional variation of pyrochlore-group minerals at the Oka Carbonatite complex, Quebec: Evidence of magma mixing? *Can. Mineral.* **2004**, *42*, 1159–1168.

69. McHone, J.G. Constraints on the mantle plume model for Mesozoic alkaline intrusions in northeastern North America. *Can. Mineral.* **1996**, *34*, 325–334.

70. Faure, S.; Tremblay, A.; Angelier, J. State of intraplate stress and tectonism of northeastern America since Cretaceous times, with particular emphasis on the New-England-Quebec igneous province. *Tectonophysics* **1986**, *255*, 111–134.

71. Crough, S. Mesozoic hotspot epeirogeny in eastern North America. *Geology* **1981**, *9*, 2–6.

72. Duncan, R.A. Age progressive volcanism in the New England seamounts and the opening of the central Atlantic Ocean. *J. Geophys. Res.* **1984**, *89*, 9980–9990.

73. Sleep, N.H. Monteregian hotspot track: A long-lived mantle plume. *J. Geophys. Res.* **1990**, *95*, 21983–21990.

74. Burke, K.; Khan, S.D.; Mart, R.W. Grenville province and Monteregian carbonatite and nepheline syenite distribution related to rifting, collision, and plume passage. *Geology* **2008**, *36*, 983–986.

75. Simonetti, A.; Bell, K. Nd, Pb and Sr isotopic data from the Napak carbonatite-nephelinite centre, eastern Uganda: An example of open-system crystal fractionation. *Contrib. Mineral. Petrol.* **1994**, *115*, 356–366.

76. Hart, S.R.; Gerlach, D.C.; White, W.M. A possible new Sr-Nd-Pb mantle array and consequences for mantle mixing. *Geochim. Cosmochim. Acta* **1986**, *50*, 1551–1557.

77. Bell, K.; Tilton, G. Probing the mantle: The story from carbonatites. *Eos Trans. Am. Geophys. Union* **2002**, *83*, 273–277.

78. Bell, K.; Simonetti, A. Carbonatite magmatism and plume activity: Implications from the Nd, Pb and Sr isotope systematics of Oldoinyo Lengai. *J. Petrol.* **1996**, *37*, 1321–1329.

79. Simonetti, A.; Bell, K.; Viladkar, S.G. Isotopic data from the Amba Dongar carbonatite complex, West-central India: Evidence for an enriched mantle source. *Chem. Geol.* **1995**, *122*, 185–198.

80. Simonetti, A.; Goldstein, S.L.; Schmidberger, S.S.; Viladkar, S.G. Geochemical and Nd, Pb, and Sr isotope data of Deccan alkaline complexes—Inferences on mantle sources and plume-lithosphere interaction. *J. Petrol.* **1998**, *39*, 1847–1864.

81. Tilton, G.R.; Bell, K. Sr-Nd-Pb isotope relationships in late Archean carbonatites and alkaline complexes: Applications to the geochemical evolution of Archean mantle. *Geochim. Cosmochim. Acta* **1994**, *58*, 3145–3154.

82. Bell, K.; Simonetti, A. Source of parental melts to carbonatites-critical isotopic constraints. *Mineral. Petrol.* **2010**, *98*, 77–89.

83. Hoernle, K.; Tilton, G.; le Bas, M.J.; Duggen, S.; Garbe-Schönberg, D. Geochemistry of oceanic carbonatites compared with continental carbonatites: Mantle recycling of oceanic crustal carbonate. *Contrib. Mineral. Petrol.* **2002**, *142*, 520–542.

84. Rondenay, S.; Bostock, M.G.; Hearn, T.M.; White, D.J.; Ellis, R.M. Lithospheric assembly and modification of the SE Canadian Shield: Abitibi-Grenville teleseismic experiment. *J. Geophys. Res.* **2000**, *105*, 13735–13754.

85. Aktas, K.; Eaton, D.W. Upper-mantle velocity structure of the lower Great Lakes region. *Tectonophysics* **2006**, *420*, 267–281.

86. Eby, G.N. Age relations, chemistry, and petrogenesis of mafic alkaline dykes from the Monteregian Hills and younger White Mountain igneous provinces. *Can. J. Earth Sci.* **1985**, *22*, 1103–1111.

87. Samson, I.M.; Liu, W.; Williams-Jones, A.E. The nature of orthomagmatic hydrothermal fluids in the Oka carbonatite, Quebec, Canada: Evidence from fluid inclusions. *Geochim. Cosmochim. Acta* **1995**, *59*, 1963–1977.

Chemical Abrasion Applied to LA-ICP-MS U–Pb Zircon Geochronology

Quentin G. Crowley [1,*], **Kyle Heron** [1], **Nancy Riggs** [2], **Balz Kamber** [1], **David Chew** [1], **Brian McConnell** [3] **and Keith Benn** [4]

[1] Department of Geology, School of Natural Sciences, Trinity College, Dublin 2, Ireland;
E-Mails: heronky@tcd.ie (K.H.); kamberbs@tcd.ie (B.K.); chewd@tcd.ie (D.C.)

[2] School of Earth Sciences and Environmental Sustainability, Northern Arizona University, Flagstaff, AZ 86011, USA; E-Mail: nancy.riggs@nau.edu

[3] Geological Survey of Ireland, Beggars Bush, Dublin 4, Ireland; E-Mail: brian.mcconnell@gsi.ie

[4] Tasiast Mauritanie Limited SA, ZRA 741, BP 5051, Nouakchott, Mauritania, West Africa;
E-Mail: keith.benn@kinross.com

* Author to whom correspondence should be addressed; E-Mail: crowleyq@tcd.ie

Abstract: Zircon ($ZrSiO_4$) is the most commonly used mineral in U–Pb geochronology. Although it has proven to be a robust chronometer, it can suffer from Pb-loss or elevated common Pb, both of which impede precision and accuracy of age determinations. Chemical abrasion of zircon involves thermal annealing followed by relatively low temperature partial dissolution in HF acid. It was specifically developed to minimize or eliminate the effects of Pb-loss prior to analysis using Thermal Ionization Mass Spectrometry (TIMS). Here we test the application of chemical abrasion to Laser Ablation Inductively Coupled Plasma Mass Spectrometry (LA-ICP-MS) by analyzing zircons from both untreated and chemically abraded samples. Rates of ablation for high alpha-dose non-treated zircons are up to 25% faster than chemically abraded equivalents. Ablation of 91500 zircon reference material demonstrates a *ca.* 3% greater down-hole fractionation of $^{206}Pb/^{238}U$ for non-treated zircons. These disparities necessitate using chemical abrasion for both primary reference material and unknowns to avoid applying an incorrect laser induced fractionation correction. All treated samples display a marked increase in the degree of concordance and/or lowering of common Pb, thereby illustrating the effectiveness of chemical abrasion to LA-ICP-MS U–Pb zircon geochronology.

Keywords: U–Pb geochronology; zircon; chemical abrasion; Laser Ablation Inductively Coupled Plasma Mass Spectrometry (LA-ICP-MS); Pb-loss

1. Introduction

U–Pb geochronology is a widely used radiometric dating tool applied to geological time-scales from 800 thousand years to >4 billion years. This is made possible due to dual uranium parent isotopes of ^{235}U and ^{238}U decaying to ^{207}Pb and ^{206}Pb with half-lives of *ca.* 704 million years and *ca.* 4.468 billion years respectively. U–Pb geochronology was initially developed in the mid-1950s [1–3] and only made possible by earlier advances in mass spectrometry [4]. Thermal Ionization Mass Spectrometry (TIMS) is the method of choice for studies requiring the highest precision U–Pb data. Micro-beam techniques such as Secondary Ionization Mass Spectrometry (SIMS) and Laser Ablation Inductively Coupled Plasma Ionization Mass Spectrometry (LA-ICP-MS) are generally used when there is a requirement for a greater number of analyses at a lower analytical precision, or where a spatially resolved analysis is beneficial. Zircon ($ZrSiO_4$) is the most commonly used mineral in U–Pb geochronology and is found in a wide compositional range of igneous, metamorphic and sedimentary rocks. In terms of crystal structure and composition, zircon has a tetrahedral Si site and a distorted octahedral Zr site [5]. Hf, U, Th, Ti, rare earth elements, Y and Sc may enter the Zr site, but only relatively small amounts of Pb are accepted. As a result of this, U is compatible but Pb is incompatible in the zircon crystal structure, so that common (or initial) Pb is preferentially excluded at the time of zircon formation, whereas ^{206}Pb and ^{207}Pb can reside within zircon following *in-situ* radioactive decay from the parent U isotopes. The exact mode of incorporation and place of residence of radiogenic Pb in the zircon crystal structure is thought to be strongly dependent on composition, such as the presence or absence of phosphate or hydrous phases [6].

Zircon commonly occurs as an accessory phase following high-grade metamorphism and partial melting and demonstrates a high closure temperature for Pb (>900 °C [7,8]), yet paradoxically the U–Pb isotope system may be disturbed at low temperatures resulting in Pb-loss. Conditions resulting in zircon being susceptible to Pb-loss may arise from the crystal lattice being damaged by the emission of alpha particles and alpha recoil processes [9]. In this manner, high-U zircons or high-U domains within zircons may accumulate radiation damage and be susceptible to Pb-loss. This effect is particularly pronounced at low temperatures where there is insufficient thermal energy to repair (anneal) radiation damage to the crystal lattice. The seemingly contradictory behavior of Pb in zircon may be explained by a general difference in valence states for ^{204}Pb, compared to ^{206}Pb and ^{207}Pb; although uncertainty exists, it is possible that ^{204}Pb is divalent, whereas ^{206}Pb and ^{207}Pb may originally exist in tetravalent states at the site of production in the zircon crystal lattice [10,11]. Prevalence of tetravalent Pb in zircon may be enhanced by a strong oxidizing environment at alpha-recoil sites and would be maintained by the continuous loss of beta-particles [12]. Radiogenic Pb however, may subsequently be reduced during alteration processes, so divalent lead which previously existed in tetravalent form in the zircon crystal lattice may have a tendency to occupy sites affected by both accumulated radiation damage and alteration. Common Pb is incorporated into the zircon crystal lattice in small amounts, or present in micro-inclusions. Considering that divalent Pb is incompatible in zircon, any fluid-induced reduction of Pb^{4+} to Pb^{2+} may result in loss of Pb. Aside from Pb-loss being related to radiation damage, it can also arise following physical distortion of the crystal lattice structure during rock deformation processes (e.g., shearing). This phenomenon results in mobilization of Pb from distorted zircons and can occur at lower metamorphic grade than in undistorted counterparts [13]. The U–Pb isotope system in zircon may therefore be disturbed by a process of Pb-loss causing erroneous younger apparent ages.

Although it is possible to obtain meaningful radiometric ages from zircons that have suffered Pb-loss, a greater number of analyses are usually required to construct a statistically valid Discordia array and intercept ages. Additionally, these intercept ages are generally less precise than ages (e.g., a Concordia age or weighted mean $^{206}Pb/^{238}U$ age) calculated from a uniform population of zircons with no Pb-loss. In order to reduce uncertainties due to Pb-loss a method of "air-abrasion" was developed to physically abrade the exterior portions of zircons [14]. As the outer portions of zircons are more susceptible to Pb-loss due to either higher-U domains, and/or contact with alteration fluids, removal of these outer domains results in less discordant or concordant zircon analyses when compared with non-abraded samples. The amount of zircon removed during air-abrasion is primarily dependent on the duration of treatment, with more zircon being removed for longer abrasion periods. It is not suited to all grain morphologies (e.g., delicate acicular needles) and runs the risk of removing narrow rims which may be of a different age compared to the bulk of the zircon grain. The air abrasion method was widely adopted for over two decades as a zircon pretreatment for TIMS, where whole grains, or multi-grain fractions are subsequently dissolved and purified prior to U–Pb isotope analysis. Although air-abrasion is still used for combating the effects of Pb-loss in minerals other than zircon (e.g., titanite), it has largely been replaced by a newly developed technique termed chemical abrasion [15]. The chemical abrasion protocol was perfected following a series of similar, but less successful attempts [16–20], to chemically pre-treat zircons and lessen, or remove, the effects of Pb-loss. Chemical abrasion of zircon involves two steps: (1) thermally annealing grains between 850 and 1000 °C for 48 h; and (2) subjecting annealed grains to HF acid attack for 12 h at temperatures below which full dissolution normally takes place. This two-step process whereby alpha track damage is thermally annealed and higher-U domains are preferentially dissolved from the entire grain also has the effect of removing more soluble forms of Pb (i.e., Pb^{2+}).

Chemical abrasion has been adapted for incrementally dissolving complex single grains to produce high-precision TIMS ages not discernible using micro-beam techniques [21]. It has also been demonstrated to be effective for SIMS U–Pb zircon geochronology [22]. This study aims to investigate if chemical abrasion can be applied to LA-ICP-MS U–Pb zircon geochronology to reduce discordance. Although the technique has previously been applied to a small number of LA-ICP-MS U–Pb zircon studies [23–25], this particular application has not been rigorously assessed. We test a modified chemical abrasion technique by applying an annealing-leaching treatment to zircon reference materials and a range of zircons of unknown age. Comparisons of data are made between treated and un-treated zircons to assess any physical and chemical effects and to evaluate the method as a pretreatment utility prior to LA-ICP-MS analysis.

2. Experimental Section

2.1. Sample Description

Untreated and/or chemically abraded zircons were analyzed from five samples, comprising three natural reference materials 91500 [26], Plešovice [27] and Penglai [28] and two unknowns TSUPB021 and 10/225. Non-treated Plešovice was used as a secondary reference material for non-treated samples only and is not described below.

The 91500 zircon reference material is thought to have come from a syenite pegmatite near Kuehl Lake in Ontario, Canada. The original sample consisted of a single 238 g crystal, part of which was split into a large number of 200 μm to 2 mm fragments and made available to geochronology laboratories worldwide. 91500 zircon has an accepted U–Pb age of 1062.4 ± 0.4 million years (Ma) [26]. Secondary electron (SEM) and cathodoluminesence (CL) images of a chemically abraded 91500 zircon fragment (Figure 1a) show pitting in the grain interior, resulting from etching by HF acid. Note that this pitting is not present in untreated 91500 zircon reference material [26]. All zircons were polished down to reveal grain interiors after chemical abrasion; this signifies that HF acid vapour penetrated into central portions of even large (*ca.* 150 μm diameter) grains.

Figure 1. Secondary electron (left) and cathodoluminesence (right) images of representative zircons from: (**a**) 91500-CA; (**b**) Penglai-CA; (**c**) TSUPB021; (**d**) TSUPB021-CA; (**e**) 10/225; and (**f**) 10/225-CA. Note that "-CA" indicates zircons have been subjected to chemical abrasion.

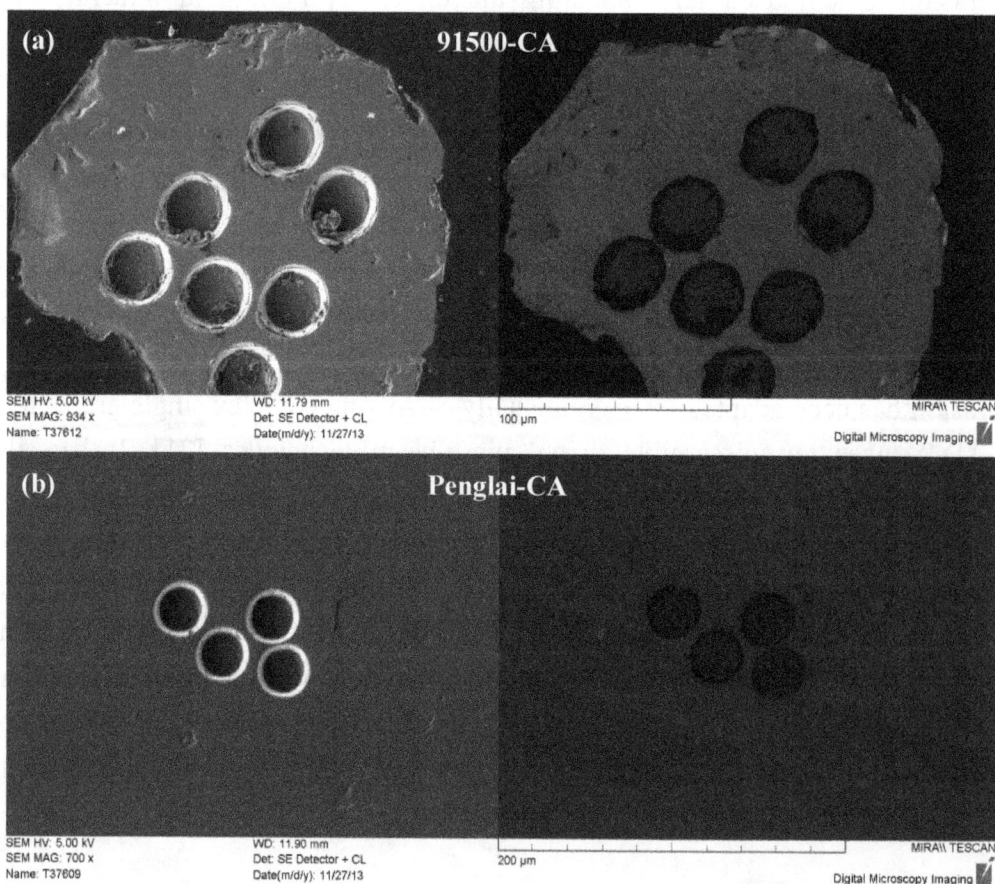

Figure 1. *Cont.*

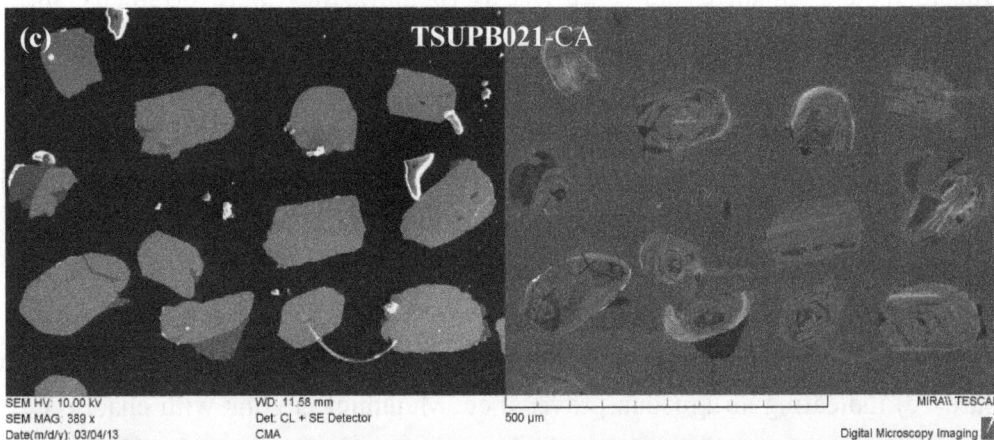

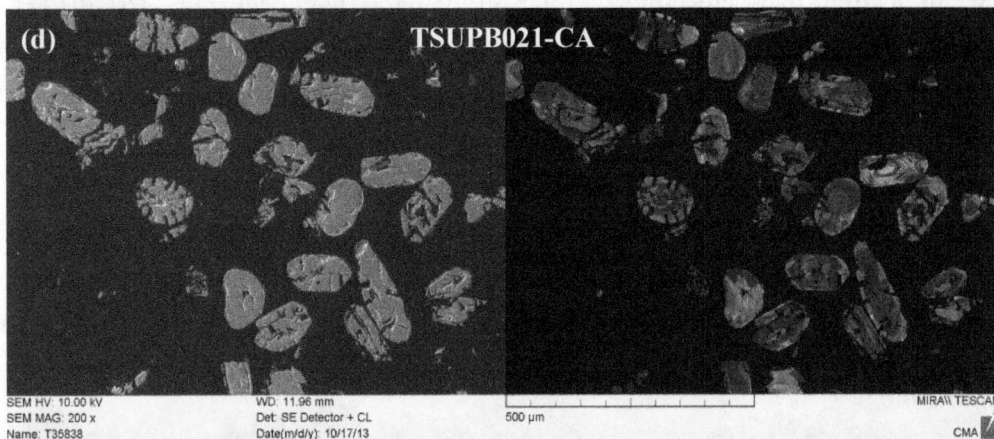

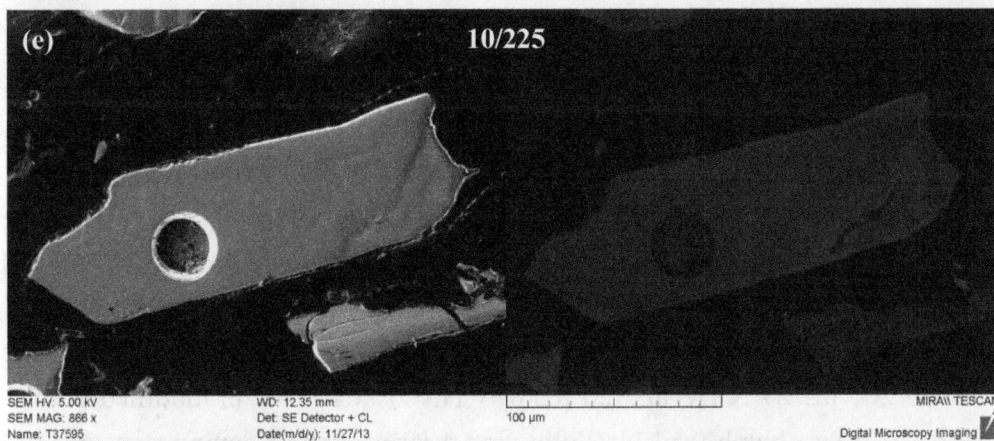

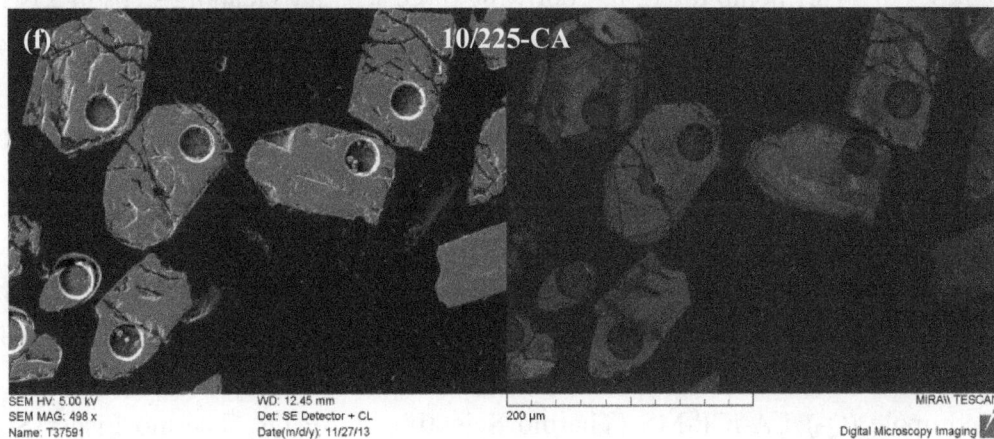

Penglai zircons are sourced from early Pliocene alkaline basalts from the northern Hainan Island of South China. Due to their abundance and large size (>10 mm), they were originally suggested as reference material for micro-beam analysis of Hf–O isotopes and for U–Pb geochronology [28]. Penglai zircons however, define a linear discordant array on a Tera-Wasserburg diagram with a "preferred" $^{206}Pb/^{238}U$ age of 4.4 ± 0.1 Ma [28]. This discordance makes Penglai unsuitable for use as a primary reference material for U–Pb geochronology, unless a common Pb correction is undertaken [29]. As with 91500, slight pitting due to chemical abrasion is evident in the SEM image. Oscillatory zoning is obvious in CL (Figure 1b), consistent with an igneous origin.

TSUPB021 is an Archean greywacke from the Tasiast terrane of Mauritania, West Africa. Zircons are sub-angular to rounded, 150–100 × 100–80 × 80–60 μm and typically display oscillatory zoning in CL images (Figure 1c) indicating an igneous provenance. Metamict zircons with chaotic CL patterns are also evident, indicating that substantial radiation damage has occurred. TSUPB021 zircons subjected to chemical abrasion have been visibly and severely affected, with widespread partial dissolution along micro-cracks, radiating fractures and dark CL (high U) domains throughout the grains (Figure 1d).

Sample 10/225 is a granite cobble from an Ordovician sedimentary sequence in South Mayo, west of Ireland. Zircons are euhedral to subhedral, 180–50 × 80–60 × 60–40 μm and typically display oscillatory zoning and metamict domains in CL (Figure 1e–f). Chemically abraded zircons from this sample display widespread severe pitting and dissolution zones along fractures through entire grains (Figure 1e).

2.2. Sample Preparation

Samples TSUPB021 and 10/225 were crushed and zircons separated by standard techniques (Rogers water table, Frantz magnetic separation, and methylene iodide heavy-liquid separation). Several hundred zircons from these samples were hand-picked under a binocular microscope, in order to ensure representative populations. Fragments of 91500 and Penglai zircon reference materials and representative zircons separated from TSUPB021 and 10/225 were thermally annealed at 850 °C in quartz glass beakers for 60 h. The annealed crystals were then ultrasonically washed in 4 N HNO_3, rinsed in ultra-pure water, then further washed in warm 4 N HNO_3 prior to rinsing with ultra-pure water to remove surface contamination. The annealed and cleaned zircon fractions were then chemically abraded in 200 μL 29 N HF and 20 μL 8 N HNO_3 at 180 °C in modified 1.5 mL Teflon micro-centrifuge tubes in a Parr bomb for 12 h. Following three separate cleaning steps in 3 N HCl and ultra-pure water, treated and untreated zircons were mounted in separate epoxy plugs, ground down to expose grain interiors and polished with 6 μm and 1 μm diamond paste. All sample preparation was undertaken at the Department of Geology, Trinity College Dublin (TCD).

2.3. Instrumentation

Untreated zircons from sample 10/225 were analyzed at the Department of Earth Science, University of California Santa Barbara, USA (UCSB), by Multi-Collector (MC) LA-ICP-MS (Nu Plasma, Nu Instruments Ltd., Wrexham, UK) [30], whereas all others presented in this study were analyzed by Quadrupole (Q-) LA-ICP-MS (Thermo Scientific iCAP Qs, Thermo Fisher Scientific, Waltham, MA, USA) at the Department of Geology, TCD, Dublin Ireland. Both laboratories utilized a

Photon Machines 193 nm ArF Excimer laser equipped with a HeLex ablation cell, 91500 zircon [26] as a primary reference material and Plešovice zircon [27] as a secondary reference material. A standard-sample bracketing approach was utilized using 91500 as the primary reference material after every *ca.* 10 analyses of unknowns. Chemically abraded samples were matched with chemically abraded reference materials in the same analytical session. For analyses conducted at TCD the LA-ICP-MS system was tuned while ablating a line raster on NIST SRM (National Institute of Standards, Standard Reference Material) 612 synthetic glass reference material in order to obtain optimal sensitivity for Pb, Th and U, while keeping ThO+/Th low and Th/U close to 1. ThO+/Th and Th/U were typically < 0.002 and > 0.9, respectively. Full instrument set-up conditions are given in Table 1.

Table 1. Running conditions for ICP-MS and laser ablation at University of California, Santa Barbara (UCSB), and Trinity College, Dublin (TCD). Note that for analyses performed at TCD ^{91}Zr was included to monitor general signal intensity and ^{88}Sr was monitored to evaluate alteration (pristine or non-hydrothermal zircon does not contain significant levels of Sr).

Parameter	UCSB	TCD
ICP instrument model	Nu Plasma HR MC-ICP-MS	Thermo-Scientific iCAP-Qs (quadrupole)
Plasma RF power (W)	1300	1450
Plasma gas flow (L/min)	0.8	0.7
Monitored masses	^{238}U, ^{232}Th, ^{208}Pb, ^{207}Pb, ^{206}Pb, ^{204}Pb/^{204}Hg	^{238}U, ^{232}Th, ^{208}Pb, ^{207}Pb, ^{206}Pb, ^{204}Pb/^{204}Hg, ^{200}Hg, ^{91}Zr, ^{88}Sr
LA instrument model	Photon Machines Analyte 193	Photon Machines Analyte 193
Laser	ATLEX-SI 193 nm ArF excimer	ATLEX-SI 193 nm ArF excimer
Fluence (J/cm^2)	3.5	*ca.* 3.3
Repetition rate (Hz)	4	4
Delay between analyses (s)	20	15
Ablation duration (s)	30	30
Carrier gas flow (L/min)	0.25 (He)	0.70 (He), 0.006 (N$_2$)
Spot diameter (μm)	30	30

2.4. Data Processing

All data reduction was performed using the Iolite software package [31–34] which operates within Igor Pro [35]. Although Iolite has been available for only a few years, it has become widely adopted for its visual and interactive approach to processing very large time-integrated datasets. Once raw data have been imported from the mass spectrometer software, Iolite uses several steps to carry out data processing. These include: (1) baseline corrections; (2) defining integrations for U–Pb primary reference materials; (3) calculating a down-hole U–Th–Pb fractionation correction; (4) defining integrations for secondary reference materials and unknowns; and (5) applying a session-wide fractionation and instrumental drift corrections. Applying a session-wide correction is a widely used approach, as it is deemed a robust way to correct for laser-induced fractionation correction [31,33]. Other methods of applying an inter-element fractionation include correction by linear regression, whereby the intercept gives a "true" ratio at the initial stage of ablation [36]. The VizualAge data reduction scheme [37] within Iolite was used to process data presented here. It has the capability of producing "live Concordia" diagrams for visualising data while selecting integration intervals during data processing. Isoplot [38] was used for plotting U–Pb data. The "TuffZirc" algorithm [39] is used to calculate an age for sample 10/225. If this algorithm can find a coherent group of at least five analyses (or 0.3 × total number of analyses, whichever is larger), the age and uncertainty of the median of the coherent group is calculated. This function can tolerate up to 70% of the data being non-cogenetic (*i.e.,* xenocrystic or suffered from Pb-loss) and produces an asymmetrical uncertainty expanded in the direction of most complexity. At best it will produce an error comparable to that of the most precise analysis used in the calculation.

3. Results and Discussion

3.1. Characterizing Ablation Profiles and Down-Hole Fractionation

Natural variation in the U and Th concentration of zircon, and hence the amount of accumulated radiation damage [9], has a direct influence on zircon ablation rates [40,41]. Generally, high U zircons (e.g., up to 3000 ppm U; Plešovice zircon reference material [27]) are expected to ablate at higher rates than those with lower U concentrations. Untreated and chemically abraded zircons from sample TSUPB021 were ablated during two separate analytical sessions. Measurement of ablation pit profiles from these zircons was carried out at the Centre for Microscopy and Analysis at TCD using an Omniscan MicroXan White Light Interferometer. This technique measures surface roughness and texture and is capable of rapid 3-dimensional scanning to capture data for off-line plotting and visualization. Optical interferometry has revealed differences in ablation profiles from non-treated and chemically abraded zircons (Figure 2). In general, untreated zircons ablated in a more uniform manner (Figure 2a,c). In terms of ablation depths, untreated zircons have pits which are *ca.* 25% deeper than chemically abraded counterparts (Figure 2b,d). This equates to *ca.* 67 nm and *ca.* 50 nm ablation depths per laser shot for untreated and chemically abraded zircons respectively.

The reasons for such a disparity in pit depths from TSUPB021 zircons are unclear but may be due to elimination of alpha-dose-induced crystal lattice damage in annealed zircons, a factor which contributes to a so-called "matrix effect" in LA-ICP-MS analyses [40,41]. A *ca.* 5%–6% difference in the ablation rate of several untreated and annealed zircon reference materials, including Plešovice, has previously been documented [41]. By virtue of their relatively high U and Th concentrations and Archean age, untreated zircons from sample TSUPB021 have a higher alpha-dose and lower density compared with most commonly used zircon reference materials. Thermal annealing of zircons results in higher densities and a corresponding shallower pit depth [41]. The considerable difference in ablation depths between pre- and post-annealed TSUPB021 zircons is presumably due to a greater disparity in crystal density when compared to similarly treated zircon reference materials [41]. At least some of the observed difference in ablation rates in untreated and chemically abraded zircons may also result from less effective laser coupling in chemically abraded zircons due to small-scale etching and creation of a 3-dimensional porous texture. This observation has important implications for the use of chemical abrasion in LA-ICP-MS analyses of zircon since down-hole fractionation increases with an increase in ablation pit depth. Assessment of down-hole fractionation of $^{206}Pb/^{238}U$ for untreated and chemically abraded 91500 zircons reveals a distinct difference (Figure 3), with a *ca.* 3% higher fractionation factor calculated for untreated analyses. This demonstrates chemical abrasion of zircons affects their physical response to ablation, which in turn influences laser-induced fractionation.

Figure 2. Ablation pit profiles from sample TSUPB021: (**a**) 3D view of untreated zircons; (**b**) cross-sectional view of untreated zircons; (**c**) 3D view of chemically abraded zircons; and (**d**) cross-sectional view of chemically abraded zircons.

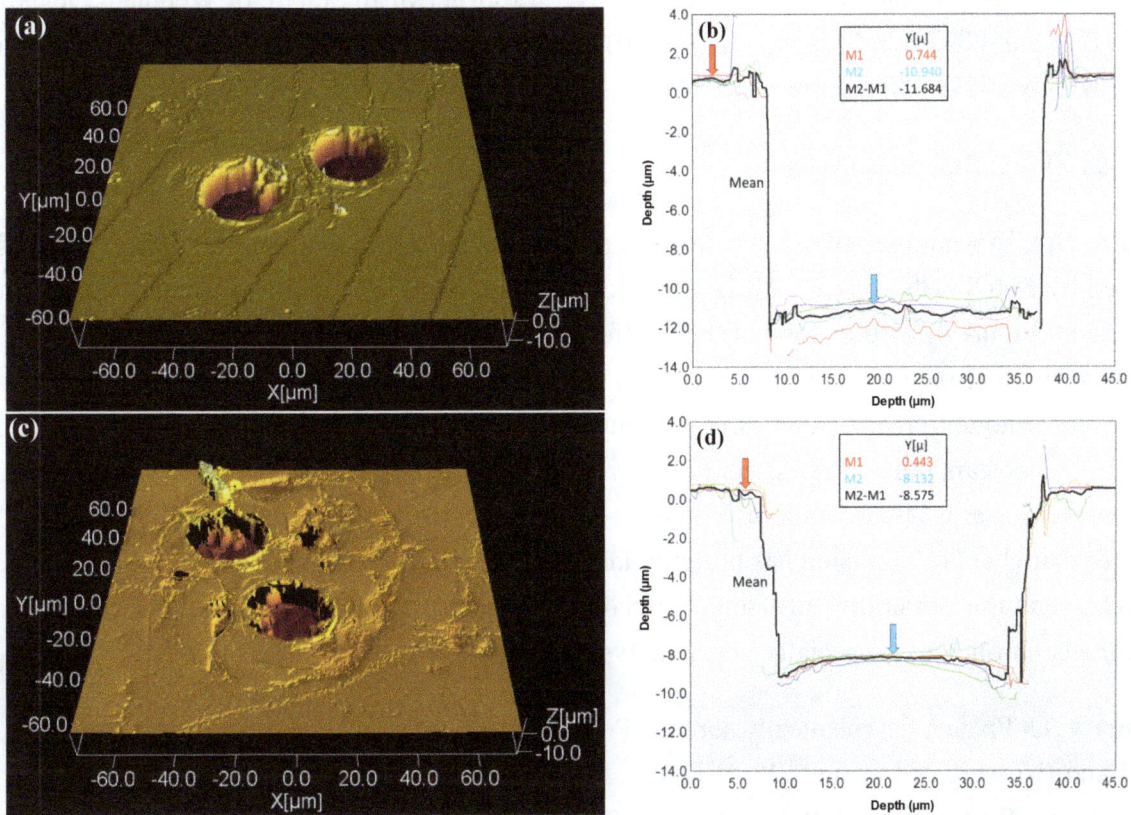

Figure 3. Comparison of time-resolved $^{206}Pb/^{238}U$ down-hole fractionation from static ablations (30 μm spot diameter) of untreated 91500 zircon (blue lines and dashed curve) and chemically abraded 91500 zircon (pink lines and solid curve) reference materials. Pale coloured lines indicate ratios from individual analyses, whereas heavy-coloured lines represent a time-resolved mean value calculated from all analyses. Black curves are calculated as a best-fit exponential spline through mean values. There is a *ca.* 3% difference between the two curves, taken as an average over the full duration of ablation.

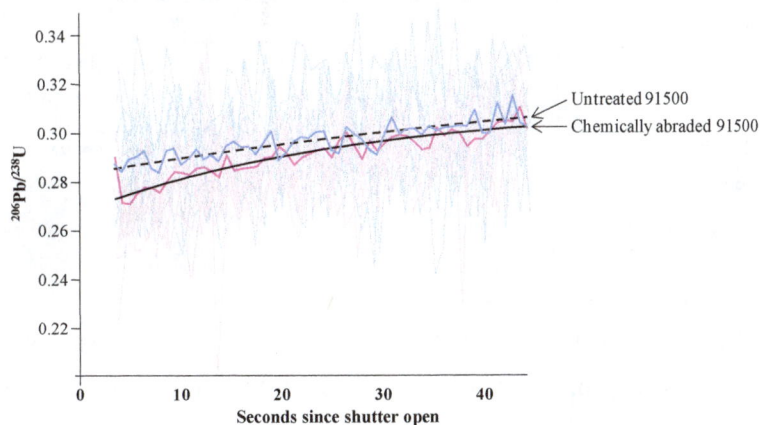

3.2. U–Th–Pb Data

U–Th–Pb data for Penglai, TSUPB021 and 10/225 are provided in the supplementary material. Errors and uncertainties are presented at a 2σ confidence level. Chemically abraded 91500 zircon was used as a primary reference material for chemically abraded secondary reference material and chemically abraded unknowns. Plešovice zircon, which was used as a secondary reference material at both UCSB and TCD gave mean $^{206}Pb/^{238}U$ ages of 342 ± 23 Ma (Mean Square of Weighted Deviation (MSWD) = 0.86) and 337.8 ± 3 Ma (MSWD = 5.0) respectively, both within error of the accepted age of 337.13 ± 0.37 Ma [27]. All reported data are non-common Pb corrected.

3.2.1. Penglai Zircon Reference Material

Representative sub-samples of several large mm-size zircon fragments were chemically abraded and analysed by Q-ICP-MS at TCD. All 20 analyses are concordant and define a mean $^{206}Pb/^{238}U$ age of 4.3 ± 0.2 Ma (Figure 4), within error of the "preferred age" of 4.4 ± 0.1 Ma [28].

For consistency between analytical protocols, a 30 µm spot diameter was used for all analyses in this study. The resultant error of our calculated mean $^{206}Pb/^{238}U$ age for Penglai zircons (4.7% 2σ, including decay constant uncertainties) could likely be reduced with a larger spot size. The concordant nature of our data is in contrast to all published U–Pb data for this reference material [28,29,37]. This demonstrates that chemical abrasion has preferentially and effectively removed common Pb from these zircons and opens the possibility of using Penglai as a primary reference material for micro-beam U–Pb zircon geochronology, especially when analysing young (<10 Ma) material.

Figure 4. U–Pb data for chemically abraded Penglai zircon reference material: (**a**) Concordia plot; and (**b**) weighted mean $^{206}Pb/^{238}U$ age. Error ellipses and box heights are 2σ. Conf.: Confidence; Wtd: weighted; Rej.: rejected; MSWD: Mean Square of Weighted Deviation.

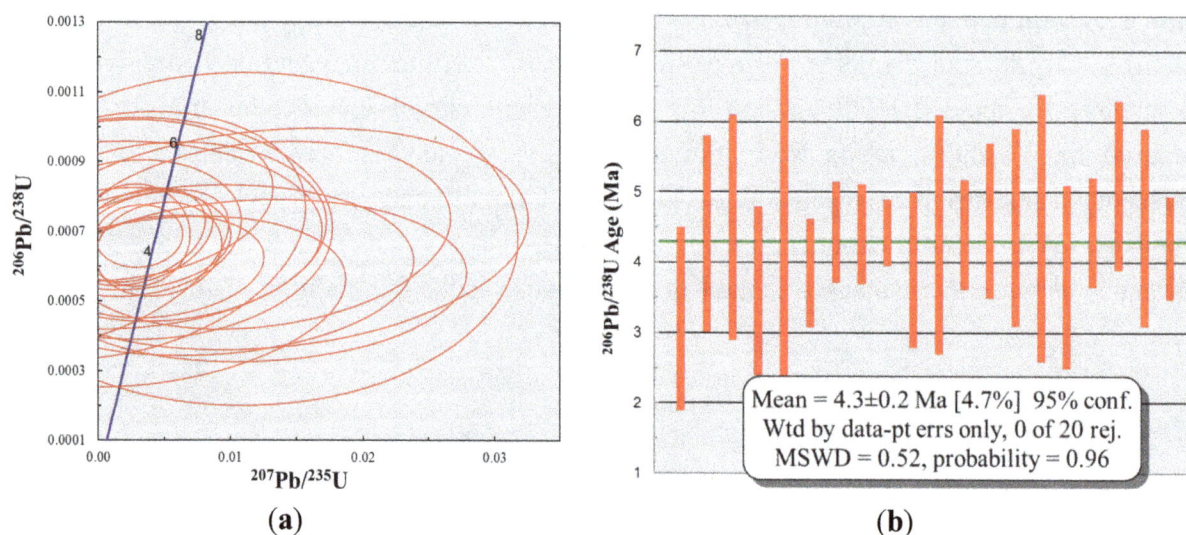

(**a**)   (**b**)

3.2.2. Sample TSUPB021

Both untreated (*n* = 77) and chemically abraded (*n* = 60) zircons from this sample were analysed by Q-ICP-MS at TCD. A significant proportion of untreated zircons (*ca.* 30%) are discordant due to severe Pb-loss (Figure 5a), which is quite typical of zircons from Archean rocks. This contrasts dramatically with chemically abraded zircons from the same sample where only 3% of analyses are >10% discordant and the remaining analyses plot close to, or overlap with, the Concordia curve (Figure 5b). Although untreated zircons from this sample display a greater range of U concentration (*ca*. 30 to 7300 ppm, but mostly within *ca*. 30 to 1000 ppm) compared to those which were chemically abraded (*ca*. 70 to 500 ppm) (Figure 6a), importantly all main age spectra present in the untreated sample are evident in the chemically abraded counterpart (Figure 6b). Minor peaks appearing in only the untreated zircons at *ca*. 2910 and 3120 are likely an artefact of the relatively small number or zircons analysed.

Figure 5. U–Pb Concordia diagrams for sample TSUPB021: (**a**) Non-treated zircons; and (**b**) chemically abraded zircons. Data-point error ellipses are 2σ. Red ellipses: analyses > 10% discordant. Black ellipses: analyses < 10% discordant.

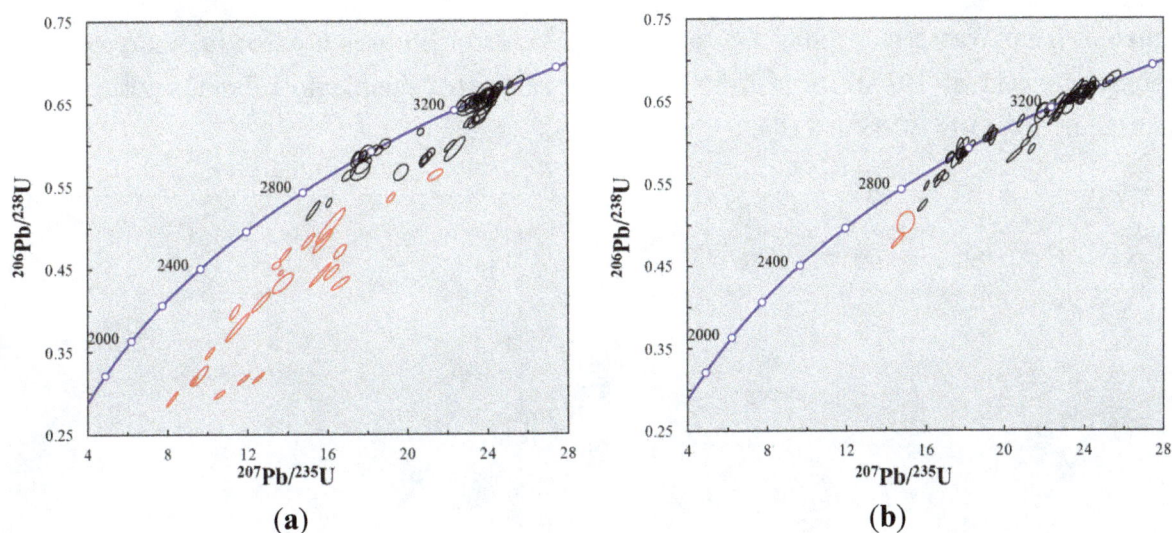

(**a**)   (**b**)

Figure 6. Chemical and age data for sample TSUPB021: (**a**) U ppm *vs*. ^{207}Pb/^{206}Pb age (Ma), note that analyses > *ca*. 1100 ppm U have not been plotted but do not define any ages not already represented in the diagram. Error ellipses are 2σ. (**b**) Probability-age plot for both untreated and chemically abraded zircons for analyses < 10% discordant.

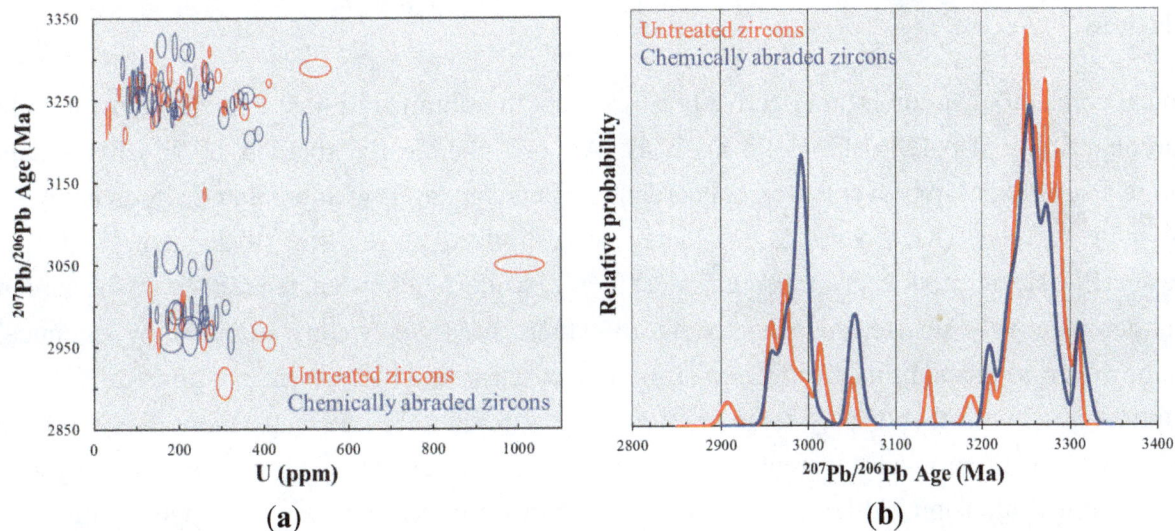

(**a**)   (**b**)

3.2.3. Sample 10/225

Untreated zircons from sample 10/225 were analysed at the UCSB facility. U–Pb data show a concordant cluster at *ca.* 480 Ma and a large number of discordant analyses with a high degree of scatter (Figure 7a). This scatter and discordance most likely result from a combination of elevated common Pb, Pb-loss and possibly some degree of inheritance. Despite the large number of analyses (*n* = 109) and the relatively high-precision nature of the U–Pb data, this scatter impedes a robust weighted mean age determination. Analysis of chemically abraded zircons from the same sample (*n* = 33) demonstrates a tight cluster of data overlapping with the Concordia curve and four discordant analyses (Figure 7b), with the latter likely reflecting "mixed ages" due to sampling of an inherited component and magmatic overgrowth. The "TuffZirc" algorithm [39] returns an age of 479.90 ± 1.2 Ma for the non-treated (*n* = 46/109) and 479.7 +5.3/−2.2 Ma for the chemically abraded (*n* = 24/33) zircon populations respectively.

Figure 7. Tera-Wasserburg plots for sample 10/225, error ellipses are 2σ: (**a**) untreated zircons analysed at UCSB by LA-MC-ICP-MS; and (**b**) chemically abraded zircons analysed by LA-Q-ICP-MS at TCD.

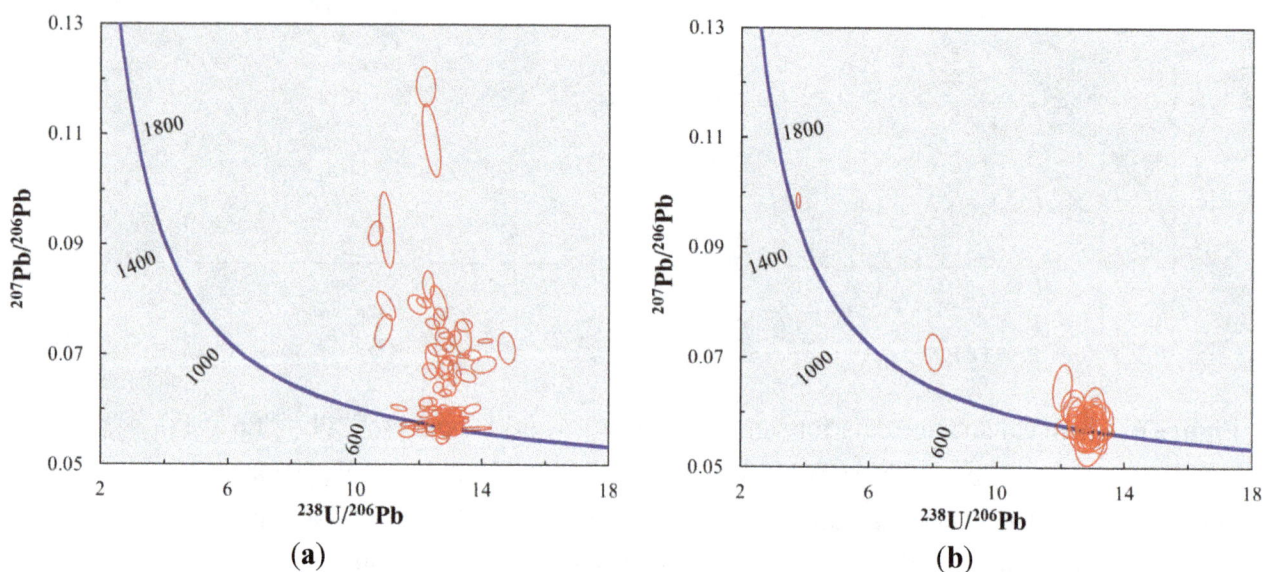

4. Conclusions

Chemical abrasion of zircon is an effective pre-treatment to minimize Pb-loss and elevated common Pb in zircon prior to LA-ICP-MS U–Pb analysis. All our chemically abraded U–Pb zircon data demonstrate a substantial improvement in concordance when compared with untreated zircon analyses from the same samples. Discordance is eliminated or minimized by removal of divalent Pb during exposure to HF acid at moderate temperatures (180 °C), while leaving the remainder of the zircon crystal lattice relatively unaffected. The extent to which zircon is visibly affected by chemical abrasion, however, seems to primarily depend on U concentration and the presence of physical defects, such as fractures. Zircons that display some or all of these attributes are inherently more likely to have suffered Pb-loss, so there is an apparent correlation between the extent of Pb-loss and the physical effects of chemical abrasion. Although this physical effect on concordant zircons appears minimal, comparison of down-hole fractionation of $^{206}Pb/^{238}U$ in untreated and chemically abraded 91500 reference material reveals a *ca.* 3% difference, with the untreated zircons displaying a greater degree of laser-induced fractionation effects. This disparity is likely depth dependant, as ablated pits in chemically abraded high alpha-dose zircons known to have suffered Pb-loss are up to 25% shallower than ablation pits in untreated zircons from the same sample. Although this may be an extreme example, it illustrates that primary reference materials should be chemically abraded when analysing chemically abraded unknowns and secondary reference materials. It does not, however, eliminate the possibility that reference materials and unknowns will respond differently to the treatment, thereby resulting in differing down-hole fractionation of U–Pb and Th–Pb isotopes. Accurate measurement and comparison of ablation depths may help to minimize a mismatch between reference materials and sample down-hole fractionation corrections.

Analysis of zircons from an Archean sample (TSUPB021) demonstrates preferential dissolution of zircon at U concentrations greater than *ca.* 1100 ppm, the majority of which are severely discordant as measured in untreated counterparts. In this particular case the main age spectra between untreated and chemically abraded zircons are identical, although it would in theory be possible to bias a sample by preferential dissolution of high-U zircons with a distinct age signature. Although higher U zircons (up to 1400 ppm) have been reported to have survived intact following chemical abrasion [22], these instances have been from considerably younger zircons. The chemical abrasion technique should therefore be used with caution when dealing with Archean zircons with elevated U concentrations.

In summary, although initially developed for TIMS U–Pb geochronology, chemical abrasion can easily and routinely be applied as a pre-treatment protocol to improve concordance for the majority of samples for LA-ICP-MS U–Pb geochronology.

Acknowledgments

Thanks to Tony Simonetti for his suggestion to write this research paper and to two anonymous reviewers for their constructive comments. Both Andrew Kylander-Clark and Cora McKenna are thanked for their assistance with mass spectrometry at UCSB and TCD respectively.

Author Contributions

Quentin G. Crowley developed the experiment design, carried out the majority of sample analyses and drafted the paper. All co-authors contributed to the writing of this paper. Kyle Heron collected sample TSUPB021, carried out rock crushing and mineral separation of samples TSUPB021 and 10/225, analyzed sample TSUPB021, performed CL imaging and measurement of pit depth profiles. Nancy Riggs carried out analysis of the non-treated sample 10/225 at UCSB. Balz Kamber supplied Penglai zircon reference material. David Chew supplied 91500 zircon reference material. Brian McConnell assisted with collection of sample 10/225. Keith Benn assisted with collection of sample TSUPB021.

Conflicts of Interest

The authors declare no conflict of interest.

References

1. Kulp, J.L.; Bate, G.L.; Broecker, W.S. Present status of the lead method of age determination. *Am. J. Sci.* **1954**, *252*, 345–365.

2. Tilton, G.R.; Patterson, C.; Brown, H.; Inghram, M.; Hayden, R.; Hess, D.; Larsen, E. Isotopic composition and distribution of lead, uranium, and thorium in a Precambrian granite. *Geol. Soc. Am. Bull.* **1955**, *66*, 1131–1148.

3. Wetherill, G.W. Discordant uranium-lead ages: 2. Disordant ages resulting from diffusion of lead and uranium. *J. Geophys. Res.* **1963**, *68*, 2957–2965.

4. Nier, A.O. Variations in the relative abundances of the isotopes of common lead from various sources. *J. Am. Chem. Soc.* **1938**, *60*, 1571–1576.

5. Finch, R.J.; Hanchar, J.M. Structure and chemistry of zircon and zircon-group minerals. *Rev. Mineral. Geochem.* **2003**, *53*, 1–25.

6. Kogawa, M.; Watson, E.B.; Ewing, R.C.; Utsunomiya, S. Lead in zircon at the atomic scale. *Am. Mineral.* **2012**, *97*, 1094–1102.

7. Cherniak, D.; Watson, E. Pb diffusion in zircon. *Chem. Geol.* **2001**, *172*, 5–24.

8. Lee, J.K.; Williams, I.S.; Ellis, D.J. Pb, U and Th diffusion in natural zircon. *Nature* **1997**, *390*, 159–162.

9. Nasdala, L.; Hanchar, J.M.; Kronz, A.; Whitehouse, M.J. Long-term stability of alpha particle damage in natural zircon. *Chem. Geol.* **2005**, *220*, 83–103.

10. Tanaka, K.; Takahashi, Y.; Horie, K.; Shimizu, H.; Murakami, T. Determination of the oxidation state of radiogenic Pb in natural zircon using X-ray absorption near-edge structure. *Phys. Chem. Miner.* **2010**, *37*, 249–254.

11. Xu, X.-S.; Zhang, M.; Zhu, K.-Y.; Chen, X.-M.; He, Z.-Y. Reverse age zonation of zircon formed by metamictisation and hydrothermal fluid leaching. *Lithos* **2012**, *150*, 256–267.

12. Kramers, J.; Frei, R.; Newville, M.; Kober, B.; Villa, I. On the valency state of radiogenic lead in zircon and its consequences. *Chem. Geol.* **2009**, *261*, 4–11.

13. MacDonald, J.M.; Wheeler, J.; Harley, S.L.; Mariani, E.; Goodenough, K.M.; Crowley, Q.; Tatham, D. Lattice distortion in a zircon population and its effects on trace element mobility and U-Th-Pb isotope systematics: Examples from the Lewisian Gneiss Complex, northwest Scotland. *Contrib. Mineral. Petrol.* **2013**, *166*, 21–41.

14. Krogh, T. Improved accuracy of U-Pb zircon ages by the creation of more concordant systems using an air abrasion technique. *Geochim. Cosmochim. Acta* **1982**, *46*, 637–649.

15. Mattinson, J.M. Zircon U–Pb chemical abrasion ("CA-TIMS") method: Combined annealing and multi-step partial dissolution analysis for improved precision and accuracy of zircon ages. *Chem. Geol.* **2005**, *220*, 47–66.

16. Krogh, T.; Davis, G. Alteration in zircons with discordant U-Pb ages. *Carnegie Inst. Wash. Yearb.* **1974**, *73*, 560–567.

17. Krogh, T.; Davis, G. Alteration in zircons and differential dissolution of altered and metamict zircon. *Carnegie Inst. Wash. Yearb.* **1975**, *74*, 619–623.

18. Mattinson, J.M. A study of complex discordance in zircons using step-wise dissolution techniques. *Contrib. Mineral. Petrol.* **1994**, *116*, 117–129.

19. McClelland, W.C.; Mattinson, J.M. Resolving high precision U–Pb ages from Tertiary plutons with complex zircon systematics. *Geochim. Cosmochim. Acta* **1996**, *60*, 3955–3965.

20. Todt, W.A.; Büsch, W. U-Pb investigations on zircons from pre-Variscan gneisses—I. A study from the Schwarzwald, West Germany. *Geochim. Cosmochim. Acta* **1981**, *45*, 1789–1801.

21. Crowley, Q.; Key, R.; Noble, S. High-precision U–Pb dating of complex zircon from the Lewisian Gneiss Complex of Scotland using an incremental CA-ID-TIMS approach. *Gondwana Res.* **2014**, doi:10.1016/j.gr.2014.04.001.

22. Kryza, R.; Crowley, Q.G.; Larionov, A.; Pin, C.; Oberc-Dziedzic, T.; Mochnacka, K. Chemical abrasion applied to SHRIMP zircon geochronology: An example from the Variscan Karkonosze Granite (Sudetes, SW Poland). *Gondwana Res.* **2012**, *21*, 757–767.

23. McConnell, B.; Riggs, N.; Crowley, Q.G. Detrital zircon provenance and Ordovician terrane amalgamation, western Ireland. *J. Geol. Soc.* **2009**, *166*, 473–484.

24. Riggs, N.R.; Barth, A.P.; González-León, C.M.; Jacobson, C.E.; Wooden, J.L.; Howell, E.R.; Walker, J.D. Provenance of Upper Triassic Strata in Southwestern North America as Suggested by Isotopic Analysis and Chemistry of Zircon Crystals. In *Mineralogical and Geochemical Approaches to Provenance*; Rasbury, E.T., Hemming, S.R., Riggs, N.R., Eds.; Special Paper Volume 487; Geological Society of America: Boulder, CO, UDA, 2012; pp. 13–36.

25. Riggs, N.; Reynolds, S.; Lindner, P.; Howell, E.; Barth, A.; Parker, W.; Walker, J. The early Mesozoic cordilleran arc and late Triassic paleotopography: The detrital record in Upper Triassic sedimentary successions on and off the Colorado Plateau. *Geosphere* **2013**, *9*, 602–613.

26. Wiedenbeck, M.; Alle, P.; Corfu, F.; Griffin, W.; Meier, M.; Oberli, F.; Quadt, A.V.; Roddick, J.; Spiegel, W. Three natural zircon standards for U-Th-Pb, Lu-Hf, trace element and REE analyses. *Geostand. Newsl.* **1995**, *19*, 1–23.

27. Sláma, J.; Košler, J.; Condon, D.J.; Crowley, J.L.; Gerdes, A.; Hanchar, J.M.; Horstwood, M.S.; Morris, G.A.; Nasdala, L.; Norberg, N. Plešovice zircon—A new natural reference material for U–Pb and Hf isotopic microanalysis. *Chem. Geol.* **2008**, *249*, 1–35.

28. Li, X.H.; Long, W.G.; Li, Q.L.; Liu, Y.; Zheng, Y.F.; Yang, Y.H.; Chamberlain, K.R.; Wan, D.F.; Guo, C.H.; Wang, X.C. Penglai zircon megacrysts: A potential new working reference material for microbeam determination of Hf–O isotopes and U–Pb age. *Geostand. Geoanal. Res.* **2010**, *34*, 117–134.

29. Chew, D.; Petrus, J.; Kamber, B. U–Pb LA–ICPMS dating using accessory mineral standards with variable common Pb. *Chem. Geol.* **2014**, *363*, 185–199.

30. Kylander-Clark, A.R.; Hacker, B.R.; Cottle, J.M. Laser-ablation split-stream ICP petrochronology. *Chem. Geol.* **2013**, *345*, 99–112.

31. Hellstrom, J.; Paton, C.; Woodhead, J.; Hergt, J. Iolite: Software for Spatially Resolved LA-(QUAD and MC) ICPMS Analysis. In *Laser Ablation ICP–MS in the Earth Sciences: Current Practices and Outstanding Issues*; Sylvester, P., Ed.; Short Course 40; Mineralogical Association of Canada: Quebec City, QC, Canada, 2008; pp. 343–348.

32. Woodhead, J.; Hellstrom, J.; Paton, C.; Hergt, J.; Greig, A.; Maas, R. A Guide to Depth Profiling and Imaging Applications of LA–ICP–MS. In *Laser Ablation ICP–MS in the Earth Sciences: Current Practices and Outstanding Issues*; Sylvester, P., Ed.; Short Course 40; Mineralogical Association of Canada: Quebec City, QC, Canada, 2008; pp. 135–145.

33. Paton, C.; Hellstrom, J.; Paul, B.; Woodhead, J.; Hergt, J. Iolite: Freeware for the visualisation and processing of mass spectrometric data. *J. Anal. At. Spectrom.* **2011**, *26*, 2508–2518.

34. The Iolite Project. Available online: http://www.iolite.org.au (accessed on 26 May 2014).

35. WaveMetrics. Available online: http://www.wavemetrics.com (accessed on 26 May 2014).

36. Janoušek, V.; Gerdes, A.; Vrána, S.; Finger, F.; Erban, V.; Friedl, G.; Braithwaite, C.J.R. Low-pressure granulites of the Lišov Massif, southern Bohemia: Viséan metamorphism of late Devonian plutonic arc rocks. *J. Petrol.* **2006**, *47*, 705–744.

37. Petrus, J.A.; Kamber, B.S. VizualAge: A novel approach to laser ablation ICP-MS U-Pb geochronology data reduction. *Geostand. Geoanal. Res.* **2012**, *36*, 247–270.

38. Isoplot Version 4.1. Available online: http://www.bgc.org/isoplot_etc/isoplot.html (accessed on 29 May 2014).

39. Ludwig, K.; Mundil, R. Extracting reliable U-Pb ages and errors from complex populations of zircons from Phanerozoic tuffs. *Geochim. Cosmochim. Acta* **2002**, *66*, A463.

40. Allen, C.M.; Campbell, I.H. Identification and elimination of a matrix-induced systematic error in LA–ICP–MS ^{206}Pb/^{238}U dating of zircon. *Chem. Geol.* **2012**, *332*, 157–165.

41. Marillo-Sialer, E.; Woodhead, J.; Hergt, J.; Greig, A.; Guillong, M.; Gleadow, A.; Evans, N.; Paton, C. The zircon "matrix effect": Evidence for an ablation rate control on the accuracy of U–Pb age determinations by LA-ICP-MS. *J. Anal. At. Spectrom.* **2014**, *29*, 981–989.

4

Mapping Changes in a Recovering Mine Site with Hyperspectral Airborne HyMap Imagery (Sotiel, SW Spain)

Jorge Buzzi [1], Asunción Riaza [1,*], Eduardo García-Meléndez [2], Sebastian Weide [3] and Martin Bachmann [3]

[1] Instituto Geológico y Minero de España (IGME), Geological Survey of Spain, La Calera 1, 28760 Tres Cantos, E-28003 Madrid, Spain; E-Mail: j.buzzi@igme.es

[2] Facultad de Ciencias Ambientales, Universidad de León, Campus de Vegazana s/n, E-24071 León, Spain; E-Mail: egarm@unileon.es

[3] Remote Sensing Data Centre, German Aerospace Research, Deustsche Zentrum für Luft und Raumfahrt (DLR), P.O. Box 1116, D-82234 Wessling, Germany; E-Mails: sebastianweide@gmx.de (S.W.); martin.bachmann@dlr.de (M.B.)

* Author to whom correspondence should be addressed; E-Mail: a.riaza@igme.es

Abstract: Hyperspectral high spatial resolution HyMap data are used to map mine waste from massive sulfide ore deposits, mostly abandoned, on the Iberian Pyrite Belt (southwest Spain). Mine dams, mill tailings and mine dumps in variable states of pyrite oxidation are recognizable. The interpretation of hyperspectral remote sensing requires specific algorithms able to manage high dimensional data compared to multispectral data. The routine of image processing methods used to extract information from hyperspectral data to map geological features is explained, as well as the sequence of algorithms used to produce maps of the mine sites. The mineralogical identification capability of algorithms to produce maps based on archive spectral libraries is discussed. Trends of mineral growth differ spectrally over time according to the geological setting and the recovery state of the mine site. Subtle mineralogical changes are enhanced using the spectral response as indicators of pyrite oxidation intensity of the mine waste piles and pyrite mud tailings. The changes in the surface of the mill tailings deserve a detailed description, as the surfaces are inaccessible to direct observation. Such mineralogical changes respond faithfully to industrial activities or the influence of climate when undisturbed by human influence.

Keywords: imaging spectroscopy; mine site recovery; change detection; pyrite weathering

---

1. Introduction

Geology-based geoenvironmental mineral deposit models point to the need for "useful geophysical techniques to identify, delineate, and monitor, environmental signatures associated with mined and unmined mineral deposits" [1]. Hyperspectral airborne HyMap data are used in this work to map the extension and evolution of pyrite weathering products in a mine site during the recovery of the mine facilities. Imaging spectroscopy using spectral libraries has been developed as a reliable technique for quick mineralogical analysis of mine wastes, which saves both time and costs *versus* conventional sample collection [2–6]. Additionally, it permits the mineralogical diagnosis of ephemeral thin crusts concentrating heavy metals on inaccessible surfaces [1], providing an invaluable tool for environmental evaluation and information.

Hyperspectral remote sensing deals with airborne and spaceborne imaging spectrometers with a spectral resolution similar to field or laboratory instruments. A large number of channels in two dimensions requires image processing procedures able to manage the high dimensionality of this data. Algorithms particularly designed to extract spectral features from a large number of data arranged in layers were developed and gathered in widely available software [7]. The new analytical techniques permit a new understanding of the geological expression on the surface of the Earth, which is not accessible through spectral broad band sensors. The spectral resolution of airborne or spaceborne hyperspectral data is similar to field or laboratory spectra. Each pixel of an image is a continuous spectra suitable to be part of a map through digital image processing.

Massive sulfide deposits have been mined in the Iberian Pyrite Belt for at least 5000 years [8]. Mining activity implies the removal of large amounts of rocks, exposing to the air unstable mineral substances that weather when exposed to the atmosphere following a well-known sequence of secondary minerals [9].

Rapid oxidation and evaporation of sulfides on mine waste produces iron-bearing sulfate and other metals as secondary minerals [10]. A sequence of salts is established from a solution of pyrite, from early to later formed [9,11]. Ferrous sulfate salts are found close to pyrite sources (melanterite $Fe^{2+}(SO_4)\cdot7(H_2O)$, rozenite $Fe^{2+}(SO_4)\cdot4(H_2O)$, szomolnokite $Fe(SO_4)\cdot H_2O$), whereas ferric-bearing minerals (rhomboclase $HFe^{3+}(SO_4)_2\cdot4(H_2O)$, voltaite, halotrichite $Fe^{2+}Al_2(SO_4)_4\cdot22(H_2O)$) can be considered hydrologic dead-ends, where most of the Fe^{2+} has had time to oxidize to Fe^{3+}.

Based on laboratory experiments of evaporating acid mine waters, the paragenetic sequence seems as below (p. 152 in [10], mentioning [9]; and "C. Maenz, written communication, 1995"):

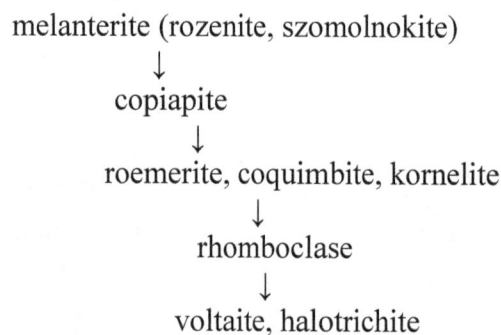

melanterite (rozenite, szomolnokite)
↓
copiapite
↓
roemerite, coquimbite, kornelite
↓
rhomboclase
↓
voltaite, halotrichite

Further oxidation leads to the formation of schwertmannite ($Fe^{3+}16O_{16}(OH)_{12}(SO_4)_2$) and the group of jarosite-alunite (($SO_4)_2KFe_3(OH)_6$-$KAl_3(SO4)_2(OH)_6$). The mineralogy of oxidized zones on gossans is dominated by hematite (Fe_2O_3), goethite ($FeO(OH)$) and jarosite (($SO_4)_2KFe_3(OH)_6$).

schwertmannite

↓

jarosite-alunite

↓

hematite, goethite

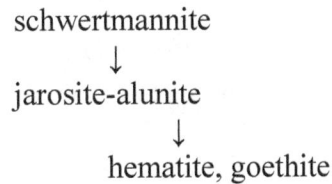

In the mine site of study, industrial activities ended in 2001, and the regional authorities began recovery in 2006. The mineralogical changes involved in mine waste weathering are controlled by climate and geomorphology [12], when undisturbed by human activity. The mineralogical, temporal evolution and spatial pattern of the secondary minerals growing on mine waste range from mud tailings to rock piles and vary also with the landforms on the waste piles and dumps. The waste removal and movement of machinery during the recovery activities has a great influence on the mineralogy of the dust throughout the area [13].

2. The Mine Site Facilities: Industrial Operation and Recovery

By now, industrial operation on most metallic mine sites in the Iberian Pyrite Belt has ceased. Most mine facilities are abandoned and under the environmental control of the authorities. There are many such facilities, and the wide regional extent of abandoned mine waste makes the Iberian Pyrite Belt an ideal field for testing environmental monitoring methods.

The operating underground mine works of the Sotiel mine site are displayed along the Odiel River (Figure 1). A conveyor belt transported the ore uphill from the mine to the processing plant. Ashes from the ore processing plant are impounded on an adjacent dam. Pyrite mud is stored in another dam draining to the Odiel River.

In the year 2001, the industrial activities ended. The rehabilitation of the mine site began in 2006 with the ash dam. The purpose of the rehabilitation is to isolate the contaminating matter, preventing its expansion into the environment. The water of the dam has been drained and the bottom protected to prevent water infiltration, which can activate the oxidation process of pyrite. The dry bottom has been sealed with a repeated sandwich of several layers, including an overlying textile sheet, a clay-rich layer, an iron oxide rich layer and an organic matter rich layer. Grass is planted with *Poaceae* over the final organic matter layer to restore ordinary soil development. The dry grass (flight 4 August 2008) growing in spring is cut in the summer to prevent fires (flight 13 August 2009).

The rehabilitation of the mill tailings dam and the dismantlement of the ore processing plant began in the autumn of 2008 and continued during the winter of 2009. By 13 August 2009, the last HyMap flight available, the mud had been removed from the dam, and the bottom was in the process of being drained and consolidated before the operation of machines. In the following years, the bottom was sealed, the dam was filled with waste from the mine dumps, and the final surface was covered with yellow sand.

Hyperspectral image processing provides a fast and accurate diagnosis of the secondary minerals growing on mine waste and their mineralogical changes mapped from HyMap flights in the test site in the summers of 2005, 2008 and 2009 [13–15]. Previously published work discusses the challenges and limitations in validating mineralogical analysis in a highly spatially heterogeneous environment, combined with fast temporal geochemical mobility [16].

Figure 1. The location of the study area in the Iberian Hesperian Massif (Spain and Portugal) (blue circle) [17]. The location of domains within the Sotiel mine site.

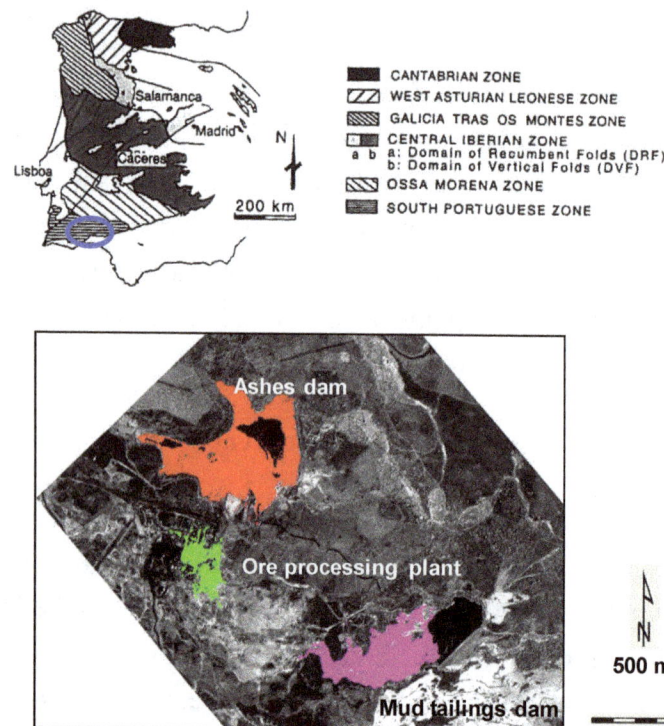

3. Data Set

HyMap airborne hyperspectral images were acquired over the Odiel River path on 17 July 2005, 4 August 2008, and 13 August 2009. HyMap is an airborne hyperspectral sensor with 128 wavebands from 436 to 2485 nm with a spectral resolution of 15 nm in the 436–125 nm wavelength range, 13 nm in the 140–180 nm wavelength range and 17 nm in the 195–248 wavelength range. Its spatial resolution is 5 m. The HyMap preprocessing was described in [12] and its use in mineral diagnosis of crusts over mine waste in [16].

Field spectral measurements were taken at 287 field locations on the mine sites with an ASD FieldSpec 3 Spectrometer [18] from 2008 to 2011 for thematic purposes, apart from calibration. An additional 45 spectral measurements were made in a dark room on selected samples of rock, mud and soil. This device measures the ground reflectance spectra covering the wavelength range from 350 nm to 2500 nm. The details of the procedure for diagnosing spectra using spectral reference libraries [14,19,20] in the area have also been previously published [21]. Eighteen existing spectra from sulfide oxidation products in public domain spectral libraries [10,22] were used as references with the Spectral Analyst [7] to get a comparative score of similarity both for spectra measured in the field and HyMap data. The Spectral Angle Mapper, Spectral Feature Fitting and Binary Encoding were taken into account in equal weighting in the final similarity score with a maximum of 1. General trends and small changes in mineralogical identification can thereby be traced in detail.

Selected representative sediment and rock samples were analyzed by X-ray diffraction (XRD). All XRD measurements were performed using a PANalytical powder diffractometer (PANanalytical, Almelo, The Netherlands) with Cu Kα radiation. Selected samples of rocks, mud and dust were also collected for conventional mineralogical analysis.

4. Geological Mapping through Image Processing

The procedure for feature extraction for thematic purposes used in this research is an interpreter oriented sequential spectral unmixing using standard algorithms, leading to a spatial pattern and spectral identification pixels within the scene displayed as a map. The dimensionality of the data is reduced through various image processing within subscenes extracted at several steps through the chain of image processing, ending in a map of the pyrite oxidation products of the abandoned mine site [12].

The first step uses a false color composite of HyMap Channels 10 (0.5719 μm), 39 (1.0063 μm) and 125 (2.4702 μm) to build a spectral library with basic land use end-members on the scene (Figure 2A). Channel 10 focuses on water properties, Channel 39 on vegetation and Channel 125 on land. Minimum Noise Fractions are used to explore the general land use features. Areas mainly covered by vegetation, clear water and urban areas were excluded from the analysis from this step on, after masks were built based on the spectral identification of the main land use regions through the Spectral Angle Mapper (Figure 2B,C). Then, the Spectral Angle Mapper produces again a first estimation of the spectral diversity of the open land, which is used as a mask for subsequent image processing steps per scene. This is the beginning of the long and iterative image processing procedure ending in a map, with the major steps shown in Figure 2.

The areas mapped as open land during the first step are analyzed using Minimum Noise Fraction Transforms, subsequent Pixel Purity Index [23] and *n*-dimensional analysis. The mine sites and main areas covered by the sulfate salts are commonly identified from bare open and cultivated land at this stage (Figure 2D). The data dimensions are reduced further, repeating the procedure to qualify areas within early-formed salts and among the more oxidized or hydrated zones (Figure 2E,F), which are isolated on corresponding masks (Figure 2G–I). A final map is compiled by gathering all the end members corresponding to pyrite oxidation products on the scene (Figure 2K). A spectral library is finally built with such end members, which are then mineralogically identified by comparison to a selection of mineral spectra from archive spectral libraries (Figure 2J), previously resampled to match the HyMap wavelength ranges.

Spectral Angle Mapper, Spectral Feature Fitting and Binary Encoding were taken into account using equal weights for each of the final similarity scores. The Spectral Angle Mapper determines the spectral similarity between two spectra by calculating the angle between spectra, treating them as vectors in a space with dimensions equal to the number of bands [24]. A maximum angle of 0.1 radians is used as the default. Spectral Feature Fitting compares the fit of image spectra to selected reference spectra using a least-squares technique [25]. The Binary Encoding algorithm encodes the data and end member spectra into zeros and ones, based on whether a band falls below or above the spectrum mean [26].

The spectrally dominant mineral was assigned to the map units (Figure 2K) using the mean spectra of the corresponding region of interest through the image processing procedure, using the spectral library with pyrite oxidation and precipitation products [7,19,20] as a reference.

Figure 2. Sequential subscenes produced by image processing (from left to right). (**A**) False color composite with HyMap Channels 9 (0.5568 μm), 28 (0.8460 μm) and 109 (2.2074 μm). (**B**) Land use map estimating vegetation cover and open soil. (**C**) Mask for vegetation. (**D**) Mask for mine waste. (**E**) Mask for hydrated tailings. (**F**) Mask for oxidated dumps. (**G**) Mask for hydrated ponds around the ore processing plant. (**H**) Mask for the hydrated mill tailings dam. (**I**) Mask for the oxidated pond of ashes. (**J**) Spectral reference library on pyrite weathering products [19,20]. Spectra are displaced vertically for clarity. (**K**) Final map of coatings of pyrite weathering products over mine waste following the reference spectral library [19]. Color coding for minerals mapped with the corresponding chemical formula are arranged in a precipitation and oxidation sequence. Increasing reddish colors suggest intense oxidation. Bluish and greenish colors concentrate on hydrated areas.

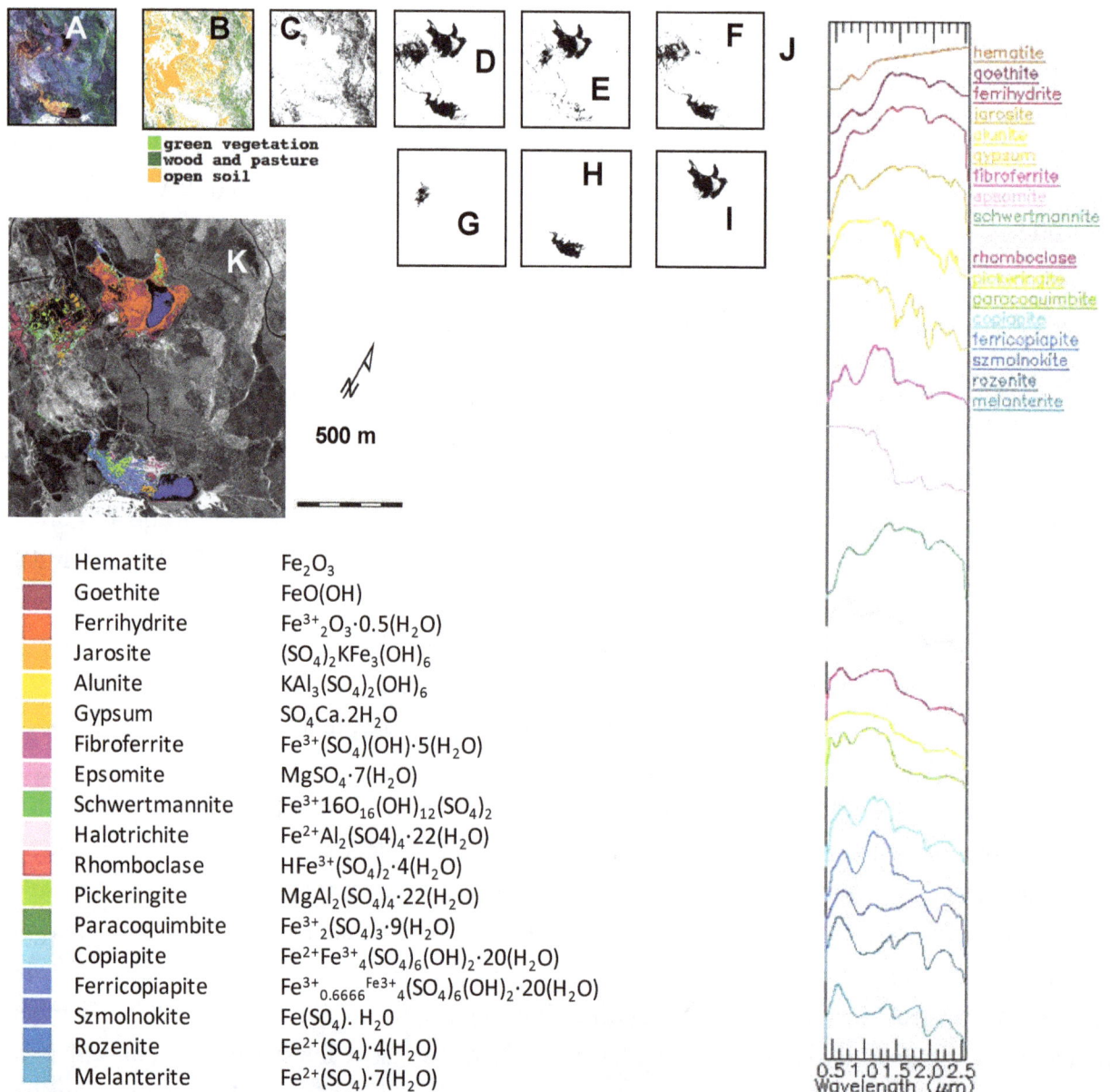

Hematite	Fe_2O_3
Goethite	$FeO(OH)$
Ferrihydrite	$Fe^{3+}_2O_3 \cdot 0.5(H_2O)$
Jarosite	$(SO_4)_2KFe_3(OH)_6$
Alunite	$KAl_3(SO_4)_2(OH)_6$
Gypsum	$SO_4Ca.2H_2O$
Fibroferrite	$Fe^{3+}(SO_4)(OH) \cdot 5(H_2O)$
Epsomite	$MgSO_4 \cdot 7(H_2O)$
Schwertmannite	$Fe^{3+}16O_{16}(OH)_{12}(SO_4)_2$
Halotrichite	$Fe^{2+}Al_2(SO4)_4 \cdot 22(H_2O)$
Rhomboclase	$HFe^{3+}(SO_4)_2 \cdot 4(H_2O)$
Pickeringite	$MgAl_2(SO_4)_4 \cdot 22(H_2O)$
Paracoquimbite	$Fe^{3+}_2(SO_4)_3 \cdot 9(H_2O)$
Copiapite	$Fe^{2+}Fe^{3+}_4(SO_4)_6(OH)_2 \cdot 20(H_2O)$
Ferricopiapite	$Fe^{3+}_{0.6666}{}^{Fe3+}_4(SO_4)_6(OH)_2 \cdot 20(H_2O)$
Szmolnokite	$Fe(SO_4).H_2O$
Rozenite	$Fe^{2+}(SO_4) \cdot 4(H_2O)$
Melanterite	$Fe^{2+}(SO_4) \cdot 7(H_2O)$

The Spectral Angle Mapper [24] was used to test the scene similarity from the second step, helping to identify end members more representative of sulfate salt coverage according to field experience and digital comparison of the shape of the corresponding spectra with archive spectral libraries. Therefore, the spectral shape information was also derived from the HyMap data.

A number of individual mineral substances precipitated from pyrite acid water were identified using public domain spectral libraries, ranging from melanterite, the first-formed sulfate, to hematite, when dehydration and oxidation is completed. Eighteen existing spectra of sulfide oxidation products in public domain spectral libraries [19,20] were used as references with the Spectral Analyst [7] to get a comparative score of similarity for spectra from every map unit. The Spectral Angle Mapper, Spectral Feature Fitting and Binary Encoding were taken into account using equal weights for each of the final similarity scores. The spectrally dominant mineral was assigned to the map units using the mean spectra of the corresponding region of interest through the image processing procedure, using the spectral library with pyrite oxidation and precipitation products.

Although mineral identification of hyperspectral pixels necessarily involves reducing high dimensionality spectral data to one compositional class, in reality, pixels are rarely pure, and their spectral signatures often reflect heterogeneous forms of mixing. Both field and laboratory spectral experience from samples collected in the mine site show that most spectral features are mixed responses from the selected minerals, which are likely to occur. Only very extended and spatially well-developed pans of intensely oxidized pyrite mud show clearly identifiable spectral signatures compared to the archive spectral libraries. The oxidation of pyrite is a progressive and heterogeneous process concerning mineralogical phases and spatial cover, as well as grain size or minor shade effects associated with small-scale geomorphology. The areas identified on the imagery as a single or double mineral presence should be interpreted as indicators of a mineralogically dominant trend displayed by HyMap imagery with a 5-m spatial resolution.

5. Mineralogical Changes Associated with Mine Site Recovery Shown by HyMap Maps

5.1. The Ash Dam

In June 2005, the northern ash pond was deeply oxidized, as can be seen in the image (Figure 3A). Hematite and goethite are widespread on the surface with reddish colors, both in the field and the maps. Comparatively, more hydrated phases occur along the rills as pickeringite or paracoquimbite, whitish in the field and bluish and greenish on the maps.

Field spectra (Figure 4A) reveal the dominant oxidized crusts in 2005 (Figure 4C), according to the results of the HyMap image processing (see Figure 3A). The reddish dominant color in the map of 2005 over the dam of ashes, suggesting an intense oxidation (Figure 3A), is due to the presence of hematite and goethite. The typical spectral features of hematite recorded in reference spectral libraries [20] are faithfully drawn by field spectra (Figure 4A).

The rehabilitation of the ash dam was begun in 2006 and was completed by the summer of 2008. All the ashes were removed, the water was drained, and the whole area was refilled, thus preventing water infiltration. In August 2008, the surface of the former ash dam was covered by high dry grass and remained so in the following years (Figure 3B,C).

Field spectra (Figure 4A) show the spectral changes from a dominant hematitic crust in 2005 (Figure 4C) to dry grass in 2008 (Figure 4D). The spectra of dry grass covering the former ash dam in 2008 measured in the field (Figure 4A) shows the typical decedent vegetation spectral response. HyMap spectra (Figure 4B) from the spectral libraries used as end members to compile the maps in Figure 3A,B in the ash dam reproduce faithfully the spectra measured in the field in 2005 and 2008.

Figure 3. (**a–c**) Maps compiled from HyMap data on the mineral products of the oxidation and dehydration of sulfide sludge, on the three available flights around the ore processing plant. (**A–C**) Maps compiled from HyMap data on the mineral products of the oxidation and dehydration of sulfide sludge over the mine site, on the three available flights. Increasing reddish colors suggest intense oxidation. Bluish and greenish colors concentrate on hydrated areas.

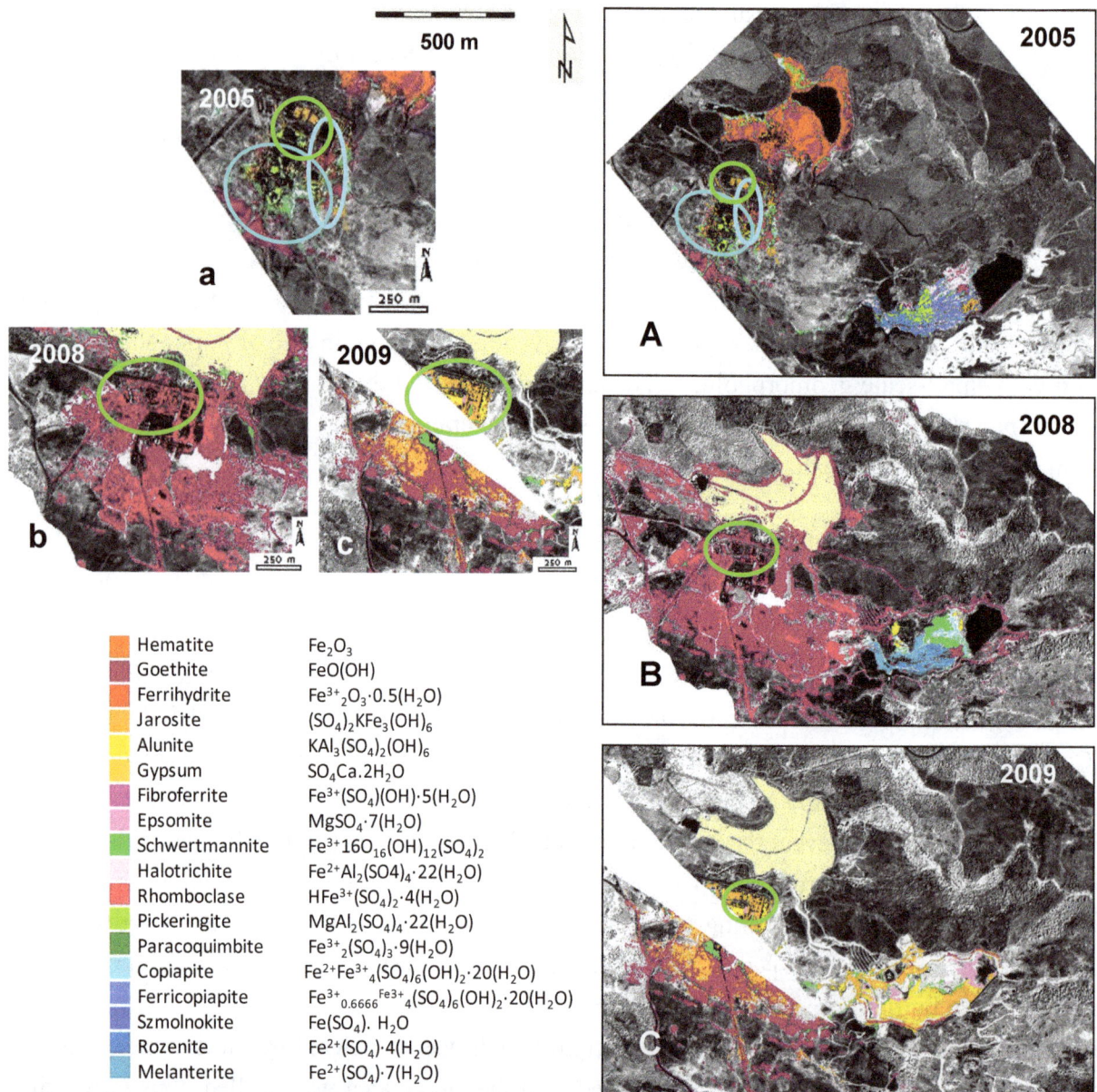

Mineral	Formula
Hematite	Fe_2O_3
Goethite	$FeO(OH)$
Ferrihydrite	$Fe^{3+}_2O_3 \cdot 0.5(H_2O)$
Jarosite	$(SO_4)_2KFe_3(OH)_6$
Alunite	$KAl_3(SO_4)_2(OH)_6$
Gypsum	$SO_4Ca.2H_2O$
Fibroferrite	$Fe^{3+}(SO_4)(OH) \cdot 5(H_2O)$
Epsomite	$MgSO_4 \cdot 7(H_2O)$
Schwertmannite	$Fe^{3+}16O_{16}(OH)_{12}(SO_4)_2$
Halotrichite	$Fe^{2+}Al_2(SO4)_4 \cdot 22(H_2O)$
Rhomboclase	$HFe^{3+}(SO_4)_2 \cdot 4(H_2O)$
Pickeringite	$MgAl_2(SO_4)_4 \cdot 22(H_2O)$
Paracoquimbite	$Fe^{3+}_2(SO_4)_3 \cdot 9(H_2O)$
Copiapite	$Fe^{2+}Fe^{3+}_4(SO_4)_6(OH)_2 \cdot 20(H_2O)$
Ferricopiapite	$Fe^{3+}_{0.6666}{}^{Fe3+}{}_4(SO_4)_6(OH)_2 \cdot 20(H_2O)$
Szmolnokite	$Fe(SO_4). H_2O$
Rozenite	$Fe^{2+}(SO_4) \cdot 4(H_2O)$
Melanterite	$Fe^{2+}(SO_4) \cdot 7(H_2O)$

Figure 4. Ash dam. (**A**) Field spectra from dominant hematite in the ash dam in 2005 and dry grass in 2008 covering the former ash dam. (**B**) HyMap spectra from dominant hematite in the ash dam in 2005 and dry grass in 2008 covering the former ash dam. (**C**) Detail of dominant hematite crusts covering the ash dam in 2005, with spectra in A. (**D**) Detail of dry grass covering the recovered former ash dam in 2008, with spectra in A. (**E**) View of dominantly hematitic crusts covering the ash dam in 2005. (**F**) View of the dry grass covering the former ash dam in 2008, with operator measuring spectra.

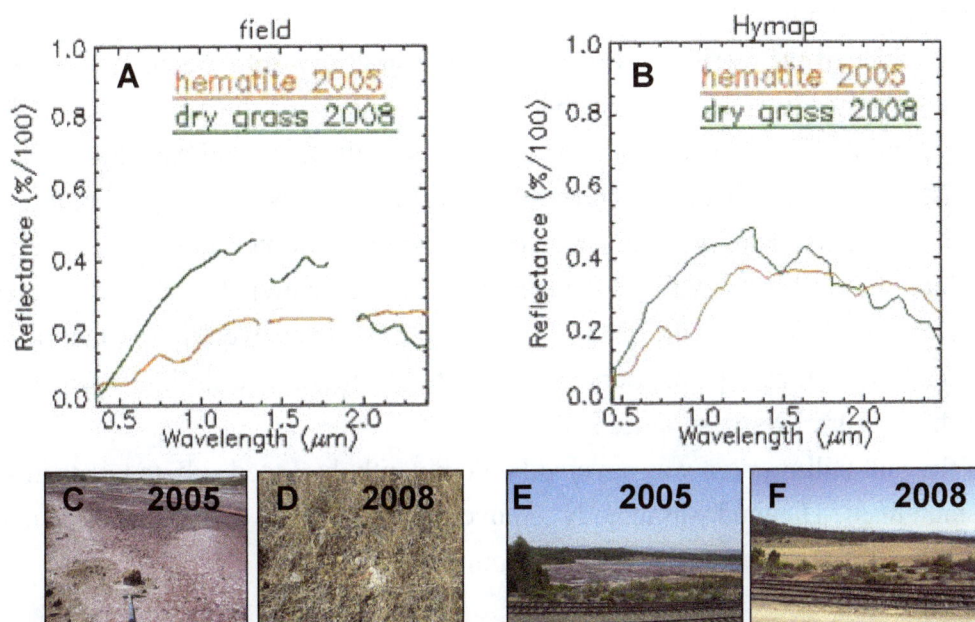

5.2. The Ore Processing Plant and Neighborhood

The ore processing plant stopped its industrial activity in 2001. All buildings and waste ponds around remained abandoned until 2006, when the recovery of the nearby ash dam began.

In 2005, the processing plant was covered by hydrated sulfate, such as paracoquimbite and pickeringite, covering the dumps around the buildings and in small ponds (Figure 3A, cyan circle) storing mill waste. Spots of oxidized iron sulfate, in the form of jarosite, and of hydroxides, as goethite (Figure 3A, green circle), cover the outer areas of the facilities.

No activity within the buildings of the plant has occurred since 2001, but some influence from moving the machinery around can be expected after 2006. The HyMap 2008 flight shows the ore processing plant and neighborhood intensely oxidized by the movement of machinery (Figure 3B), cleaning of small mud ponds and cleaning and smoothing of slopes on the dumps around the plant and the ash dam. Goethite and ferrihydrite cover the entire area. The small ponds around the plant storing mud in 2005 have disappeared by 2008.

The dismantling of buildings and machinery began during the fall of 2008 and progressed through the winter and spring of 2009. In the summer of 2009, the movement of machinery and the removal of parts in the processing plant was still active. The map of August 2009, shows the surfaces of the buildings in the plant covered by jarosite, alunite and some schwertmannite (Figure 3C, green circle), while the neighborhood of the processing plant is covered by goethite over the already cleaned dumps and rock outcrops. In 2009, the intensity of oxidation recedes from its level in 2008, both around the ore processing plant and the roads around the mill tailings plant, a level which had been caused by the movement of machines involved in the recovery.

5.3. The Mill Tailings Dam

Mapping the surface of the mill tailings dam with hyperspectral data is crucial, as there is a large mass of mud which is inaccessible to direct observation. The mineralogy of the thin crusts on the surface of the mass of mud is sensitive to water or mud supplies, whether by industrial operations, maintenance or climate variability.

The industrial activities of the ore processing plant ceased in 2001. Both the pond of ashes and the mill tailings dam kept a steady waste storage, but the maintenance of the facilities included water pumping and flowing through the water bodies. The water from the ash pond was drained to the mill tailings dam until 2006 through a channel (Figure 5A, red arrow). A mud cleaning plant at the foot of the wall of the mill tailings dam prevents the contamination of water downstream and returns processed mud behind the wall to the tailings dam (Figure 5B, red arrow). The artificial water supply at known channel outputs influences the pattern of mineral coatings over the surface of the mill tailings, as much as do the climate parameters and the natural progressive drying up of the mass of mud tailings after the industrial waste supplies stop.

The surface of the mill tailings dam was covered by dominantly hydrated sulfate crusts in 2005 and 2008, covering a mud mass. The mud was already removed in the summer of 2009, and the map shows a comparatively oxidized bottom, already drying and compacting.

5.3.1. The Influence of Water and Mud Operational Supplies: From 2005 to 2008

In 2005, the minerals coating the surface of the mill tailings were intensely hydrated, dominated by rozenite and szomolnokite (Figure 5A, green circle). A fan of uniform pickeringite spreads from the output of the draining channel (Figure 5A, blue circle) supplying water from the dam of ashes (Figure 5A, red arrow). The mud is comparatively shallower at this point, and the outer coating is comparatively more oxidized than the general mass of mud tailings around it. Oxidized crusts, goethite and jarosite, are located far from the water supply, close to the border of the water of the pond (Figure 5A, red circle), surrounded by comparatively more hydrated halotrichite (Figure 5A, orange circle).

The ash dam was cleaned in 2006, and the draining of water to the mill tailings dam stopped. A water cleaning plant at the foot of the wall of the dam prevents the spread of contamination to runoff. In 2008, the only water was supplied by two pipes at both ends of the wall of the mill tailings dam, pumping water from the water cleaning plant at the foot of the wall (Figure 5A,B, red arrows). The area covered by the most hydrated sulfate is strongly dependent on the location of the water supplied by the maintenance of facilities.

The lack of water supply from the ash dam in 2008 erased the pickeringite surface visible in 2005 where there was a water floodgate (Figure 5B, blue circle), it was replaced by a copiapite and alunite tail in 2008, when the topographically most elevated area by the former water floodgate was covered by a uniform surface of alunite (Figure 5B, cyan circle).

Figure 5. The mill tailings dam: (**A–C**) Maps compiled from HyMap data on the minerals product of oxidation and dehydration of sulfide sludge, on the three available flights on the southern mill tailings dam. Color coding for minerals mapped with the corresponding chemical formula are arranged in a precipitation and oxidation sequence. Bluish and greenish colors concentrate on comparatively hydrated areas. Orange and dark red color corresponds to oxidized end members. Acid water stored behind the wall of the dam is in black. Red arrows are the locations of the mud supply. (**D**) Panoramic view of the mill tailings dam.

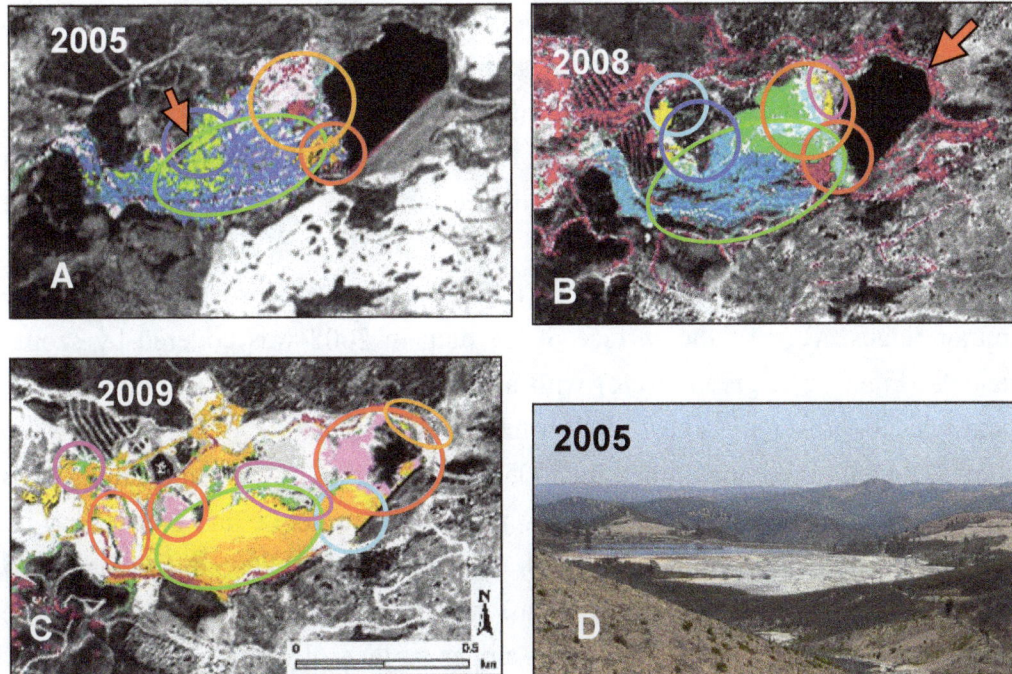

Mineral	Formula
Hematite	Fe_2O_3
Goethite	$FeO(OH)$
Ferrihydrite	$Fe^{3+}_2O_3 \cdot 0.5(H_2O)$
Jarosite	$(SO_4)_2KFe_3(OH)_6$
Alunite	$KAl_3(SO_4)_2(OH)_6$
Gypsum	$SO_4Ca.2H_2O$
Fibroferrite	$Fe^{3+}(SO_4)(OH) \cdot 5(H_2O)$
Epsomite	$MgSO_4 \cdot 7(H_2O)$
Schwertmannite	$Fe^{3+}16O_{16}(OH)_{12}(SO_4)_2$
Halotrichite	$Fe^{2+}Al_2(SO4)_4 \cdot 22(H_2O)$
Rhomboclase	$HFe^{3+}(SO_4)_2 \cdot 4(H_2O)$
Pickeringite	$MgAl_2(SO_4)_4 \cdot 22(H_2O)$
Paracoquimbite	$Fe^{3+}_2(SO_4)_3 \cdot 9(H_2O)$
Copiapite	$Fe^{2+}Fe^{3+}_4(SO_4)_6(OH)_2 \cdot 20(H_2O)$
Ferricopiapite	$Fe^{3+}_{0.6666}{}^{Fe3+}_4(SO_4)_6(OH)_2 \cdot 20(H_2O)$
Szmolnokite	$Fe(SO_4).H_2O$
Rozenite	$Fe^{2+}(SO_4) \cdot 4(H_2O)$
Melanterite	$Fe^{2+}(SO_4) \cdot 7(H_2O)$

5.3.2. Effects of Mud Removal and Drying Up: 2009

The recovery of the mill tailings dam began in the autumn of 2008, through 2011. The mill tailings were actively drained and dried during the winter and spring of 2009. In the summer of 2009 (Figure 5C), the surface of the mill tailings dam was dry and accessible by foot, in the process of consolidation to allow the operation of machinery.

The area formerly flooded behind the wall in 2005 and 2008 was in 2009 largely covered by halotrichite (Figure 5C, red circles). Halotrichite patches occur where mud remains behind the walls at the bottom, which were covered by mud in 2005 and 2008 (Figure 5C, red circles). Drying due to recovery activities during 2009 caused the generalized oxidation of the mineral coatings to jarosite and alunite (Figure 5C, green circle) with schwertmannitic outer fringes (Figure 5C, magenta circles), replacing the hydrated szomolnokite, rozenite and melanterite dominating the surface of the tailings in 2005 and 2008 (Figure 5A,B, green circles).

5.3.3. The Influence of Climate: From 2005 to 2008

The progressive dehydration and oxidation of the secondary minerals of the mill tailings dam from 2005 to 2008 is evidenced by the increasing extent of oxidized crusts. The year 2005 was dryer and warmer than 2008 [27]. The effects of the increasing temperature and less moisture available in 2005 are visible in the mineralogy of the coatings on the surface of the mill tailings dam in the areas away from channel outputs.

Rainwater and mass movements inside the mill tailings developed a micro-geomorpho-geology with rills and minor ridges. Most of the surface of the dam in 2008 was covered by szomolnokite grooved by channels (Figure 5B, green circle) with a bottom of rozenite, a slightly more hydrated sulfate. The topographically lower areas host small patches of copiapite.

The surface of the main mud mass that in 2005 was covered by rozenite and szomolnokite (Figure 5B, green circle), in 2008, was dominated by melanterite, rozenite and copiapite (Figure 5B, green circle). This surface of hydrated minerals shrunk its extension from 2005 due to the lack of water supplies from the ash dam. Since the mud tailings mass had been moving without mud supplies since 2006, the rills and ridges are better developed in the upper surface of coatings over the mud tailing mass in 2008 than in 2005. The pattern of mineralogical crusts shows a strong topographical control both in 2005 and 2008. In 2008, melanterite dominates the western tail, and rozenite draws the more hydrated rills, while copiapite traces the less humid rills.

Halotrichite in 2005 changed its pattern of secondary minerals in 2008 (Figure 5A,B, orange circle) near the water pool behind the wall of the dam. There is a wide uniform surface of schwertmannite in comparatively topographically elevated areas over a lower copiapitic bottom drawing a topographic valley draining to the pond of water. This schwertmannite and copiapite topographically controlled area in 2008 replaces the szomolnokite and halotrichite from 2005, showing an increasing oxidation process.

In the border close to the water, there is a stripe of alunite in 2008 (Figure 5B, magenta circle). This spatial pattern of schwertmannite and copiapite extends southwards, replacing a small jarosite area from 2005 (Figure 5A,B, red circle). This jarositic area was the most oxidized spot in 2005, the far end of the water supply at the time, and by 2008 has moved to the northern border of the water pond, a comparatively topographically high spot concentrating increasing oxidation processes.

5.3.4. Spectral Analysis

The spectra from the end members used to map the mineral crusts with the HyMap data are mineralogically diagnosed using a reference spectral library and specific algorithms (see Section 3). The HyMap spectra of the crusts mapped in the mud tailings dam (Figure 6) illustrate the mineralogical evolution following both the industrial operation changes and climate effects. The spectra in 2008 and 2009 (Figure 6A,B) reflect the dominant presence of hydrated sulfate crusts covering the mass of the mill tailings dam (alunite, pickeringite, copiapite and rozenite). In 2005, a warm year, goethite and gypsum occur (Figure 6A), which were absent in the spectra of the crust units mapped in 2008 (Figure 6B).

Most crusts present in 2008 are diagnosed as hydrated sulfate in the HyMap data (Figure 6B). The spectra from the hydrated sulfate, rozenite and melanterite show enhanced absorption features compared to those in 2005, indicating a wider extent and increased availability of water in 2008.

The HyMap spectra of the identified units in 2009 (Figure 6C), when the mud was already removed, show the presence of comparatively oxidized mineral crusts (goethite and jarosite). The more hydrated sulfate units in 2009 are diagnosed as alunite and schwertmannite as the first choice. However, the second diagnosed minerals are oxidized mineralogical phases (jarosite and goethite), indicating a progressing oxidation, even on the most hydrated ends.

Figure 6. The mill tailings dam. Spectra from HyMap areas mapped in (**A**) June 2005, (**B**) August 2008, and (**C**) August 2009. Spectra are arranged on a precipitation and oxidation sequence. The color coding follows the trend of the key from Figure 3, but changes are made for clarity.

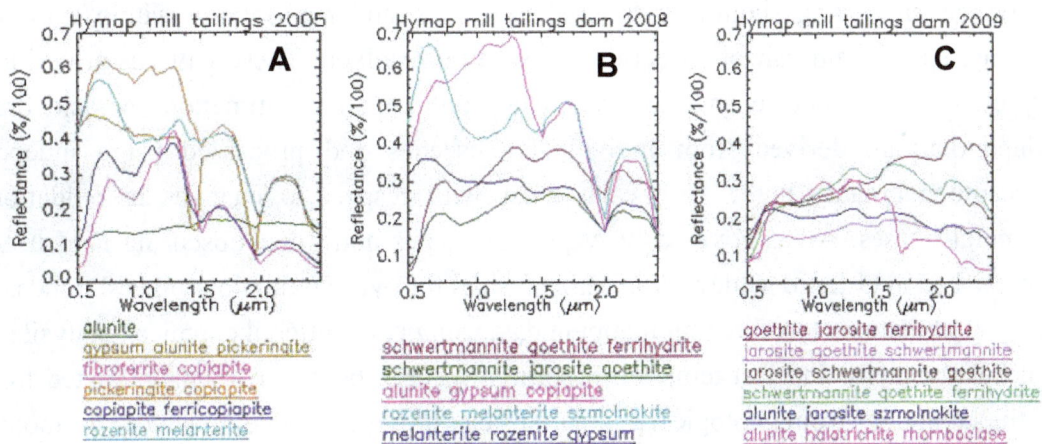

6. Discussion

Although the mineral identification of hyperspectral data necessarily involves reducing the spectral data, with its high dimensionality, to one compositional class, in reality, pixels are rarely pure, and their spectral signatures often reflect heterogeneous forms of mixing. Both field and laboratory spectral experience from samples collected at the mine site show that most spectral features are mixed responses from the selected minerals, which are likely to occur. Only extended and spatially well-developed pans of intensely oxidized pyrite mud show clearly identifiable spectral signatures compared to the archive spectral libraries. The oxidation of pyrite is a progressive and heterogeneous process concerning sulfate crusts and spatial cover, as well as grain size or minor shade effects associated with small-scale geomorphology. The areas identified in the imagery as a single or double mineral presence should be interpreted as indicators of a mineralogically dominant trend displayed by the HyMap imagery with a 5-m spatial resolution.

Prior work monitored the mineralogical changes on mine waste within this mine site considering geomorphological influences under climate variability [15]. HyMap flights from 1999 to 2005 were used to support the observations. This work, in contrast, concentrates on the mineralogical effects induced in the mine site by an intense recovery activity running from 2006 to 2009 [7–9], which is still ongoing.

The complete recovery of the ash dam opens up the possibility of monitoring future changes by the long-term spectral study of the vegetation growing over the recovered premises.

Both the progressing work on the mill tailings dam and the ore processing plant illustrate the mineralogical trends associated with changes due to the activities of recovery. The active cleaning of crusts over dumps is translated spectrally to an intense oxidation evident in 2008 around the ore processing plant, which recedes partly with the removal of buildings and machinery in 2009.

On the mill tailings dam, the changes on the locations of water supply are critical for the pattern of mineralogical coatings on the upper crust of the mud mass. The depth of mud is also partly responsible for the mineralogy of the crusts on the surface. The lack of water and mud supplies lets the mud mass movement display a minor geomorphology in the surface, which is directly translated to mineralogical changes obvious from 2005 to 2008.

This work confirms that imaging spectroscopy fills the need of "useful geophysical techniques to identify, delineate, and monitor, environmental signatures associated with mined and unmined mineral deposits" [1]. Image processing should be tailored per scene and per domain within a mine site; otherwise, complex spatial mineralization patterns cannot be resolved. These patterns are critical for accurate interpretation of climate change trends, metal contamination estimations or acid drainage prediction. Maps that are derived from individual subscenes and processed using independent processing procedures broadly agree with each other with respect to changes in oxidation and dehydration mineral phases. Areas covered by vegetation types and other obscuring land use/cover types had to be masked in order to achieve reasonable results for pyrite oxidation mineral trends.

There are strong limitations to validating mapping data using conventional chemical analysis, due to its high spatial heterogeneity and fast temporal mobility. Spectral behavior trends extracted from the images and in the laboratory from geological evaluations can provide reliable indicators for monitoring contamination from mine wastes. The challenges and limitations mapping pyrite oxidation products over mine waste dealing with preprocessing procedures, heterogeneous mineral mixtures and spectral diagnostic methods are already widely discussed [11].

7. Conclusions

The industrial activity at the mine of Sotiel-Almagrera (Huelva) stopped in the year 2000. The recovery by the regional government began in 2006 with the closure of the ash dam and was still active in 2011. The HyMap 2005 flight pictures the state of the facilities before any recovery activity, with an intensely oxidized ash dam, a comparatively hydrated mud tailings dam and oxidized buildings of the ore processing plant surrounded by crusts of hydrated sulfate on small mud ponds.

The HyMap 2008 flight shows the ash dam covered by dry grass, completely recovered, and the surroundings clean, free from any pyrite weathering product, which has been mechanically cleaned over the dumps. The ore processing plant and neighborhood are intensely oxidized by the movement of the machinery and the cleaning of the small mud ponds. The mud tailings dam is still dominatingly hydrated, although some progress in oxidation can be seen at the mud supply locations, which have ceased operation since the closure of the ash dam.

In 2009, the mud from the mill tailings dam was removed, and the state of oxidation in the crusts is intermediate, as much as the ore processing plant and surroundings, where buildings and machinery are already in the process of being dismantled. The movement of machinery throughout the facilities makes the oxidation state of the whole surface appear uniform in the imagery.

The micro-geomorphology controls the mineralogical coatings on the surface of masses of mill tailings, when there are no artificial supplies of water or mud. The same is true on rock piles of larger dimensions, where minor slopes and topographically low areas are more sensitive and display the mineralogical changes associated with climate variability.

This proves that hyperspectral imagery is an invaluable witness to the changes experienced by abandoned mine sites, whether under the influence of climate and the availability of environmental humidity or under intense human activity.

Acknowledgments

The National Research Plan of Spain (CGL2007-60004/CLI, BES-2008-003648) has funded this work. Both the German Space Agency (Oberpfaffenhofen, Bayern) and the University of Nantes (France) provided training in advanced spectral analytical techniques (Short Stages SEST2009010937 and SEST1000I001495XVO). Jose Manuel Moreira contributed with information about mine inventories and regional environmental databases. The Andalusian Regional Government, the Junta de Andalucía, permitted entry to the abandoned mine site of Sotiel at the border of the river. Local industries also allowed access to the river through their facilities.

Author Contributions

Jorge Buzzi, Asunción Riaza and Eduardo García-Meléndez shared all the interpretative image processing, field geological work and geological interpretation. Sebastian Weide and Martin Bachmann were involved in the flight survey planning and the advanced pre-processing of all HyMap data, and supported the image analysis.

Conflicts of Interest

The authors declare no conflict of interest.

References

1. Plumlee, G.S. The environmental geology of mineral deposits. Part A: Processes, techniques, health issues. *Rev. Econ. Geol.* **1999**, *6*, 71–116.
2. Swayze, G.A.; Smith, K.S.; Clark, R.N.; Sutley, S.J.; Pearson, R.M.; Vance, J.S.; Hageman, P.L.; Briggs, P.H.; Meier, A.L.; Singleton, M.J.; *et al.* Using imaging spectroscopy to map acidic mine waste. *Environ. Sci. Technol.* **2000**, *34*, 47–54.

3. Choe, E.; van der Meer, F.; van Ruitenbeek, F.; van der Werff, H.; de Smeth, B.; Kim, K. Mapping of heavy metal pollution in stream sediments using combined geochemistry, field spectroscopy, and hyperspectral remote sensing: A case study of the Rodalquilar mining area, SE Spain. *Remote Sens. Environ.* **2008**, *112*, 3222–3233.

4. Goetz, A.F.H. Three decades of hyperspectral remote sensing of the Earth: A personal view. *Remote Sens. Environ.* **2009**, *113*, 5–16.

5. van der Meer, F.D.; van der Werff, H.M.A.; van Ruitenbeek, F.J.A.; Hecker, C.A.; Bakker, W.H.; Noomen, M.F.; van der Meijde, M.; Carranza, E.J.M.; de Smeth, J.B.; Woldai, T. Multi- and hyperspectral geologic remote sensing: A review. *Int. J. Appl. Earth Obs. Geoinf.* **2012**, *14*, 112–128.

6. Quental, L.; Sousa, A.J.; Marsh, S. Identification of materials related to acid mine drainage using multi-source spectra at S. Domingos Mine, southeast Portugal. *Int. J. Remote Sens.* **2013**, *34*, 1928–1948.

7. Exelis. *ENVI User's Guide*; Exelis Visual Information Solutions: Boulder, CO, USA, 2011.

8. Leblanc, M.; Morales, J.A.; Borrego, J.; Elbaz-Poulichet, A. 4,500-year-old mining pollution in Southwestern Spain: Long-term implications for modern mining pollution. *Econ. Geol.* **2000**, *95*, 655–662.

9. Buurman, P. *In vitro* weathering products of pyrite. *Geol. Mijnvouw* **1975**, *54*, 101–105.

10. Nordstrom, D.K.; Alpers, C.N. Geochemistry of Acid Mine Waters. In *The Environmental Geochemistry of Mineral Deposits. Part A: Processes, Techniques, and Health Issues*; Plumlee, G.S., Logsdon, M.J., Eds.; Society of Economic Geologists: Littleton, CO, USA, 1999; pp. 133–160.

11. Alpers, C.N.; Nordstrom, D.K.; Spitzley, J. Extreme Acid Mine Drainage from a Pyritic Massive Sulfide Deposit: The Iron Mountain Endmember. In *Environmental Aspects of Mine-Wastes*; Jambor, J.L., Blowes, D.W., Ritchie, A.I.M., Eds.; Mineralogical Association of Canada: Quebec City, QC, Canada, 2003; pp. 407–430.

12. Riaza, A.; Müller, A. Hyperspectral remote sensing monitoring of pyrite mine wastes: A record of climate variability (Pyrite Belt, Spain). *Environ. Earth Sci.* **2010**, *61*, 575–594.

13. Buzzi, J. Imaging Spectroscopy to Evaluate the Contamination from Sulphide Mine Waste in the Iberian Pyrite Belt Using Hyperspectral Sensors (Huelva, Spain). Ph.D. Thesis, Universidad de León, León, Spain, 14 December 2012.

14. Buzzi, J.; Carrère, V.; Riaza, A.; García-Meléndez, E.; Bachmann, M. Modified Gaussian Modelization Applied to Hyperspectrtal Data in an AMD-Contaminated Area: Case of Odiel River (Huelva, SW Spain). In Proceedings of the 7th EARSeL Workshop on Imaging Spectroscopy, Edinburgh, UK, 11–13 April 2011.

15. Buzzi, J.; Riaza, A.; García-Meléndez, E.; Holzwarth, S. Change Detection in Sediments of a River Affected by Acid Mine Drainage Using Airborne Hyperspectral Hymap Data (River Odiel, SW Spain). In Proceedings of the 4th EARSeL Workshop on Remote Sensing and Geology, Mykonos, Greece, 24–25 May 2012; pp. 134–158.

16. Riaza, A.; Buzzi, J.; García-Meléndez, E.; Carrère, V.; Müller, A. Monitoring the extent of contamination from acid mine drainage in the Iberian Pyrite Belt (SW Spain) using hyperspectral imagery. *Remote Sens.* **2011**, *3*, 2166–2186.

17. Gibbons, W.; Moreno, T. *The Geology of Spain*; Geological Society: London, UK, 2002.

18. Analytical Spectral Devices, Inc. *ASD, FieldSpec® 3 User Manual, ASD Document 600540 Rev. F*; Analytical Spectral Devices, Inc.: Boulder, CO, USA, 2006.

19. Crowley, J.K.; Williams, D.E.; Hammarstrom, J.M.; Piatak, N.; Chou, I.-M.; Mars, J.C. Spectral reflectance properties (0.4–2.5 μm) of secondary Fe-oxide, Fe-hydroxide, and Fe-sulphate-hydrate minerals associated with sulphide-bearing mine wastes. *Geochem. Explor. Environ. Anal.* **2003**, *3*, 219–228.

20. Clark, R.N.; Swayze, G.E.; Wise, R.; Livo, E.; Hoefen, T.; Kokaly, R.; Sutley, S.J. *USGS Digital Spectral Library Splib06a*; U.S. Geological Survey: Reston, VA, USA, 2007.

21. Riaza, A.; Garcia-Melendez, E.; Mueller, A. Spectral identification of pyrite mud weathering products: A field and laboratory evaluation. *Int. J. Remote Sens.* **2011**, *32*, 185–208.

22. Clark, R.N.; Swayze, G.A.; Gallagher, A.; King, T.V.V.; Calvin, W.M. *The U.S. Geological Survey Digital Spectral Library: Version 1: 0.2 to 3.0 μm*; U.S. Geological Survey: Reston, VA, USA, 1993.

23. Boardman, J.W. Automated Spectral Unmixing of AVIRIS Data Using Convex Geometry Concepts. In Proceedings of the Summaries of the 4th JPL Airborne Geoscience Workshop, Washington, DC, USA, 25–29 October 1993.

24. Kruse, F.A.; Lebkoff, A.B.; Boardman, J.B.; Heidebrecht, K.B.; Shapiro, A.T.; Barloon, P.J.; Goetz, A.F.H. The Spectral Imaging Processing System (SIPS)—Interactive Visualization and Analysis of Imaging Spectrometer Data. *Remote Sens. Environ.* **1993**, *44*, 145–163.

25. Clark, R.N.; Gallagher, A.J.; Swayze, G.A. Material Absorption Band Depth Mapping of Imaging Spectrometer Data Using the Complete Band Shape Least-Squares Algorithm Simultaneously Fit to Multiple Spectral Features from Multiple Materials. In Proceedings of the 3rd Airborne Visible/Infrared Imaging Spectrometer (AVIRIS) Workshop, Pasadena, CA, USA, 20–21 May 1991.

26. Goetz, A.F.H.; Vane, G.; Solomon, J.E.; Rock, B.N. Imaging spectrometry for earth remote sensing. *Science* **1985**, *228*, 1147–1153.

27. Agencia Estatal de Meteorología, Spain (AEMET). Resumen Anual Climatológico de los años 2004, 2005, 2006, 2007, 2008 y 2009. Available online: http://www.aemet.es (accessed on 11 April 2014).

5

Evaluation of Uranium Concentration in Soil Samples of Central Jordan

Ned Xoubi

Nuclear Engineering Department, King Abdulaziz University, P.O. Box 80204, Jeddah 21589, Saudi Arabia; E-Mail: nxoubi@kau.edu.sa

Academic Editor: Mostafa Fayek

Abstract: Naturally occurring radionuclides such as uranium, thorium and their decay products (^{226}Ra, ^{222}Rn) are present in a number of geological settings in Jordan. Motivated by the existence of uranium anomalies coupled with its lack of conventional energy resources, Jordan decided that the development of this indigenes resource (uranium) is the first step in introducing nuclear power as part of its energy mix. Uranium deposits in Central Jordan were perceived not only as a secured resource that will fulfill Jordan's energy needs, but also as an economic asset that will finance Jordan's nuclear program. The average uranium concentration of 236 soil samples using ICP-Mass (inductively coupled plasma mass spectrometry) was found to be 109 parts per million (ppm). Results analysis revealed a wide range of 1066 ppm for uranium concentration, and a median of 41 ppm uranium. The measurements frequency distribution indicates that 72% of samples measured had a uranium content of less than 100 ppm, a concentration that characterizes overburden and tailings quality, rather than minable reserves. This paper presents and evaluates the concentration of uranium in central Jordan, being the most promising area with the highest radioactive anomalies in Jordan.

Keywords: uranium; exploration; Jordan; nuclear energy; ICP-Mass

1. Introduction

Worldwide most countries rely on imports to fulfill their energy needs. Jordan is no exception, with the lack of conventional energy resources and motivated by the existence of uranium occurrences coupled with official studies claiming proven reserves of high-grade uranium. The government of Jordan declared in 2007 its intention to introduce nuclear power as part of its energy mix [1].

The development of this indigenous fuel (uranium) resource was presented by Officials not only as a secured resource that will help fulfill Jordan's energy needs, but also an economic asset that will finance Jordan's nuclear program [2,3].

In 2008, Jordan Atomic Energy Commission (JAEC) was established, and its Chairman stated that, "The first step in Jordan's nuclear program is the mining and extraction of uranium, which constitute a strategic wealth being available in commercial quantities in Jordan" [4]. He added, "Uranium reserves in central Jordan are estimated at about 70,000 metric tons, with an average concentration of 500 ppm" [5].

In 1980, a countrywide Gamma airborne survey was carried out revealing 11 areas with high background radiation intensity of up to 2000 counts per second (cps) as shown in Figure 1 [6]. Although most areas are related to known phosphorite formations (outlined in black), where Phosphate is being mined and exported, five of the areas were identified as potential uranium deposits (outlined in red); the most promising of the five is an area located the central part of the Jordan.

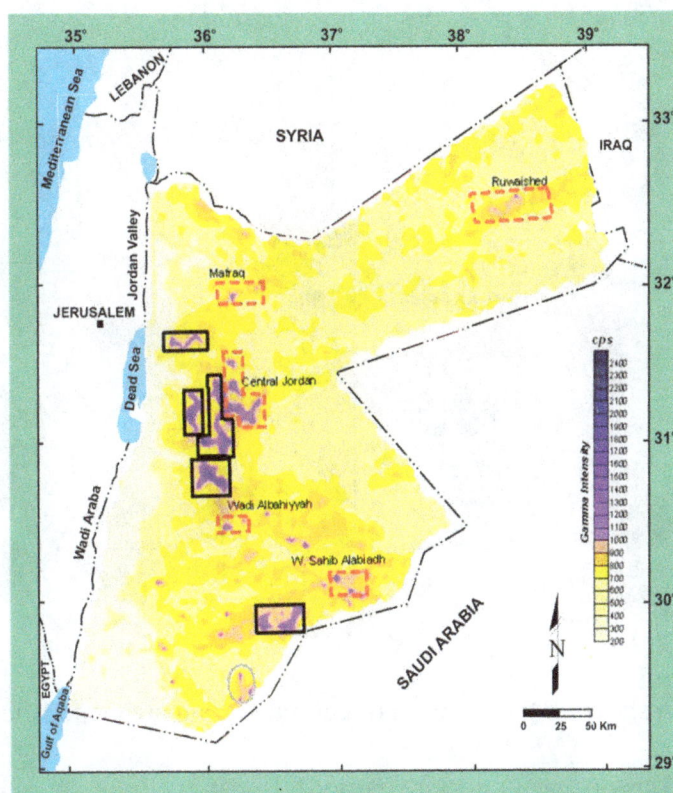

Figure 1. Jordan countrywide Gamma intensity airborne survey, showing areas of high radiation. Known phosphate mining areas are outlined in black, areas with uranium anomalies are outlined in red, and one area with thorium anomalies is outlined in gray (modified after [6]).

Naturally occurring radionuclides such as uranium, thorium and their decay products (^{226}Ra, ^{222}Rn) are present in a number of geological settings in Jordan. Uranium occurrences in phosphorite, oil shale, limestone, marble, sandstone have been studied over the last 40 years primarily during geological mapping and exploration projects [7–14]. Jordan does not have any assured uranium resources, or uranium mining.

The aim of this work is to present and evaluate the concentration of uranium in central Jordan, and to analyze the results in the framework of international standards for uranium ore, resources and reserves classification.

2. Geological Setting

Jordan is a small country situated to the north west of the Arabian Peninsula, south of Syria and east of the Great Rift Valley home of the Dead Sea and Jordan River, between latitudes 29° and 33° north and longitudes 34° and 39° east. The study area is located in central Jordan, approximately 65 km southeast of the capital Amman, and 40 km East of the Dead sea, and consists of two zones, the Fertile Zone (FZ) and Khan Az-Zabib (KZ) (Figure 2).

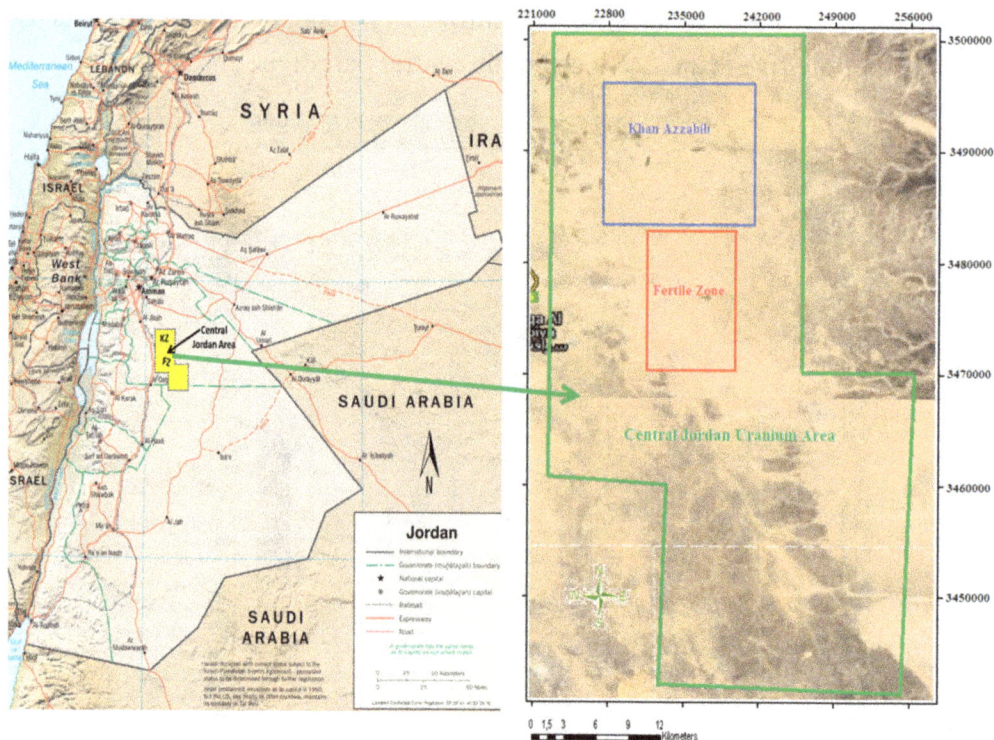

Figure 2. Map of Jordan showing the location of central Jordan area (modified after [15,16]).

During the Late Cretaceous to early Eocene period, Jordan was situated in a shallow marine environment of the Tethys Sea. In the Cenomanian times, transgression took place in relation to a global warming event, which resulted in the deposition of mostly calcareous marine sediments.

During the Late Eocene, a regression led to a period of uplifting, folding and faulting in the region, which are mostly related to the continued tectonic movement along the Great Rift. This period is also characterized by left lateral trans-tensional events, which were responsible for the Dead Sea rifting, extensive basalt flows, that extend to Syria and Saudi Arabia [17–19].

The central Jordan area is located near the opening area of the Dead Sea, a sector that is strongly structured by fault system; the main faults are the Zerqa Main, Daba and Siwaqa fault systems [20,21].

Central Jordan area was mapped by the Natural Resources Authority (NRA) [20,21] and its geology was described in detail as illustrated in Figure 3 [13,18,19,22].

In 2008, Jordan Atomic Energy commission (JAEC) delineated an area of 1469 km^2 in central Jordan for uranium exploration and mining, the area was licensed to Areva and was closed to all mining activities except uranium. The central Jordan uranium area map is shown in Figure 1, and its coordinates are tabulated in Table 1.

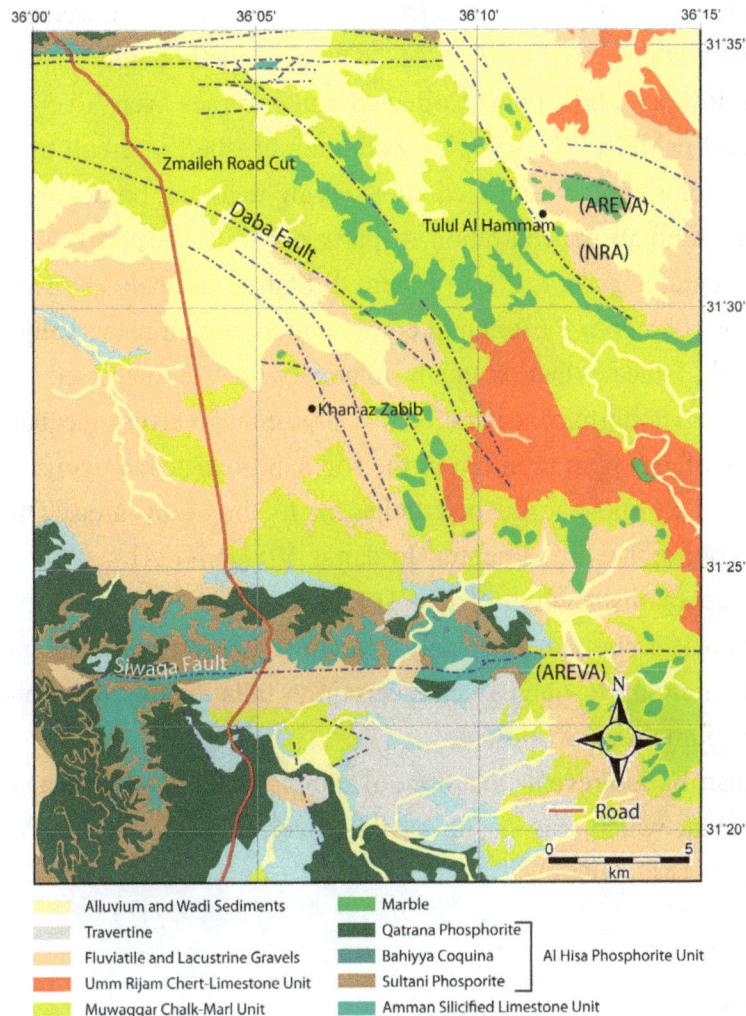

Figure 3. Geological map of central Jordan (modified after [13]).

Table 1. Coordinates of the Central Jordan uranium area, in UTM (Universal Transverse Mercator) 37° north, and in GPS.

North (UTM)	East (UTM)	Latitude	Longitude
3,500,685	246,010	31°36'	36°19'
3,470,281	245,863	31°20'	36°19'
3,470,281	256,291	31°20'	36°26'
3,441,306	255,339	31°04'	36°26'
3,442,076	232,782	31°04'	36°11'
3,460,200	233,403	31°14'	36°12'
3,460,734	222,363	31°14'	36°05'
3,500,685	222,510	31°36'	36°04'

Two zones in central Jordan (Figure 1) were identified as having the highest concentration of uranium anomalies in the country [15]; the Fertile Zone with an area of 100 km^2, and Khan Azzabib with an area of 178 km^2. JAEC claim that the average uranium concentration in the two zones is 500 ppm [5].

3. Results and Discussion

The uranium concentration of 236 soil samples from central Jordan was measured using inductively coupled plasma mass spectrometry (ICP-MS) [23], a precise method that is capable of detecting uranium concentration as low as 0.001 ppm. Collected soil samples were oven dried for 24 h and sieved to remove stones and pebbles. The samples were then crushed in a ball mill to pass through a fine mesh sieve (0.5 mm), each sample was mixed after sieving to assure homogeneity.

Measurement results of uranium concentration in parts per million (ppm) *vs.* sample number are tabulated in Table 2, and depicted graphically in Figure 4. Statistical analysis revealed that the results exhibit a very wide range of 1066 ppm for uranium concentration, with a mean value of 109 ppm, and a median of 41 ppm. Which indicate that uranium mineralization is dispersed over wide areas in small-localized areas rather than uniform concentration in a large enough layers adequate for exploitation.

The mean average value of 109 ppm uranium in the soil samples, is analogous to the concentration of secondary uranium found in phosphorite beds adjacent to the study area. The average uranium concentration in Jordanian phosphate ore of central Jordan is 105 ppm [24].

The measurements frequency distribution (Table 3) indicates that 72% of samples measured had a uranium content of less than 100 ppm, an extremely low concentration that is characteristic of natural radioactivity rather than uranium resources, and certainly falls below the cut-off grade of any commercially operated uranium mines worldwide.

In fact, the frequency distribution reveals that 86% of samples measured have a concentration below the cut-off grade of 250 ppm, which was set-up by the government of Jordan for uranium resource estimation in Central Jordan [15].

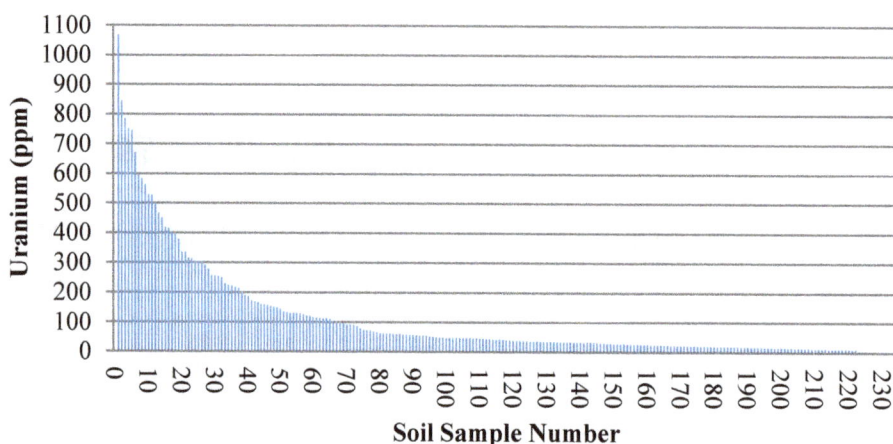

Figure 4. Uranium concentration in parts per million (ppm) in soil samples from Central Jordan using inductively coupled plasma mass spectrometry (ICP-MS).

Table 2. Uranium concentration in soil samples using ICP-MS, in parts per million (ppm).

Sample Number	Uranium (ppm)	Sample Number	Uranium (ppm)	Sample Number	Uranium (ppm)
1	1068	36	219	71	93
2	845	37	215	72	93
3	785	38	198	73	92
4	752	39	191	74	87
5	747	40	188	75	80
6	671	41	174	76	75
7	595	42	170	77	72
8	585	43	167	78	72
9	564	44	162	79	69
10	530	45	160	80	67
11	529	46	159	81	63
12	499	47	155	82	63
13	467	48	152	83	62
14	452	49	150	84	62
15	419	50	147	85	62
16	416	51	136	86	60
17	404	52	134	87	60
18	396	53	132	88	58
19	380	54	132	89	58
20	336	55	132	90	57
21	336	56	131	91	57
22	317	57	128	92	56
23	314	58	123	93	55
24	306	59	122	94	55
25	305	60	118	95	55
26	305	61	116	96	52
27	294	62	116	97	51
28	280	63	114	98	50
29	259	64	114	99	49
30	257	65	111	100	49
31	256	66	105	101	48
32	251	67	101	102	48
33	231	68	98	103	48
34	226	69	96	104	48
35	224	70	93	105-236	<48

Results were grouped into eight categories depending on the uranium content of the sample as illustrated in Figure 5, the lowest category is for uranium of less than 50 ppm of which 139 samples fallen into this category. The highest category is for uranium of more than 500 ppm of which only 11 samples fallen into this category. The rest of our measurements, 86 samples fell in the six categories in between with the majority of result closer to the lower concentration categories.

The US Nuclear Regulatory commission (NRC) has established a threshold of 500 ppm (0.05%) uranium concentration for ore grade to be deemed as source material under its regulations 10 CFR (Code of Federal Regulation) 40.4; NRC has carefully taken into account current technology and economics in selecting this threshold [25]. Applying the NRC threshold, we found that only 5% of the samples have an adequate amount of uranium concentration to pass the threshold and maybe considered with potential for exploration.

Table 3. Measurements frequency distribution.

Uranium Concentration (ppm)	Number of Samples	Frequency (%)	Cumulative Frequency (%)
Less than 50	139	59%	59%
50–99	30	13%	72%
100–149	18	8%	79%
150–199	12	5%	84%
200–249	5	2%	86%
250-299	6	3%	89%
300-499	15	6%	95%
More than 500	11	5%	100%

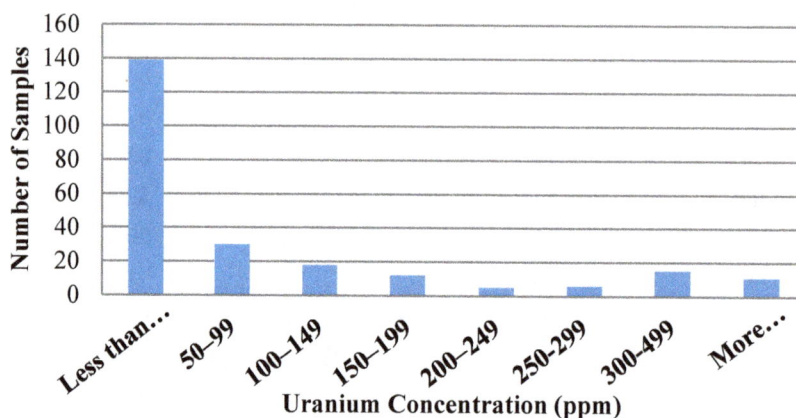

Figure 5. Uranium concentration in parts per million (ppm) *vs.* number of samples.

4. Conclusions

Countrywide Gamma airborne survey of Jordan revealed the presence of high radiation intensity areas, some associated with the presence of uranium mineralization. Two zones in central Jordan (Fertile zone and Khan Az-Zabib) were identified as having the highest concentration of uranium anomalies in the country.

The average uranium concentration of 236 soil samples from central Jordan was found to be 109 ppm. Farther analysis showed that more than 72% of the samples measured had a uranium content of less 100 ppm, an extremely low concentration that is characteristic of natural radioactivity rather than uranium resources.

Statistical analysis revealed that the results exhibit a very wide range of 1066 ppm for uranium concentration, with a median of 41 ppm. Results that are normally indicative of scattered mineralization rather than uniform concentration in large enough layers adequate for exploitation.

We can finally conclude, based on our findings, that the central Jordan area offers little or no uranium prospectively. Furthermore, that apart from perhaps by-product uranium arising from phosphate mining, Jordan does not seem to have any prospect for radioactive minerals.

Acknowledgement

The author would like to extend special thanks to the three anonymous reviewers for their valuable and constructive comments, which have helped improving the quality of the article.

Conflicts of Interest

The authors declare no conflict of interest.

References

1. Eldar, A. Jordan Aims to Develop Nuclear Power. Available online: http://www.haaretz.com/news/king-abdullah-to-haaretz-jordan-aims-to-develop-nuclear-power-1.210546 (accessed on 20 March 2015).

2. Dawood, A.; Sayidh, W.; Nofal, I. Head of the Atomic Energy Commission in a Comprehensive Dialogue, Ad-Dustour. Available online: http://www.addustour.com/2009/6/16/18/ (accessed on 20 March 2015).

3. Toukan, K. Jordan's Nuclear Reactor as an Alternative Source of Energy, Alrai Studies Center. Available online: http://alraicenter.com/alraicenter.com/User_Site/Site/View_Articlear.aspx?type=2&ID=254 (accessed on 20 March 2015).

4. Toukan, K. Jordan and France Sign a Protocol on the Use Nuclear Energy for Peaceful Purposes. Alarab Alyawm: Amman, Jordan, 28 August 2008. (In Arabic)

5. Haddad, S.; Hgazeen, F. No Real Obstacles in Front of Jordan to Implement its Nuclear Program. Alarab Alyawm: Amman, Jordan, 19 July 2009. (In Arabic)

6. Phoenix Corporation. *Comprehensive Airborne Radiation Survey of the Hashemite Kingdom of Jordan*; Natural Resources Authority (NRA): Amman, Jordan, 1980.

7. Abu-ajamieh, M. *Uranium Reserves in Jordan, Geological Survey and Bureau of Mines*; Natural Resources Authority (NRA): Amman, Jordan, 1974.

8. Healy, R.; Young, J. *Mineralogy of U-Bearing Marls from the Jordanian Desert*; Unpublished Report; Cameco Corporation: Saskatoon, SK, Canada, 1998.

9. Smith, B.; Powell, J.; Bradley, A.; Gedeon, R.; Amro, H. Naturally occurring uranium pollution in Jordan. *Int. J. Rock Mech. Min. Sci. Geomech. Abstr.* **1996**, *33*, 96A.

10. Abed, A.M.; Khalid, H. Uranium distribution in the Jordanian phosphorite. *Dirasat* **1985**, *12*, 91–103.

11. Helmdach, F.; Khoury, H.; Meyer, J. Secondary uranium mineralization in the Santonian-Turonian, near Zarqa, North Jordan. *Dirasat* **1985**, *12*, 105–112.

12. Xoubi, N. Jordan's Recent & Ongoing Activities in Uranium Exploration. In Proceedings of the 43rd Meeting of Joint Uranium Group, Vienna, Austria, 17–19 June 2009.

13. Khoury, H.; Salameh, E.; Clark, I. Mineralogy and origin of surficial uranium deposits hosted in travertine. *Appl. Geochem.* **2014**, *43*, 49–65.

14. Fleurance, S.; Cuney, M.; Malartre, F.; Reyx, J. Origin of the extreme polymetallic enrichment (Cd, Cr, Mo, Ni, U, V, Zn) of the Late Cretaceous—Early Tertiary Belqa Group, central Jordan. *Palaeogeogr. Palaeoclim. Palaeoecol.* **2013**, *369*, 201–219.

15. Uranium Mining Agreement. *Government of Jordan and Areva, Jordan Official Gazette No 5037*; Prime Ministry–Directorate of the Official Gazette: Amman, Jordan, 2010; pp. 3424–3551.

16. Central Intelligence Agency. *The World Fact Book*; Base 803051AI (C00697)4-04; Central Intelligence Agency: Washington, DC, USA, 2013.

17. Bender, F. *Geology of Jordan: Contribution of the Regional Geology of the Earth*; Gebrüder Borntraeger: Berlin, Germany, 1974.

18. Powell, J.H.; Moh'd, B.K. Evolution of Cretaceous to Eocene alluvial and carbonate platform sequences in central and south Jordan. *GeoArab. Middle East Pet. Geosci.* **2011**, *16*, 29–82.

19. Powell, J.H. *Stratigraphy and Sedimentology of the Phanerozoic Rocks in Central and Southern Jordan, Bull. 11*; Part B: Kurnub, Ajlun and Belqa Group; Geology Directorate, Natural Resources Authority (NRA): Amman, Jordan, 1989.

20. Jaser, D. *The Geology of Khan Ez Zabib, Bull. 3*; Natural Resources Authority (NRA): Amman, Jordan, 1986.

21. Barjous, M. *The Geology of Siwaqa, Bull. 4*; Natural Resources Authority (NRA): Amman, Jordan, 1986.

22. Quennell, A. Geological Map of Jordan (East of the Rift Valley 1:250,000). Department of Lands and Survey: Amman, Jordan, 1956.

23. Toukan, K. Strategy of Nuclear Energy in Jordan Abdul Hameed Shoman Foundation. Available online: http://www.shoman.org.jo/lecture/GetMaterial.aspx?MID=478 (accessed on 20 March 2015).

24. Abed, A. Review of uranium in the Jordanian phosphorites: Distribution, genesis and industry. *Jordan J. Earth Environ. Sci.* **2012**, *4*, 35–45.

25. Environmental Protection Agency (EPA). *Technologically Enhanced Naturally Occurring Radioactive Materials from Uranium Mining*; US Environmental Protection Agency: Washington, DC, USA, 2008.

6

The Distribution, Character, and Rhenium Content of Molybdenite in the Aitik Cu-Au-Ag-(Mo) Deposit and Its Southern Extension in the Northern Norrbotten Ore District, Northern Sweden

Christina Wanhainen [1,*], Wondowossen Nigatu [2], David Selby [3], Claire L. McLeod [4], Roger Nordin [5] and Nils-Johan Bolin [6]

[1] Division of Geosciences and Environmental Engineering, Luleå University of Technology, SE 971 87 Luleå, Sweden

[2] Gunnarn Exploration AB, Blåvagen 207 Box 149, SE 923 23 Storuman, Sweden; E-Mail: wondossenbekele@yahoo.com

[3] Department of Earth Sciences, University of Durham, Durham DH1 3LE, UK; E-Mail: david.selby@durham.ac.uk

[4] Department of Earth and Atmospheric Sciences, University of Houston, Houston, TX 77204-5007, USA; E-Mail: clmcleod@central.uh.edu

[5] Exploration Department, Boliden Mineral AB, SE 936 81 Boliden, Sweden; E-Mail: roger.nordin@boliden.com

[6] Division of Process Technology, Boliden Mineral AB, SE 936 81 Boliden, Sweden; E-Mail: nils-johan.bolin@boliden.com

* Author to whom correspondence should be addressed; E-Mail: chwa@ltu.se

External Editor: Thomas N. Kerestedjian

Abstract: Molybdenite in the Aitik deposit and its southern extension was studied through mineralogical/chemical analysis and laboratory flotation tests. It is demonstrated that molybdenite varies considerably in grain size, ranging from coarse (>20 μm) to very fine (<2 μm) and occurs predominantly as single grains in the groundmass of the rocks, as grain aggregates, and intergrown with chalcopyrite and pyrite. The dominating molybdenite-bearing rocks are the mica schists, the quartz-monzodiorite, and the Salmijärvi biotite-amphibole gneiss, the latter containing mostly medium-coarsegrained molybdenite. Later geological features, such as garnet-magnetite-anhydrite-K feldspar alteration and pegmatite dikes appear to be responsible for a significant part of the distribution pattern of molybdenite. Molybdenite grains contain up to 1587 ppm Re, with an average of 211 ± 10 ppm in Aitik molybdenite and 452 ± 33 ppm in Salmijärvi molybdenite. The higher Re concentrations are found in molybdenite associated with sericite- and quartz-amphibole-magnetite altered rocks, whereas low Re values occur in rocks in which potassic alteration is prominent. Molybdenite recovery is influenced by the mineralogy of the host rock and the alteration grade; hence both of these factors will have an impact on potential recoveries. The recovery of molybdenite was lower from flotation feeds with significant amounts of Mg-bearing clay-micas.

Keywords: Aitik deposit; northern Sweden; molybdenite; rhenium; QEMSCAN; dilution ICP-MS; mineral processing; ore characterization

---

1. Introduction

The Aitik porphyry Cu-Au-Ag-(Mo) deposit is located 60 km north of the Arctic Circle at latitude 67°04'N and longitude 20°57'E in the Gällivare area in northern Sweden (Figure 1). The mineralization extends for ~5 km with a maximum width of 400 m, and is known to extend to a maximum depth of 800 m in the deepest investigated parts (Figure 2). The Aitik mine is by size the largest open-pit metal mine in Europe and the main pit measures over 3 km in length, 1.1 km in width and 435 m in depth (Figure 2). Production in the Salmijärvi open-pit, which is situated in the southern strike prolongation of the Aitik main pit and planned to be 800 m long, 400 m wide and 275 m deep, started at the end of 2010 (Figure 2). The Aitik deposit was discovered in 1932 and has been in production since 1968 when it started as a 2 Mt/year open-pit operation. Since then over 632 Mt of ore has been mined averaging 0.35% Cu, 0.18 ppm Au and 3.4 ppm Ag. Production in 2013 was 37 Mt of ore with an output of 70,927 tonnes of Cu, 1765 kg of Au, and 53,612 kg of Ag. Ore reserves at the announcement of the Aitik 45 Mt expansion project in May 2014, were 1085 Mt grading 0.22% Cu, 0.14 ppm Au, 1.5 ppm Ag and 26 ppm Mo, with remaining measured and indicated mineral resources totaling 1716 Mt.

A major mine expansion program for Aitik has recently been finished to reach 36 Mt of annual ore production, and a decision to go for 45 Mt capacity was taken in early summer 2014. Future plans for the Aitik mine also include a study which will assess the possibility of commencing molybdenum extraction. To date, no studies have been performed on the occurrence of molybdenite within the ore body, a detailed characterization of this mineral is essential in order to maximize future recovery of molybdenum. With current metal prices it is important to characterize the ore body of its minor components and to map and characterize metallurgical parameters that could affect the ore value, regardless if the minor component constitutes a penalty metal or a bonus metal. Since rhenium occurs naturally in molybdenite [1], extraction of molybdenite through flotation would yield a rhenium-bearing molybdenite concentrate that has a potentially higher market value.

Figure 1. Geology of the Gällivare area (modified from [2]). Inset shows the location of the Gällivare area (rectangle) and the Aitik deposit (star) within the Fennoscandian Shield.

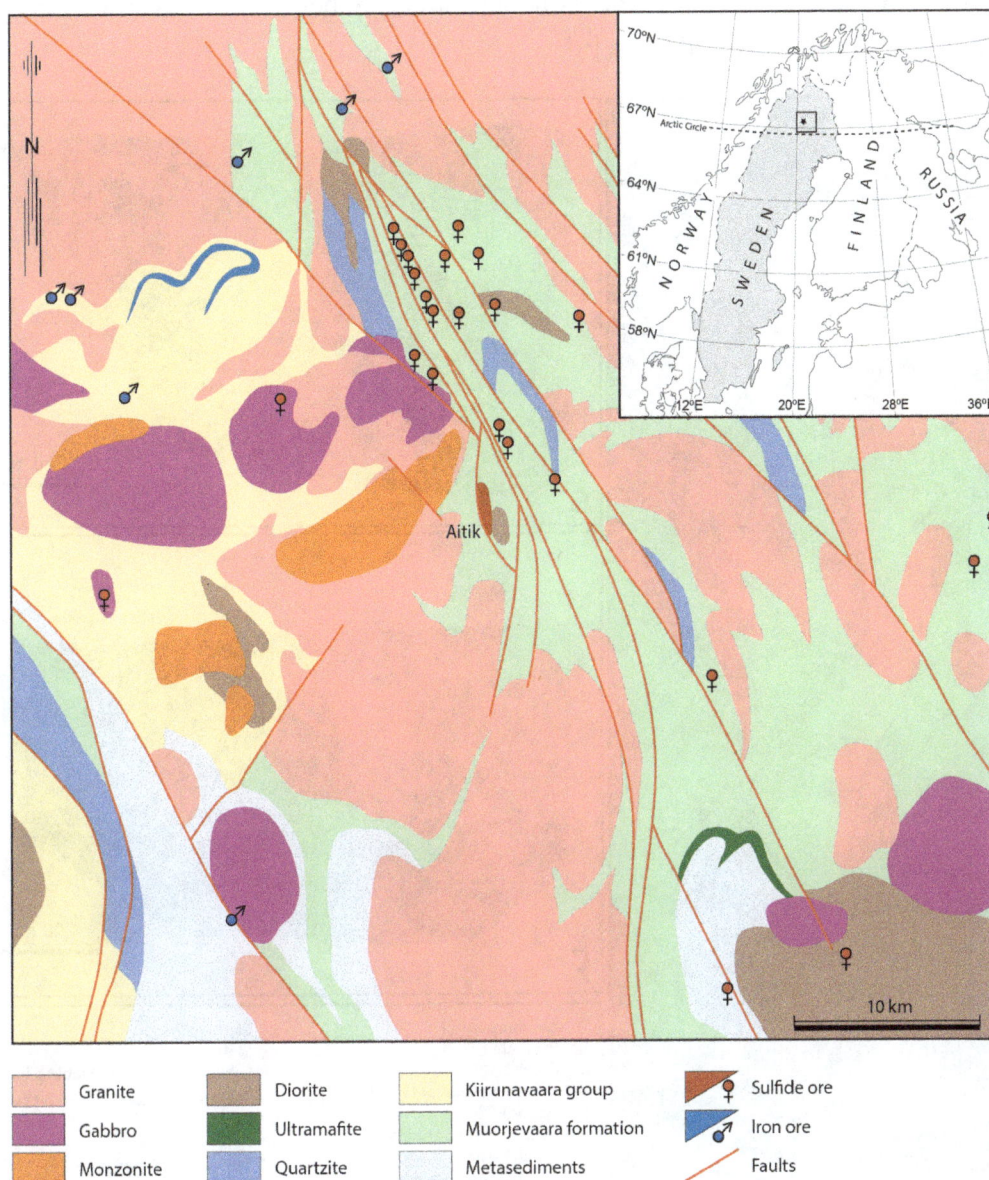

Granite		Diorite		Kiirunavaara group	Sulfide ore
Gabbro		Ultramafite		Muorjevaara formation	Iron ore
Monzonite		Quartzite		Metasediments	Faults

The Aitik and Salmijärvi ore bodies that are presented by this study are Palaeoproterozoic deposits which have been modified by multistage metamorphic-tectonic and magmatic-hydrothermal events. Redistribution of metals and possible addition of ore minerals from a later mineralizing event has led to a wide variety of mineralization styles and a complex mineralization pattern [3]. This pattern has been studied in detail regarding copper, and to some extent gold [4,5].

In this paper, we describe the character of molybdenite in the Aitik deposit, and demonstrate the application of qualitative and quantitative mineralogical data derived from optical microscope, QEMSCAN (Quantitative evaluation of minerals by scanning electron microscopy) and ICP-MS (Inductively coupled plasma mass spectrometry) to genetic and mineral processing issues. The results will form an integral part of the input required for a future blockmodel where the geological, mineralogical and metallurgical character of the ore is combined, and the true value of this deposit assessed accurately.

Figure 2. Geology and open pit contours of the Aitik and Salmijärvi deposits and their close surroundings. Local coordinates in meter; (**a**) Horizontal section at 100–200 m depth; (**b**) Schematic E-W vertical section of the Aitik deposit; (**c**) Schematic E-W vertical section of the Salmijärvi deposit. Modified from [6].

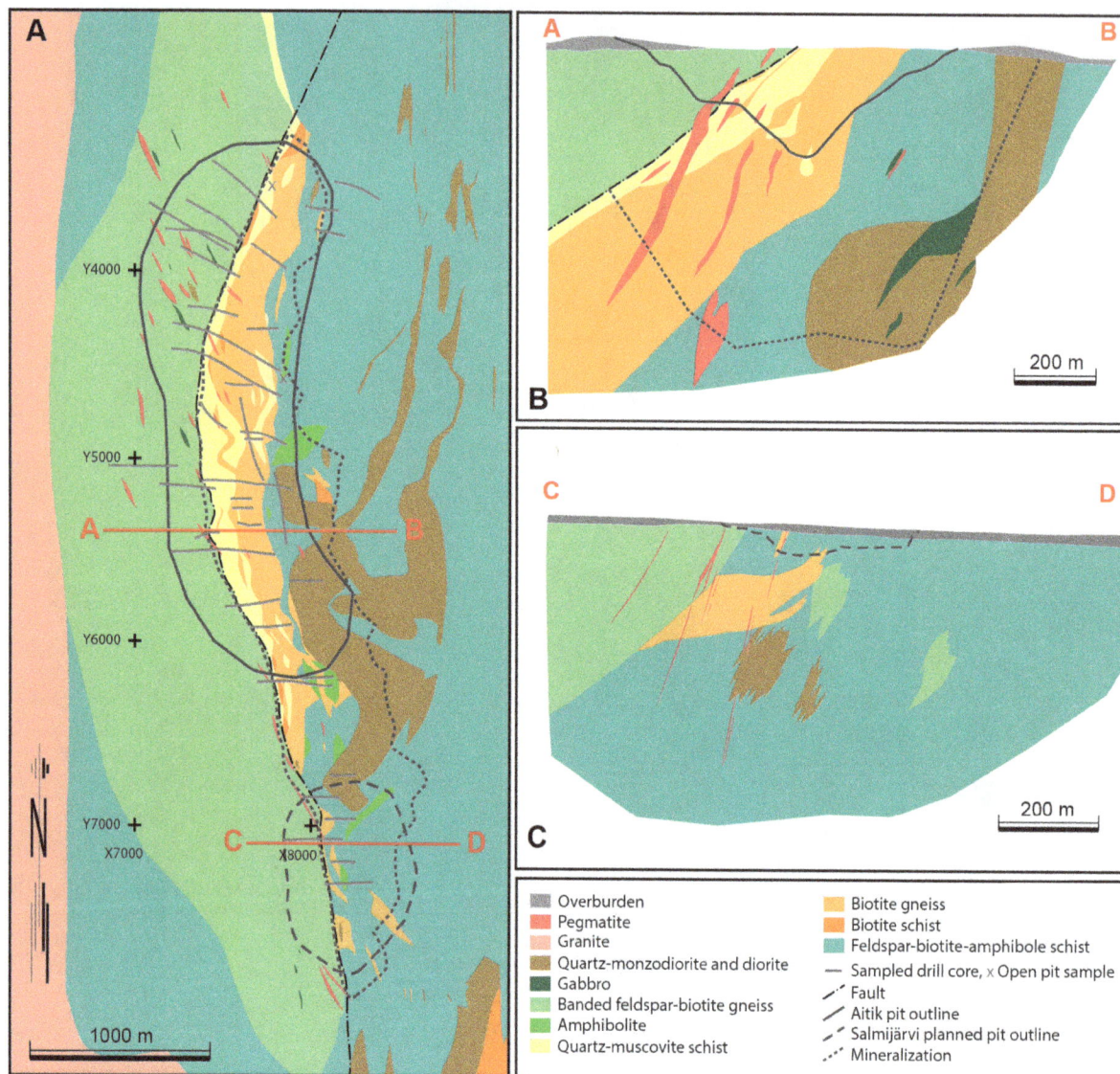

2. Geology

The Aitik porphyry Cu-Au-Ag-(Mo) deposit and its host rocks, situated ~200 km north of the Archaean-Proterozoic palaeoboundary in the Fennoscandian shield, are understood to have formed in a volcanic arc environment related to subduction of oceanic crust beneath the Archaean craton at *ca.* 1.9 Ga [7]. A quartz monzodioritic intrusion related to the formation of porphyry copper mineralization is situated in the footwall of the deposit (Figure 2). This source intrusion and related volcaniclastic rocks are mafic-intermediate in composition and belong to the regionally widespread Haparanda suite of intrusions and the Porphyrite group of volcanic rocks, respectively [8]. The deposit is overprinted by a hydrothermal event of Iron Oxide Cu-Au (IOCG) type at *ca.* 1.8 Ga [3] and strongly deformed and metamorphosed at amphibolite facies, resulting in at least 160 Ma of post-ore modification [9].

Deposit Geology

The Aitik quartz monzodiorite intrusion is comprised of younger, comagmatic, intrusions of micro-quartz monzodiorite and diorite, and is suggested to represent a cupola protruding from a pluton at depth [7]. It is typically grey and plagioclase-porphyritic with a medium grained groundmass of plagioclase, biotite, K-feldspar and quartz, and trace amounts of titanite, apatite and zircon [7]. It should be noted that the quartz monzodiorite intrusion is metamorphosed, and should accordingly be called gneiss. However, in order to stress the intrusion-related origin of the deposit this has not been applied below.

The surrounding volcaniclastic rocks constitute feldspar-biotite-amphibole gneisses in the footwall and hanging wall, and quartz-muscovite schist and biotite schist/gneiss in the ore zone (Figure 2). The feldspar-biotite-amphibole gneiss in the footwall to the ore is composed of large (6000 to 20,000 μm) amphibole phenocrysts in a groundmass of feldspar, biotite and amphibole, with accessory apatite, titanite, zircon and opaque phases. The hangingwall feldspar-biotite-amphibole gneiss is grey-greenish to grey-brownish in color and fine-grained (~200 μm). Abundant accessory magnetite and titanite occur in a groundmass of plagioclase, biotite, amphibole and quartz. The hangingwall gneiss is separated from the ore zone by a thrust (Figure 2).

The dominant rock type in the ore zone is the light grey, garnet-bearing biotite gneiss with approximately 1% to 2% garnet porphyroblasts. The microcrystalline groundmass contains K-feldspar, plagioclase, biotite, amphibole, quartz, and opaque minerals, and accessory minerals that include tourmaline, apatite, fluorite and titanite. The quartz-muscovite schist is commonly sericite-altered and mainly comprises muscovite, K-feldspar, quartz, sericite and tourmaline, with accessory apatite and opaque phases. Pyrite is common. Lenses of micro-quartz monzodiorite can be distinguished within less foliated sections of the ore zone schist and gneiss.

Several generations of pegmatitic dykes, barite veins, and quartz-sulfide veins occur in the Aitik mining area. The pegmatites strike either north-south or east-west and range in thickness from 0.5 to 20 m. Pegmatites within the ore zone and footwall frequently contain pyrite and chalcopyrite, and occasionally molybdenite, in addition to plagioclase, quartz, microcline, muscovite, biotite and tourmaline. Pegmatite dykes in the hangingwall are predominantly barren. Pegmatites striking north-south, *i.e.*, along the main foliation, in places display deformation features such as folding, while those crosscutting the foliation in an east-west direction, are always underformed. Ages of pegmatite dykes range between 1850 to 1730 Ma [9].

The main copper-bearing mineral at Aitik is chalcopyrite, which is often disseminated. Bornite and chalcocite are present in trace amounts. Other ore minerals include pyrite, magnetite, pyrrhotite, ilmenite, molybdenite, gold, and silver. Molybdenite is sporadically observed in the ore zone and footwall of the deposit, usually in association with chalcopyrite and/or pyrite, in quartz veins of varying composition, and as coarse aggregates within pegmatite dykes. Gold occurs as native metal in amalgam and in electrum [5], in close association with groundmass minerals such as K feldspar, biotite, plagioclase and quartz, but also with chalcopyrite and pyrite [10]. To date, the distribution of silver and its mode of occurrence has not been studied in detail, but the majority of the silver is likely to be hosted within the chalcopyrite [11].

Salmijärvi constitutes the southern extension of the Aitik ore body and comprises of strongly altered and metamorphosed biotite gneiss, amphibole-biotite gneiss and amphibole gneiss within the ore zone (Figure 2). These gneisses seem to be of slightly higher metamorphic grade and/or more strongly hydrothermally altered than the rocks in the Aitik main ore zone, although show similar folding styles and well developed foliations. As with the Aitik ore body, the Salmijärvi ore is crosscut by weakly mineralized pegmatite dykes and occasional quartz dykes. Quartz monzodioritic bodies occur on the footwall side of the ore (Figure 2). The hanging wall rocks consist of the same hornblende-banded garnet-porphyroblastic amphibole gneiss. In the ore zone, westward, the biotite gneiss becomes more schistose [12]. Commonly there are mm- to cm-sized green hornblende veinlets, surrounded by albite haloes within the Salmijärvi area. These veins often have a central line of chalcopyrite and are rimmed by quartz and feldspar, *i.e.*, very similar to the footwall mineralization in the Aitik main pit. The hornblende veins are often short (dm-size), sometimes with a vuggy appearance [13]. The contact between the ore zone and the hanging wall rocks is sharp and dipping 60° to the west, whereas the western ore body contact is grade controlled and thus gradational. The hanging wall contact is curved and the ore body pinches out towards the south. The contact can be seen cut by thin pegmatite dykes. The muscovite schist with common pyrite in Aitik is absent in Salmijärvi. The majority of the molybdenite is associated with quartz veins/blebs of slightly varying composition. Molybdenite occurs as small grains and clusters within mm to cm sized quartz veinlets, within mm to cm away from the quartz veins, and within dm sized quartz blebs. The quartz veins form a weak stockwork which often contains pyrite and chalcopyrite as well. The quartz stockwork forms 1%–5% of the host rock. Molybdenite also occurs as trace dissemination in the host rock.

3. Sampling and Methods

Sixty-four samples were taken from 33 drill cores from the Salmijärvi and Aitik deposits, and from the Aitik open pit. Samples were taken from different lithologies and alteration- and mineralization assemblages. Samples contained molybdenite that was both visible and not visible by the naked eye. Polished thin sections were prepared by Vancouver Petrographics Ltd, Langley, Canada, and were examined in detail to document wall rock, alteration- and mineralization phases. In order to document and characterize the different molybdenite mineralization phases, a detailed optical examination of polished thin sections in reflected and transmitted light using a standard petrographic microscope (Nikon ECLIPSE E600 POL, Nikon Instruments, Amsterdam, The Netherlands) was conducted before and after QEMSCAN, ICP-MS, and flotation analyses.

3.1. QEMSCAN Analysis

In order to assess all size fractions, and thereby obtain a reliable compilation of the overall molybdenite content at Aitik and Salmijärvi, and to be able to search a large number of samples, QEMSCAN analyses were performed. Identification of molybdenite in 30 polished thin sections by X-ray and image analysis was performed at the QEMSCAN laboratory of Intellection UK in North Wales. Polished thin sections were measured using a fieldscan operating mode with a beam-stepping interval of 25 μm. Fieldscan analyses are based on between 578,797 and 767,112 individual EDS (Energy-dispersive X-ray spectroscopy) analyses per thin section. Samples were then analyzed using the trace mineral search (TMS) mode. The backscatter electron threshold was set at 115 so that any grains "brighter" than chalcopyrite (e.g., molybdenite) were identified, and the selected grains and surrounding area were mapped using a beam-stepping interval of 1 μm. Thirteen complementary samples were analyzed at the QEMSCAN laboratory of LKAB MetLab in Luleå, Sweden. Polished thin sections were measured using the trace mineral search (TMS) mode with a beam-stepping interval of 5 μm. The backscatter electron threshold was set at 121. Fieldscan analyses were performed on specific areas that were re-mapped using a beam stepping interval of 1 μm. Some of the molybdenite grains (ca. 100–150) detected by QEMSCAN were re-located in the optical microscope in order to study mineral associations and related textures in more detail.

3.2. Molybdenite Re Abundance Determination

Seventeen drill core samples with visible molybdenite were selected for dilution ICP-MS analysis in order to determine the rhenium content of individual molybdenite grains (Table 1). Localized variation in Re abundance in the molybdenite grain [14–16] was avoided by isolating the entire area of molybdenite. Selection of molybdenite grains was done in polished thin sections together with their copy of rock chips using a high magnification Nikon Microscope at the Division of Geosciences at Luleå University of Technology, Sweden. Grains were chosen so that a variety of sizes, mineral associations and host rocks would be represented. Sampled molybdenite were extracted from the rock chips using a New Wave™ (Fremont, CA, USA) micro drill at Arthur Holmes Isotope Geology Laboratory (AHIGL) which is part of the Durham Geochemistry Centre at Durham University, UK.

The Micro drill instrument contains three basic parts; a Binocular Microscope, a high speed drill with adjustable tungsten carbide and an XYZ stage. All the components are integrated with a workstation allowing for high precision (±1 μm) movement of the drill. See [17] for further details of the microdrill setup.

Rock chips containing grains of molybdenite were first attached and mounted with double sided tape on the sample plate stage. Molybdenite was sampled using ca. 30 μm diameter drill bit and a 40 μm drill depth. Depending on grain size, several samples were taken from each grain in order to maximize the recovery of molybdenite. After each grain was sampled, the rock chip was cleaned in an ultrasonic bath in ethanol [17] and the drill bit replaced in order to avoid contamination.

Recovery of the molybdenite separate was achieved using a flotation technique with high purity water (MilliQ, Darmstadt, Germany) by a 0.5–10 μL micropipette which sucked up the floated molybdenite grains or dry separation technique (to minimize impurities such as quartz, biotite, amphibole, chalcopyrite, pyrite and magnetite) using a high magnification microscope. A total of 27 molybdenite grains were individually weighed on a gold boat using a Mettler Todeo UMT2 balance (Mettler-Toledo Ltd., Leicester, UK). The weight of the samples extracted for analysis ranged from 0.001 to 0.088 mg (Table 1). The molybdenite was washed from the gold boat in a 3.5 mL savillex vial with 0.25 mL of concentrated HCl at 80 °C overnight (>12 h). Following this stage the gold boat was removed from the vial. The molybdenite in 0.25 mL of concentrated HCl was then digested in a known amount of 185Re tracer solution together with 0.5 mL concentrated HNO_3 at 130 °C for ~24 h. Following digestion the sample was evaporated to dryness at 80 °C and then prepared for ICP-MS analysis by the addition of 1 mL 0.8 N HNO_3.

3.3. Laboratory Flotation Tests

To investigate the flotation response of different ore types (with variable molybdenum content) in the Aitik deposit, laboratory flotation tests were performed on 14 drill core samples after 7 and 11 min of grinding. Samples consist of approximately 10–15 kg of drill core each. Metallurgical testing was performed at the mineral laboratory of the Processing department (TMP) at Boliden Mineral.

Samples were obtained from representative, homogenous sections in drill cores. Drill cores were chosen so that sampling would cover the major lithologies and alteration- and mineralization types present in Aitik (except pegmatite dykes), and all samples were studied and documented in detail in hand specimen and polished thin sections before grinding. Molybdenite was not visible by the naked eye in any of the samples. Each sample was passed through a 3 mm screen and the oversize was crushed to below 3 mm. The sample was then split into 1 kg posts and stored for further treatment.

At each test, 1 kg of material was placed in a laboratory rod mill with an 8.0 kg rod charge and 550 mL of water. A Φ 195 mm × 245 mm stainless steel mill was used, rotating with 48 rpm. The diameters of the stainless rods are varying continuously from a maximum of 25 mm, down to about 5 mm. 7 and 11 min grinding time were chosen, generating particle size distributions (PSD: s) similar to a distribution from the Aitik plant. The laboratory rod mill produces less coarse particles than the autogenous grinding circuits in the plant but this does not normally affect the results.

After grinding, the pulp was transferred to a 2.5 L flotation cell. Flotation trials were conducted in a Wemco laboratory flotation unit, with a mixing speed of 1200 rpm. Tap water was added and the pH noted varied between 8.0 and 8.7 in the different tests. After conditioning for 5 min at pH 10.5 with slaked lime, Cu rougher flotation tests were run in 4 sequential steps with 1, 2, 3, and 3 min flotation times. Before each flotation step, a 1-min conditioning time was allowed for adding 5 g/tonnes of collector, frother was added if needed. Potassium xanthate (KAX) is used as collector in Aitik. The frother is a glycol, ether/glycol mixture (Nasfroth 350). All flotation products were filtered and dried and the dry weight was noted.

Each test consisted of three parallel flotation tests for which a total of 3 kg material was needed. Through this methodology enough amounts of products were produced for the assaying. Flotation products were analyzed with X-ray for a multitude of elements including Cu, Mo and S. This work was done at the Aitik process laboratory and required 12 g of each product.

4. Results

By performing petrographic, QEMSCAN, ICP-MS, and flotation laboratory test studies on molybdenite, the diversity of its character, occurrence, and rhenium content, and its response to the standard copper flotation procedure used for the Aitik ore has been thoroughly characterized. In the studied area, molybdenite grain size varies from coarse (>20 µm) to very fine (<2 µm) and occurs mainly as single grains in the groundmass of the rocks, as grain aggregates, and as intergrowths with chalcopyrite and pyrite. The molybdenite-bearing rocks are the biotite- and muscovite schists in the Aitik ore zone, the quartz monzodiorite in the Aitik footwall, and the Salmijärvi biotite-amphibole gneiss, the latter containing mostly medium-coarse grained molybdenite. In summary the rhenium abundance of molybdenite varies between 20–1587 ppm, with the higher contents obtained from Salmijärvi molybdenite samples.

4.1. QEMSCAN Analysis of the Distribution and Character of Molybdenite

Of the 36 samples from Aitik analyzed by QEMSCAN, 26 contain molybdenite and a total of 8336 molybdenite grains were found. The majority of molybdenite grains occur as 10–100 μm rectangular or subrectangular flakes lying individually or intergrown with each other (Figure 3a). Molybdenite grains in Aitik are dominantly associated with anhydrite, chalcopyrite, pyrite, biotite and K feldspar in mica schists and in quartz monzodiorite (Figures 3a–c and 4), and occasionally in pegmatite. The richest sample, containing 3047 molybdenite grains, is garnet-magnetite-quartz-anhydrite-altered biotite schist dominated by biotite, anhydrite, tourmaline and feldspar with minor sulfides. A similar sample, containing 2311 molybdenite grains, is an amphibole-quartz-sulfide-altered quartz monzodiorite, with amphibole, zeolite, quartz and anhydrite. The overall grain size distribution of molybdenite grains in Aitik is shown in Figure 5.

Figure 3. Photomicrographs of characteristic molybdenite in Aitik and Salmijärvi. Reflected light; (**a**) Scattered molybdenite grains in quartz monzodiorite in Aitik (sample Agm751-246.7). Majority of grains are 10–100 μm; (**b**) Molybdenite aggregate in anhydrite vein in homogeneous quartz monzodiorite in Aitik (sample Agm751-246.7); (**c**) Coarse-grained molybdenite intergrown with chalcopyrite in brittle mica schist in Aitik (sample Am652-518.4); (**d**) Coarse-grained molybdenite associated with magnetite and chalcopyrite in biotite-amphibole gneiss in Salmijärvi (sample Sm1022-138.3).

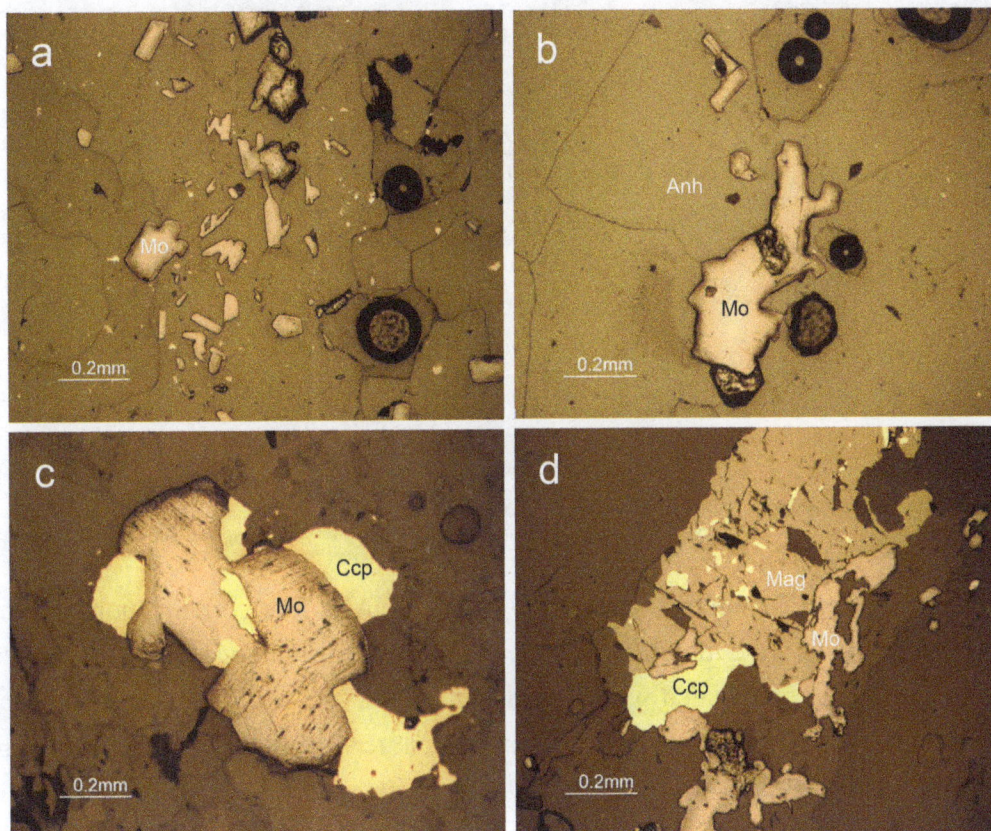

Figure 4. Average number of molybdenite grains displayed for each rock type and for three different size fractions. Graphs are based on number of grains detected by QEMSCAN in 31 polished thin section samples of 26 mm × 46 mm size.

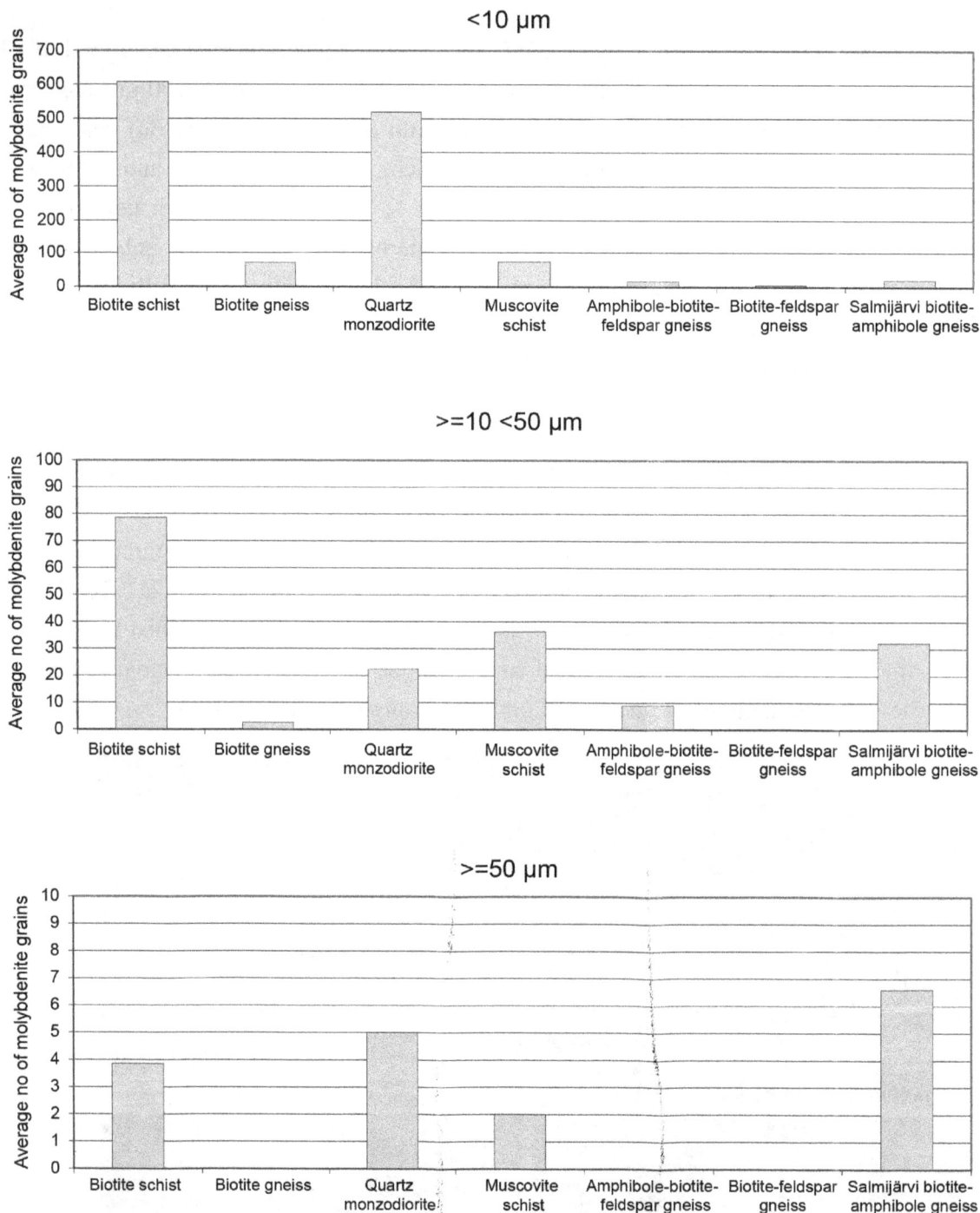

Of the seven samples from Salmijärvi analyzed by QEMSCAN, five contain molybdenite and a total of 549 molybdenite grains were found. Molybdenite grains occur mainly as 7–185 µm rectangular and rounded flakes dominantly associated with K feldspar, quartz, plagioclase, and sulfides in biotite gneiss and bioite-amphibole gneiss, and occasionally in pegmatite. The richest sample, containing 538 molybdenite grains, is a K feldspar-altered biotite gneiss with sporadic sulfide-magnetite-apatite-alteration patches. A grain size distribution built upon the type of alteration present is clearly outlined in this sample, with grains ≤10 μm mainly associated with K feldspar, quartz, and biotite, and grains 10–185 μm associated with chalcopyrite, pyrite, magnetite, K feldspar, quartz, biotite, and apatite (Figure 3d). The overall grain size distribution of molybdenite grains in Salmijärvi is shown in Figure 5.

Figure 5. Size distribution diagram for molybdenite grains in the Aitik (27 samples) and Salmijärvi (5 samples) deposits. Circles symbolize one grain.

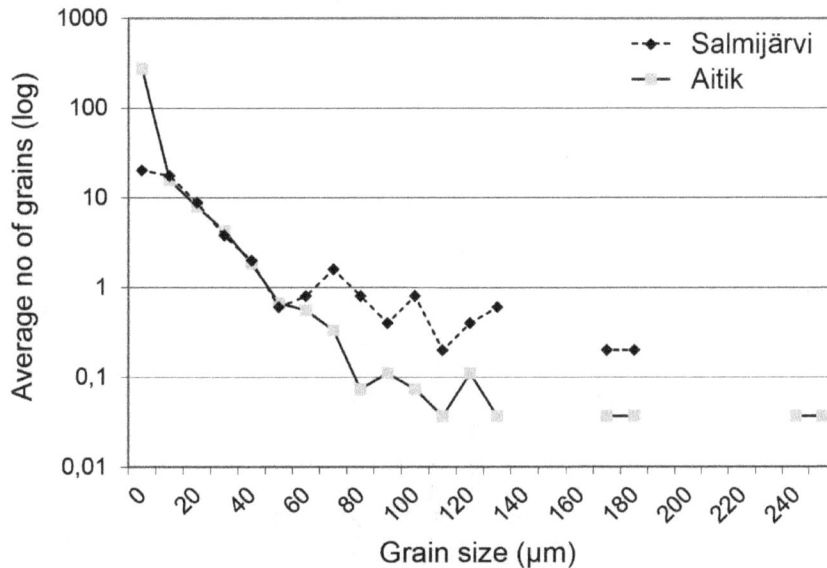

4.2. Rhenium Abundance of Molybdenite

The rhenium content of molybdenite grains from the two studied deposits show significant variation (20 to 1587 ppm) as summarized in Table 1. In Figure 6, the concentration of rhenium measured in each grain is presented in relationship to the setting of the grain (associated host rock, and alteration- and mineralization type).

In Aitik, an average Re content of 211 ± 10 ppm (4.7% 1σ, $n = 15$) is obtained. Samples of garnet-bearing biotite gneiss and pegmatite generally contain molybdenite with low concentrations of Re (on average 116 and 117 ppm, respectively) while samples of quartz monzodiorite and quartz-muscovite-(sericite) schist generally contain higher concentrations of Re in molybdenite (on average 334 and 310 ppm, respectively, Figure 6a). Molybdenite from two generations of pegmatite dykes were analyzed and revealed a low-Re concentration for molybdenite within the old (*ca.* 1.85 Ga) and deformed pegmatite dyke, and a medium-Re concentration for molybdenite within the young (*ca.* 1.73 Ga) and cross-cutting pegmatite dyke (Table 1). In Salmijärvi, an average Re content of 452 ± 33 ppm (7.3% 1σ, $n = 10$) is obtained. Samples of amphibole-(biotite) gneiss generally contain low-Re molybdenite (on average 155 ppm) while samples of biotite-(amphibole) gneiss contain high-Re molybdenite (on average 650 ppm) (Figure 6a).

Table 1. Rhenium concentration of molybdenite grains. Mineral abbreviations listed in [18].

Sample code (Mo)	Host rock	Sample description	Sample code (Re)	MoS$_2$ weight (mg)	Calc Re abundance (ppm)
Agm806-61.6	Qtz-ms schist	Qtz-ep-alteration with rich dissem of ccp, py, mag, and with scattered Mo grains assoc with matrix-qtz.	1	0.012	20 ± 1
Agm1041-111.5	Qtz-ms schist	4 m wide qtz-(cal) vein. Mo, ccp, py scattered in qtz	2	0.037	289 ± 5
	Bt gneiss	1m wide vein of qtz, fsp, bt, ms, chl, tur, and few ccp grains. Mo between fsp and bt-ms-chl clot	3	0.041	120 ± 2
Am969-493.0	Bt gneiss	1m wide vein of qtz, fsp, bt, ms, chl, tur, and few ccp grains. Mo within bt-ms-chl clot	4	0.018	67 ± 2
Sgm1021-119.5	Am-(bt) gneiss	Sporadic am-fsp-qtz-cal-chl-mag-py-(ccp)-clots with Mo. Mo assoc with am-fsp	5	0.054	148 ± 2
P-EW	Pegmatite in bt-fsp-am gneiss	6 cm wide, undeformed, E-W trending (cutting main foliation). Composed of fsp-qtz-(bt) with accessory py, ccp, mag, Mo. Mo assoc with fsp-qtz	6	0.026	172 ± 4
Am1042-542.3	Qtz monzo-diorite	5 m section of strong ser-ep-am-alteration with dissem py, ccp, (mag). Mo assoc with ser (pl)-ep-am	7	0.058	784 ± 8
	Bt-(am) gneiss	5 cm wide qtz-fsp-ep-zeo-vein rich in mag-py-ccp and with inclusions of ccp and Mo in coarse py	8	0.010	92 ± 5
Sm1022-140.8	Bt-(am) gneiss	5 cm wide qtz-fsp-ep-zeo-vein rich in mag-py-ccp and with inclusions of ccp and Mo in coarse py	15	0.005	467 ± 53
Sm1012-128.9	Am-bt gneiss	Qtz-fsp-altered with 3 mm wide kfs-qtz-vein containing sparse ccp, mag, Mo. Mo assoc with kfs	9	0.006	60 ± 6
	Bt-(am) gneiss	1 dm of qtz-am-ep-mag-ccp-py-alteration rich in fine grained Mo. Mo assoc with py	10	0.037	1587 ± 25
Sm1022-138.3	Bt-(am) gneiss	1 dm of qtz-am-ep-mag-ccp-py-alteration rich in fine grained Mo. Mo assoc with am-qtz and intergrown with ccp	11	0.086	1468 ± 119

Table 1. *Cont.*

Sample code (Mo)	Host rock	Sample description	Sample code (Re)	MoS₂ weight (mg)	Calc Re abundance (ppm)
	Qtz monzo-diorite	Am-ep-qtz-alteration with mag-py-ccp dissem Mo assoc with ep-(am-qtz-ccp)	12	0.006	133 ± 13
Am947-352.0	Qtz monzo-diorite	Am-ep-qtz-alteration with mag-py-ccp dissem Mo assoc with qtz-(am-ep-py)	13	0.034	284 ± 5
	Qtz monzo-diorite	Am-ep-qtz-alteration with mag-py-ccp dissem Mo assoc with am-(ep-qtz-ccp)	14	0.002	43 ± 13
	Bt-ms schist	Qtz-kfs-ser-(tur)-alteration with a weak dissem of ccp-(py,mag), sporadic cm-wide qtz-fsp-veins (brittle, drusy), and a rich Mo-impreg. Mo assoc with qtz-kfs-ser	18	0.007	191 ± 15
Am652-518.4	Bt-ms schist	Qtz-kfs-ser-(tur)-alteration with a weak dissem of ccp-(py,mag), sporadic 0.5 cm wide qtz-fsp-veins (brittle, drusy), and a rich Mo-impreg. Mo assoc with qtz-kfs-ser	19	0.018	742 ± 23
Agm751-246.7	Qtz monzo-diorite	3 cm wide am-zeo-ccp-py-Mo-bearing qtz-clot. Mo assoc with am-qtz-zeo-ccp	20	0.018	133 ± 4
Am947-356.9	Qtz monzo-diorite	Am-kfs-(ser)-alteration with ccp, py, Mo. Mo assoc with am-(kfs)	21	0.001	84 ± 49
Agm878-436.6	Bt-(am) schist	0.5 cm wide qtz-ccp-py-mag-Mo-vein. Mo grain situated between qtz and py	22	0.005	48 ± 6
	Am-bt gneiss	1 dm am-qtz-ccp-py-mag-clot with Mo. Mo assoc with am-(ccp?)	23	0.001	367 ± 208
Sgm1021-153.2	Am-bt gneiss	1 dm am-qtz-ccp-py-mag-clot with Mo. Mo assoc with am-(mag?)	24	0.003	46 ± 9
P-NS	Pegmatite in ms schist	Folded, N-S trending (parallel to main foliation). Composed of kfs-qtz-ms-(tur) with accessory ccp, Mo. Mo assoc with coarsegrained ms and kfs	25	0.088	61 ± 0.4
	Bt gneiss	Kfs-qtz-chl-alteration with dissem and intergrown ccp, mag, py, dissem Mo, and ccp-mag-bearing qtz-veins. Mo assoc with kfs in matrix	26	0.039	223 ± 3
Sg1025-112.4	Bt gneiss	Kfs-qtz-chl-alteration with dissem and intergrown ccp, mag, py, dissem Mo, and ccp-mag-bearing qtz-veins. Mo assoc with bt in matrix	27	0.013	63 ± 3

Figure 6. Rhenium concentrations within Aitik (A) and Salmijärvi (S) molybdenite grains. Sample numbers are given at the top of each bar; **(a)** Rhenium concentration of molybdenite *versus* molybdenite-bearing host rock; **(b)** Rhenium concentration of molybdenite *versus* associated alteration- and ore mineral assemblage. Abbreviations: qtz-quartz, bt-biotite, am-amphibole, ms-muscovite, kfs-K feldspar, chl-chlorite, cal-calcite, ser-sericite, tur-tourmaline, fsp-feldspar, ep-epidote, zeo-zeolite, pl-plagioclase, ccp-chalcopyrite, mag-magnetite, py-pyrite, pegm-pegmatite.

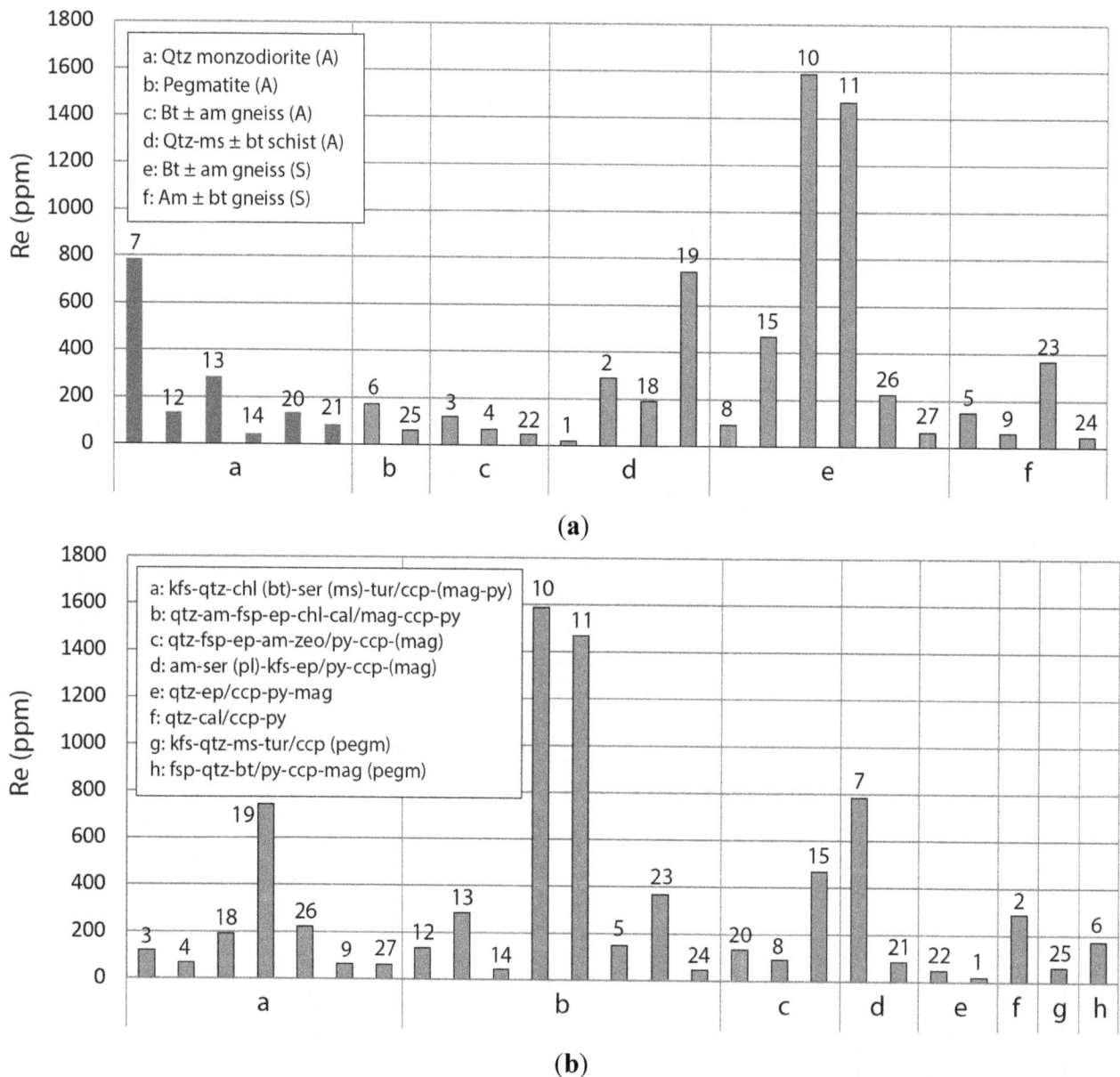

(a)

(b)

In Aitik, the Re content of all molybdenite grains associated with sericite (either sericite replacing plagioclase in the quartz monzodiorite or sericite replacing muscovite in the quartz-muscovite schist) are high (commonly above average) and ranges from 191 to 784 ppm (alteration assemblage groups a and d in Figure 6b), whereas the rhenium content of molybdenite grains associated with quartz-(kf-am) ranges from 20 to 289 ppm (alteration assemblage groups a–c, e and f in Figure 6b). In Salmijärvi, the rhenium content of all molybdenite grains associated with a quartz-amphibole-magnetite alteration assemblage, with or without other minerals such as feldspar, chlorite, calcite, epidote, and zeolite, are high (commonly above average) and ranges from 367 to 1587 ppm (alteration assemblage groups b and c in Figure 6b), whereas the rhenium content of molybdenite grains associated with K feldspar, biotite, and amphibole ranges from 46 to 223 ppm (alteration assemblage groups a–c in Figure 6b).

4.3. Molybdenum Flotation

The Mo recovery after 11 min grinding time did not substantially differ from that achieved after 7 min grinding time. The laboratory test Mo total recoveries resemble the plant recoveries to the Cu concentrate. The laboratory total recoveries are presented in Figure 7. Production results representing daily Mo recovery to Cu concentrate from nearly 2.5 years of production is presented in the same figure as limitation curves for high and low results. The expected Mo recovery is shown also. Three drill core samples with low Mo recoveries are marked out in the figure with the two last digits in the TMP sample number. These are nos. S-6792, S-6783 and S-6786. Three drill core samples with high Mo recoveries are also marked out (sample nos. S-6780, S-6785 and S-6789). The flotation recoveries together with a geological description of each test sample are presented in Table 2.

Figure 7. Molybdenum recoveries at 6% Cu rougher concentrate grade and Aitik normal (thick line), high (medium line) and low (thin line) production results. Laboratory data consists of mean values of the tests at 7 and 11 min of grinding. Plant data represents daily Mo recovery to copper concentrate from nearly 2.5 years of production (11 May 2006 to 15 October 2008). Numbers represent the two last digits in the sample numbers presented in Table 2 below.

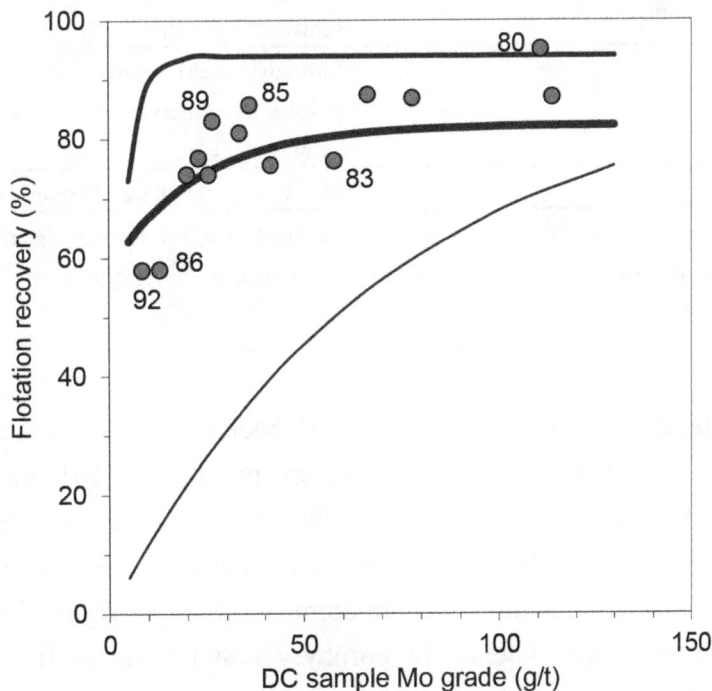

Table 2. Drill core flotation test samples, listed after Mo flotation recovery, difference from average.

Sample		Mo DC-grade	Mo recovery	Sample description
Name	TMP No.	g/t	difference	
Aan953-1B	S-6780	111	13%	Slightly am-qtz-ep-altered qtz-monzodiorite with ccp-py-mag, ccp sometimes intergrown with py
Aan1055-3A	S-6785	36	8%	Unaltered bt gneiss with ccp-mag
Aan641-3B	S-6789	26	8%	Strongly ep-chl-ttn-ser-altered am-fsp-bt gneiss with mag-py-(ccp), ccp often as inclusions in py, which is bordered by mag
Aan1110-1B	S-6781	66	6%	Strongly kf-ser-chl-altered qtz-monzodiorite with thin qtz-veins and ccp-mag-py, commonly intergrown with each other
Aan953-1A	S-6779	78	5%	Slightly ser-altered qtz-monzodiorite with ccp-mag-py, ccp commonly intergrown with mag
Aan1080-3C	S-6791	114	5%	Slightly qtz-tur-altered bt gneiss with ccp-py-mag
Aan990-3C	S-6790	33	4%	Slightly kfs-altered bt gneiss with rich ccp-py-po dissemination
Aan1045-3B	S-6787	23	3%	Strongly kfs-qtz-ep-tur-altered bt schist with po-ccp-(py)
Aan1064-2A	S-6782	20	2%	Moderately grt-mag-qtz-altered ms-(bt) schist with mag-ccp-py
Aan1064-3B	S-6788	26	−1%	Moderately am-ep-qtz-altered bt schist with py-po-ccp
Aan1069-2B	S-6784	42	−3%	Moderately chl-qtz-ser-altered ms-bt-qtz schist with py-ccp-po, ccp intergrown with po
Aan1186-2B	S-6783	58	−4%	Strongly ser-chl (1) and kfs-qtz-ep (2) altered ser-bt-chl schist with py-ccp
Aan1084-4	S-6792	8	−7%	Strongly ser-chl-kfs-qtz-altered fsp-bt gneiss with py-ccp-mag, intergrown py + ccp commonly rimmed by mag
Aan732-3A	S-6786	13	−11%	Moderately qtz-altered bt gneiss with mag-py-ccp

qtz: quartz; ms: muscovite; bt: biotite; ser: sericite; chl: chlorite; am: amphibole; fsp: feldspar; ep: epidote; ttn: titanite (sphene); kfs: K feldspar; tur: tourmaline; grt: garnet; ccp: chalcopyrite; mag: magnetite; py: pyrite; po: pyrrhotite.

5. Discussion

Porphyry systems are the most important source for molybdenum and rhenium in the world. These deposits alone account for more than 95% of molybdenum production [19]. Based on the relative content of copper and molybdenum, the porphyry deposits are divided into two categories: Cu ± Mo deposits and Mo ± Cu deposits [20]. In the copper dominant systems the molybdenum grade generally ranges from 0.005%–0.03%, while in the molybdenum dominant it ranges from 0.07%–0.3% [19,21].

The rhenium concentration in molybdenite in porphyry systems varies from the ppb level to thousands of ppm [22,23]. In Figure 8, the average rhenium content in the Aitik and Salmijärvi ore bodies are compared with those reported from other porphyry Cu and Cu-Mo deposits in the world. These deposits exhibit a relatively low quantity of molybdenite, and generally contain higher concentrations of rhenium than porphyry Mo-Cu deposits with abundant molybdenite. This may relate to the fact that small amounts of molybdenite will consume the limited rhenium budget from the ore-forming fluids whereas rhenium within large volumes of molybdenite-rich magma will be relatively diluted [14,24,25]. The Aitik and Salmijärvi deposits show similar trends in Re grades as Cu-Mo deposits in general (Figure 8).

Figure 8. Rhenium grades in porphyry Cu and Cu-Mo systems worldwide. Modified from [22].

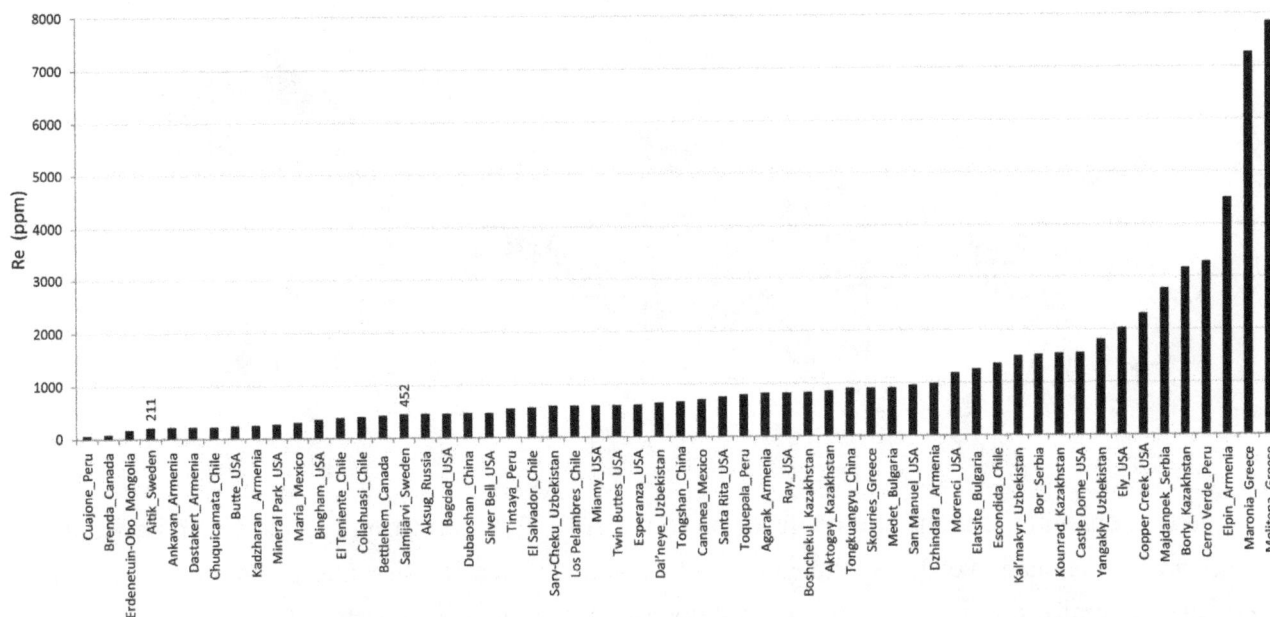

In Aitik, with its complex multi-magmatic/hydrothermal history [3], molybdenum has probably been introduced by several generations of magma, as indicated by Re-Os ages of 1.87, 1.85, and 1.73 Ga obtained for molybdenite in barite and pegmatite dykes in the Aitik ore zone [9]. These ages correspond to the Haparanda, Jyryjoki, and TIB2 magmatism, respectively, which are well-constrained magmatic events in northern Fennoscandia [26]. However, some molybdenite formation might also be the result of late hydrothermal activity and metamorphism. The authors of [27] describe a metamorphic dehydration process active in northern Sweden during the Svecofennian orogeny which produced small Mo-(W) occurrences with characteristic low-Re molybdenite formation. The Liikavaara Cu-Au occurrence, situated only 4 km east of Aitik, contains molybdenite- and scheelite-bearing quartz- and aplitic veins [28]. Its close spatial relationship to Aitik implies that fluids responsible for the Liikavaara Mo-W mineralization might also have affected the Aitik deposit and its southern extension, with possible addition of (low-Re) molybdenite as a result.

Regarding the distribution of molybdenite within the Aitik ore body, molybdenum grades are higher in the northern part of the ore body than in the southern part (Figure 9). One explanation to this pattern may be the distribution of variably altered host rocks, since certain alteration types and associated veining, rather than a specific rock type, seems to be an important factor controlling the distribution of molybdenite in Aitik.

The majority of studied molybdenite grains are found within mica schists and quartz monzodiorite, and coarser molybdenite grains (>50 µm) are exclusively found in samples of these rock types (Figure 4). The mica schists, and especially the quartz-muscovite-(sericite) schist, are nearly absent towards the south and towards depth (Figure 2), thus partly explaining the large-scale distribution pattern (Figure 9).

Figure 9. Contoured grade map of Mo at the 300 m level in the Aitik deposit. Grades are taken from the block model of the Aitik mine, and based on *ca.* 15,000 Mo analysis of drill core sections.

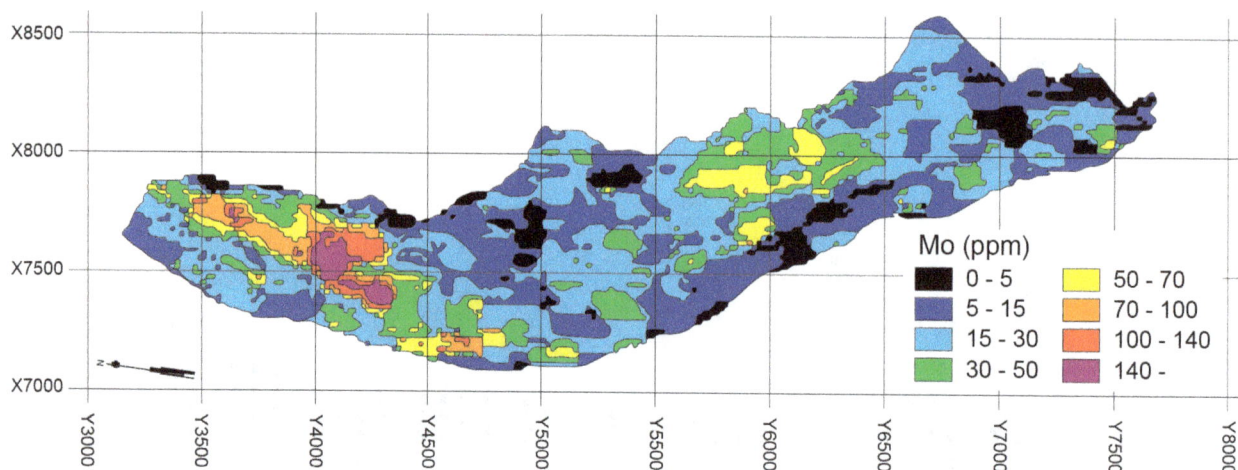

The quartz monzodiorite is the source intrusion to the Aitik mineralization [7] and the mica schists and gneiss are co-magmatic volcanosedimentary rocks that have been strongly altered and deformed [8]. Molybdenite mineralization in porphyry ore systems often occurs after the main Cu-Au mineralizing event [29]. In a recent study of the Bingham porphyry deposit, [30] describes a process of selective metal deposition from a magmatic fluid due to changes of redox and pH conditions associated with a more reducing environment in the fluid source. The molybdenite mineral associations revealed in this study however indicate opposite conditions (e.g., anhydrite, magnetite, see discussion below). Contrary to the more common Phanerozoic porphyry systems the Aitik and Salmijärvi ore bodies have been subjected to at least two metamorphic events, varying degrees of multiple deformation, and an overprinting hydrothermal event of IOCG-type [3], and the distribution and character of ore minerals as seen today is probably more the result of a combination of post-ore processes including recrystallization, remobilization, and mineral/element differentiation, rather than primary features. These processes are known to lead to both concentration and dispersion of valuable minerals and also to considerable variations in ore grade within a single deposit [31].

As mentioned above, molybdenite in Aitik is found in a wide variety of mineralogical associations, although an association with anhydrite dominates together with chalcopyrite, pyrite, biotite, K feldspar, pegmatite dykes, and barite- and quartz veins. The surprisingly strong association with anhydrite instead of expected quartz indicates that there might be more anhydrite in the deposit than previously thought, *i.e.*, that anhydrite is mistaken for quartz/feldspar-alteration and veining when drill cores are being logged.

Molybdenite in Salmijärvi is dominantly associated with pegmatite dykes and late K feldspar-epidote-quartz ± magnetite-alteration of the gneisses. The richest sample, containing 538 molybdenite grains, is a K feldspar-altered biotite gneiss with sporadic sulfide-magnetite-apatite-alteration patches of typical IOCG-character (iron-oxide copper gold). The size of molybdenite grains in Salmijärvi is significantly larger than those found in Aitik (Figure 5), a feature that might be caused by the slightly higher grade of metamorphism implied for the Salmijärvi area [13]. In Salmijärvi, a coarsening of the molybdenite grain size seems to follow the degree of alteration, with the overprinting sulfide-magnetite-apatite-alteration containing most of the coarser grains. The relationship between grain size and alteration type, with grains ≤10 μm mainly associated with K feldspar, quartz, and biotite, and grains 10–185 μm associated with chalcopyrite, pyrite, magnetite, K feldspar, quartz, biotite, and apatite, indicates that even if the latter mineral association is not the most common one, it probably constitutes the largest mass of molybdenite due to the larger grain size, an observation with relevance for future processing of the ore. The strong alteration is of oxidized character, occasionally rich in goethite and magnetite, which gives the rock an easily recognized rusty red color. This alteration type has not been observed in the Aitik ore body.

It is clear that strong alteration and post-ore veins of variable composition, often in combination with patches and clots of remobilized Cu-Fe sulfides appear to favor the occurrence of molybdenite in Aitik and Salmijärvi. This is probably related to the hydrothermal fluid that causes the alteration, in combination with deformation, both of which would allow molybdenum to migrate [32]. Thus, less molybdenite is found in the biotite gneiss, which is less altered and less tectonically affected than the mica schists, an observation also made in the Archaean Cu-Au-Mo occurrences of the Chibougamau district, Québec [33].

In Aitik molybdenite, with an average rhenium concentration of 211 ppm, molybdenite hosted by quartz monzodiorite and quartz-muscovite-sericite schist generally contain higher concentrations of rhenium (on average 244 and 310 ppm, respectively) than those hosted by pegmatites and gneisses (on average 116 and 157 ppm, respectively). In fact, the rhenium content of all studied molybdenite grains associated with sericite, either sericite replacing plagioclase in the quartz monzodiorite or sericite replacing muscovite in the quartz-muscovite schist, are high (commonly above average) and ranges from 191 to 784 ppm, whereas the rhenium content of molybdenite grains associated with quartz-(K feldspar-amphibole) ranges from 20 to 289 ppm. The variation of rhenium in molybdenite from the Sar Cheshmeh porphyry Cu-Mo deposit, Iran, shows an increased rhenium concentration in molybdenite that is intimately associated with sericitic altered veins of low quartz content, whereas a low rhenium concentration is obtained for molybdenite in quartz veinlets showing strong silicification [34]. Other observations note high-rhenium in quartz-sericite-altered (phyllic) rocks and low-rhenium in K-feldspar/biotite-altered (potassic) rocks, and suggest an inverse relationship between the temperature of formation and the Re content of molybdenite [22]. Similar observations were made of molybdenite from the Bingham deposit [24]. As such, high-temperature, magmatic fluid may form low-rhenium molybdenite, similar to those grains analyzed from Aitik pegmatites and quartz-(-feldspar-amphibole)-clots, and a low-temperature, hydrothermal/metamorphic fluid could generate high-rhenium molybdenite such as the grains analyzed from Aitik muscovite-sericite schist and strongly sericite-altered quartz monzodiorite.

A similar pattern, although with molybdenite associated with a different mineral assemblage, is found in Salmijärvi where the rhenium content of all molybdenite grains associated with quartz-amphibole-magnetite alteration, with or without other minerals such as feldspar, chlorite, calcite, epidote, and zeolites, is high (commonly above the average 452 ppm) and ranges from 367 to 1587 ppm, whereas the rhenium content of molybdenite grains associated with K feldspar, biotite, and amphibole ranges from 46 to 223 ppm. This magnetite-rich alteration is common in Salmijärvi and Aitik and in many other metal-rich occurrences in the northern Norrbotten ore district (e.g., Nautanen and Tjårrojåkka) and is linked to the overprinting hydrothermal event of IOCG-type observed in the region [3]. Oxidizing fluids, which are favorable for transport of rhenium [35], might thus be partly responsible for deposition of high-rhenium molybdenite in Salmijärvi and Aitik, and just as with the distribution of molybdenite, certain alteration types rather than specific rock types can be of guidance when documenting the overall rhenium abundance in molybdenite-bearing deposits.

Molybdenite from two generations of pegmatite dykes analyzed in this study revealed a much lower rhenium concentration for molybdenite within the old (*ca.* 1.85 Ga) and deformed pegmatite dyke, than for molybdenite within the young (*ca.* 1.73 Ga) and cross-cutting pegmatite dyke (Table 1, Figure 6). The concentrations obtained (61 and 172 ppm, respectively) are in reasonable agreement with rhenium concentrations measured for Re-Os geochronological studies of the same pegmatite dykes by [9] giving values of 54 and 215 ppm, respectively. The concentration of rhenium in molybdenite is known to be a combination of several interplaying factors, e.g., composition of parent magmas and host rocks [14,22,36], degree of oxygen and chloride fugacity of the ore fluid [37], and changes in chemical and physical conditions during crystallization [22]. Reference [38] in their study of molybdenite in the Archaean Boddington Cu-Au deposit, Western Australia, also suggest that even though rhenium can remain locked in molybdenite under deformation, dissolution and reprecipitation may occur giving molybdenite a different rhenium concentration and grain size.

Processing Implications

Molybdenum is commonly recovered from copper concentrates in a process that starts with dewatering prior to the molybdenum circuit to get rid of excess reagents, followed by conditioning with a depressor for copper. The most common depressor is NaHS, sometimes in combination with cyanide. Finely ground activated carbon can be used to further remove excess of reagents and higher temperatures are sometimes used to promote the removal of collectors. Oxygen destroys the depressor to some extent and is avoided by using, e.g., nitrogen as the flotation gas. Regrinding of intermediate concentrates is normal to get better liberation and to make the froth easier to pump. Conventional flotation cells are used for rougher and scavenger flotation and quite often also for the first cleaners. Since the introduction of columns in the early 80's, more or less every molybdenum circuit has columns for the last cleaning steps. Due to the flaky character of molybdenite grains, with strong hydrophobic and inert faces and hydrophilic and reactive edges, additions of fuel oils to enhance the flotation is common practice. Molybdenum extraction from the molybdenite concentrate is performed by roasting of molybdenite at a temperature of 500–700 °C. During this process the molybdenum transforms to molybdenum trioxide and the rhenium stored within molybdenite transforms to rhenium heptoxide gas (Re_2O_7) [39]. The molybdenum recovery obtained by this process is less than 93%, generally ranging between 25%–85% [40].

The volatilized rhenium can be caught using a wet electrostatic system combined with dry apparatus, and transferred to a solution containing sulfuric acid, which leaches the rhenium from the enriched flue dust [39]. The separation of rhenium from this solution is achieved by sorption using ion-exchanges resins and/or liquid extraction [39,41] followed by addition of potassium chloride and ammonium chloride in order to precipitate or isolate rhenium elution in the form of perrhenate salts. The non-hazardous and stable ammonium perrhenate (NH_4ReO_4) is the common market product of rhenium. The rhenium recovery obtained by this process is generally less than 60% [40].

Mo Extraction

The possibility to recover molybdenum from the copper concentrate at Aitik has been investigated several times during the lifetime of the mine. The normal procedure with depression of chalcopyrite and other sulfides with NaHS as described in the previous section, has been proven to work for the Aitik concentrate [42].

However, even if Mo assays are obtained fast enough to be of guidance in process regulations (which is often not the case) they might not provide an accurate indication of recoverable molybdenum as long as factors influencing the extraction process are not defined and taken into account. In the current study, where one focus has been to identify lithological and mineralogical factors in the incoming ore which might affect molybdenum recovery, flotation laboratory tests reveal that three drill core samples resulting in Mo-recoveries lower than expected (−4% to −11%) considering the drill core Mo grades are from two different rock types (Table 2): feldspar-biotite gneiss and biotie gneiss. These samples have three characteristics in common (1) they contain 30%–50% Fe-Mg micas (2) they contain rare quartz veinlets, and (3) they contain molybdenite grains of dominantly less than 10 μm size. QEMSCAN data of sample Aan1186-2B reveal that most mica is phlogopite, the Mg-rich end member of the biotite group. Chlorite in the same sample is of chamosite-composition (a Mg-Fe-rich member of the chlorite group), and a significant part of the minerals end up as unidentified Mg-minerals (Figure 10). Furthermore, 63% of detected molybdenite grains are less than 10 μm. QEMSCAN data from four samples taken only 16–17 m downhole from Aan732-3A contain 445 molybdenite grains where 93%–100% are less than 10 μm.

In sample Aan1064-2A, giving a Mo-recovery almost as expected (+2%), approximately the same amount of mica is present (45%), although in the form of annite (Fe-biotite) and muscovite (Figure 10). It is known that fine-grained mica minerals can disturb the flotation process by interfering in the separation of chalcopyrite and molybdenite [43]. Results from this study indicate that the mica composition may also be of significance. According to [44], there are four non-sulfide minerals in the Bingham Mine, Utah, that are negatively correlated with molybdenum recovery: talc, andradite, calcite and amphibole. These minerals contain Ca^{2+} or Mg^{2+} or both. Zanin et al. (2009) studied the effect of Ca^{2+} and Mg^{2+} in solution and gangue minerals on the recovery of molybdenite by performing laboratory flotation tests on coarse molybdenite particles from two different host rocks. They came to the conclusion that the adsorption of Ca^{2+} and Mg^{2+} to the edges of molybdenite particles may have the combined effect of reducing the hydrophobicity of molybdenite grains, and bridging to specific, negatively charged, gangue minerals in the slurry, possibly leading to the formation of slime coatings. The bridging effect was also shown to be dependent on the type of host rock [45].

Reference [44] furthermore identifies biotite/phlogopite as being positively correlated with molybdenum recovery, a finding that are not discussed further by [44], and that is in contrast to results presented in this paper and to the work by [43] where a number of clay-micas (e.g., chamosite) are identified as causing poor molybdenum recovery.

Some drill core samples produced higher Mo-recoveries than expected. Looking at the three top samples (+8% to +13%), these also represent different rock types: quartz monzodiorite, biotite gneiss, and amphibole-feldspar-biotite gneiss, but they are all characterized by common quartz veins and quartz clots containing chalcopyrite and pyrite, a feature that is missing in the samples discussed above. The amount of mica minerals is limited in these samples (Figure 10). Furthermore, no molybdenite was found when analyzing thin sections from two of these flotation samples by QEMSCAN. Since molybdenum grades are normal to high (26–111 ppm) in these samples, this probably indicates that the molybdenite is coarse grained and therefore difficult to trap in a thin section.

Figure 10. QEMSCAN diagrams showing the mineralogical composition for Mg-rich (Aan1186-2B), Fe-rich (Aan1064-2A), and mica-poor samples (Aan641-3B).

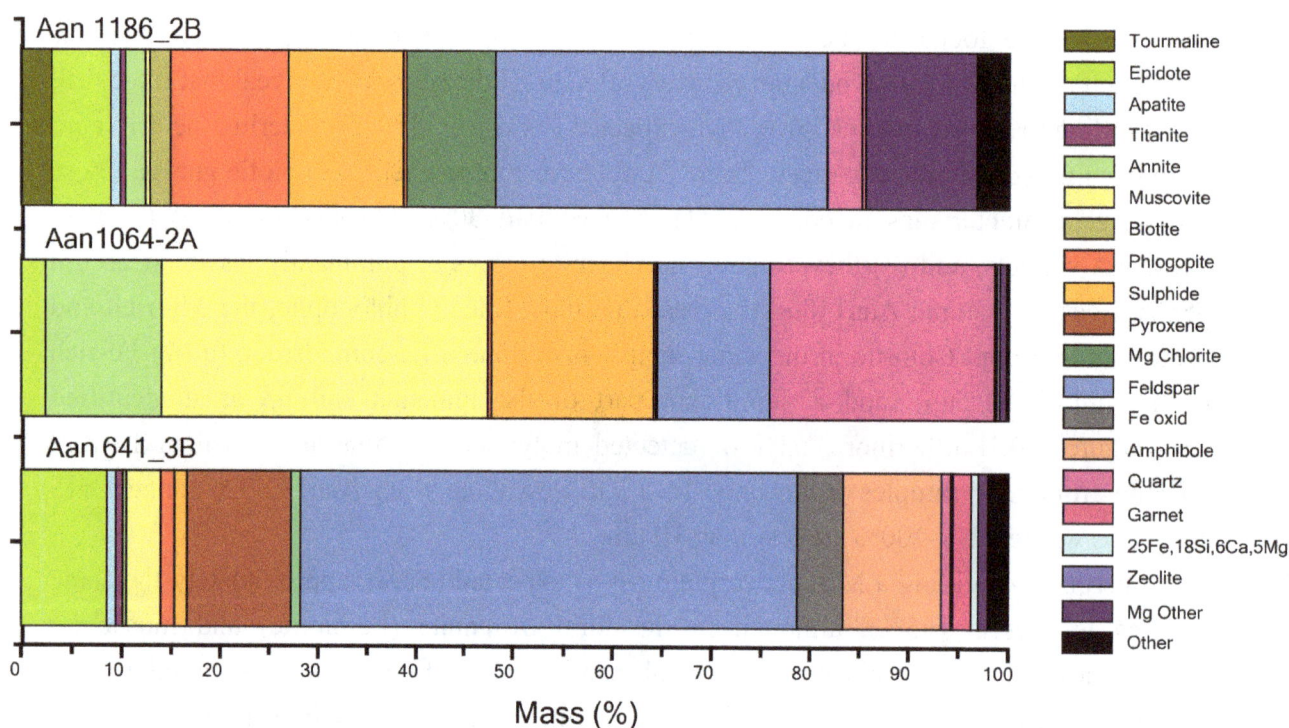

One reason for the obtained results thus seems to be found in the geological and mineralogical character of the rock, *i.e.*, the host rock lithology, the vein- and alteration types present, and the size of molybdenite grains, thus emphasizing the importance of careful characterization of ores, including the distribution and texture of both ore and gangue minerals, in order to develop optimized treatment processes.

Re Extraction

The possibility to recover rhenium from the molybdenite concentrate at Aitik has never been tested or investigated. As can be seen from the result of this work, the Aitik molybdenite contains on average 213 ppm Re, but this varies considerably between different particles (20–1587 ppm). High-Re molybdenite is mainly found in quartz monzodiorite and quartz-muscovite-sericite schist. The majority of molybdenite grains in Aitik are less than 10 microns though, and the quartz-muscovite-sericite schist disappears towards depth and towards south. Regarding the pegmatite dykes, these are due to their hardness occasionally separated from the ore to avoid low throughput. Their molybdenite and rhenium contents, although limited, could form an added value and thus motivate more of such handling, and perhaps change the mine cut-off grade. In Salmijärvi, however, the average Re content is higher (452 ppm), the size of molybdenite grains is larger, and the high-Re molybdenite is found within the common biotite-(amphibole) gneiss.

A mass balance prognosis for molybdenum concentrate production at Aitik is presented in Table 3. A Mo content of 0.0025% in the Aitik ore, and a recovery of *ca.* 85% to the copper concentrate can be assumed. Furthermore, a 60% recovery in the Mo circuit from the Cu concentrate can be assumed based on laboratory and plant scale tests [42]. An annual ore production of 36 Mt will give a production of 1047 tonnes MoS_2 concentrate with 53% Mo, thus a Mo production of 544 tonnes/year. Together with a rough estimation of the rhenium content of the Aitik deposit based on the average rhenium concentration obtained in this study (213 ppm), the current reserve of the deposit (747 Mt), the MoS_2 concentrate (1047 tonnes), and a general Re recovery (60%), it is obvious that these new products have the potential of increasing the value of the Aitik ore and extend the life of the mine.

Table 3. Mass balance prognosis for molybdenum concentrate production at Aitik.

Product	Weight ktonne	Grade Cu	[%] Mo	Distribution Cu	[%] Mo
Ore feed	36,000	0.25	0.0025	100	100
Cu conc.	324	25	0.2479	90	85
Mo conc.	0.849	1.6	53	0.02	50
Cu final conc.	323	25.06	0.1093	89.98	35
Final tailing	35,676	0.025	0.00027	10	15

These are good examples demonstrating the importance of transforming qualitative and quantitative micro-scale information about the ore to a larger scale of direct importance for future production. By studying in-situ samples, the important link between analytical data from carefully chosen and examined samples to the character of specific rock volumes within the mine can be established, even if selective mining seldom is used in porphyry ores, information of this kind helps to improve predictions and mine planning.

6. Conclusions

Molybdenite in the Aitik area varies considerably in grain size, between the two studied ore bodies and within each ore body, and is dominantly very fine-grained. Most molybdenite grains are found as single grains, as aggregates, or intergrown with chalcopyrite, and are associated with late, post-ore geological features, such as overprinting garnet-magnetite-anhydrite-quartz-K feldspar-apatite-alteration, quartz-veining, and pegmatite dykes. The rhenium concentration of molybdenite varies considerably, with the higher concentrations obtained from Salmijärvi molybdenite samples. The large variation is probably the result of several magmatic/hydrothermal and metamorphic episodes that have been active in the region. The Aitik and Salmijärvi deposits exhibit similar molybdenum and rhenium contents as Mo-Re-producing porphyry Cu-Mo deposits worldwide (e.g., El Teniente and Chuiqicamata in Chile, and Bingham and Copper Creek in USA), and pilot plant tests have shown that recovery for molybdenum at Aitik is good compared to other deposits in the world. If molybdenum and rhenium circuits are installed, the price of molybdenum and rhenium in combination with the calculated amount of metal present in the Aitik ore body indicates that the two metals would constitute significant byproducts of copper production in Aitik. The data presented in this study describes the character and distribution of molybdenite in order to help predicting the outcome of this future large investment. The distribution, character, and rhenium concentration of molybdenite within the two investigated ore bodies seems to be related to specific alteration and vein features rather than to rock types, and these features also seem to have an impact on the final recovery of molybdenum. They are thus important to estimate quantitatively for processing and mine planning.

Acknowledgments

Boliden AB is acknowledged for the financial support of this study, and for the permission to publish this paper. We would like to thank Björn Johansson, Boliden AB, for the flotation test setup and performance, and Jenny Wikström, LKAB, for assisting with complementary QEMSCAN data. We also thank the external editor and two anonymous reviewers for their constructive comments on the manuscript. Support from the ProMinNET (Nordic Researcher Network in Geometallurgy and Process Mineralogy) researchers is acknowledged. This is a CAMM (Centre of Advanced Mining and Metallurgy) publication.

Author Contributions

C.W. conceived the study and did the field sampling in Aitik and Salmijärvi. C.W., D.S. and N.-J.B. designed the experiments. C.W., W.N., C.M. and N.-J.B. performed the experiments and analyzed the data together with D.S. and R.N. All authors have contributed to the writing of the manuscript and with the production of figures and tables.

Conflicts of Interest

The authors declare no conflict of interest.

References

1. Fleischer, M. The geochemistry of rhenium, with special reference to its occurence in molybdenite. *Econ. Geol.* **1959**, *54*, 1406–1413.

2. Martinsson, O.; Wanhainen, C. Fe oxide and Cu-Au deposits in the northern Norrbotten ore district. In Proceedings of the 12th Biennial SGA Meeting, Uppsala, Sweden, 12–15 August 2013.

3. Wanhainen, C.; Broman, C.; Martinsson, O.; Magnor, B. Modification of a Palaeoproterozoic porphyry system in northern Sweden: Integration of structural, geochemical, petrographic, and fluid inclusion data from the Aitik Cu-Au-Ag deposit, northern Sweden. *Ore Geol. Rev.* **2012**, *48*, 306–331.

4. Wanhainen, C.; Kontturi, M.; Martinsson, O. Copper and gold distribution at the Aitik deposit, Gällivare area, northern Sweden. *Appl. Earth Sci. Trans. Inst. Min. Metall. B* **2003**, *112*, 260–267.

5. Sammelin-Kontturi, M.; Wanhainen, C.; Martinsson, O. Gold mineralogy at the Aitik Cu-Au-Ag deposit, Gällivare area, northern Sweden. *GFF* **2011**, *133*, 1–12.

6. Jakobsson, P.; Joslin, G.; Knipfer, S.; Nordin, R.; Wasström, A.; Wanhainen, C. The Aitik porphyry Cu-Au-Ag-(Mo) deposit in Sweden. In Proceedings of the Eleventh Biennal SGA Meeting, Antofagasta, Chile, 26–29 September 2011; pp. 346–347.

7. Wanhainen, C.; Billström, K.; Martinsson, O. Age, petrology and geochemistry of the porphyritic Aitik intrusion, and its relation to the disseminated Aitik Cu-Au-Ag deposit, Northern Sweden. *GFF* **2006**, *128*, 273–286.

8. Wanhainen, C.; Martinsson, O. Geochemical characteristics of host rocks to the Aitik Cu-Au deposit, Gällivare area, northern Sweden. In Proceeding of the Fifth Biennial SGA Meeting and the Tenth Quadrennial IAGOD Meeting, London, UK, 22–25 August 1999; pp. 1443–1446.

9. Wanhainen, C.; Billström, K.; Martinsson, O.; Stein, H.; Nordin, R. 160 Ma of magmatic/hydrothermal and metamorphic activity in the Gällivare area: Re-Os dating of molybdenite and U-Pb dating of titanite from the Aitik Cu-Au-Ag deposit, northern Sweden. *Miner. Deposita* **2005**, *40*, 435–447.

10. Wanhainen, C.; Johansson, B. Character of gold within the Aitik Ore body: Preliminary results from a geometallurgical study. In Proceedings of the Conference in Minerals Engineering, Luleå University of Technology, Luleå, Sweden, 5–6 February 2008; pp. 143–150.

11. Monro, D. The Geology and Genesis of the Aitik Copper-Gold Deposit, Arctic Sweden. Ph.D. Thesis, University of Wales, College of Cardiff: Cardiff, United Kingdom, 1988.

12. Nordin, R. Geologisk beskrivning. In *Ansökan om Bearbetningskoncession Aitik K nr 4*; Gällivare Kommun: Norrbottens län, Sweden, 2005; p. 7. (In Swedish)

13. Sarlus, Z. Geology of the Salmijärvi Cu-Au Deposit. Master's Thesis, Luleå University of Technology, Luleå, Sweden, 2012, p. 75.

14. Stein, H.; Markey, R.; Hannah, J.; Schersten, A. The remarkable Re-Os Chronometer in molybdenite: How and why it works. *Terra Nova* **2001**, *13*, 479–486.

15. Kosler, J.; Simonetti, A.; Sylvester, P.; Cox, R.; Tubrett, M.N.; Wilton, D. Laser ablation ICP-MS measurements of Re/Os in molybdenites and implications for Re-Os geochronology. *Can. Mineral.* **2003**, *41*, 307–320.

16. Selby, D.; Creaser, R. Macroscale NTIMS and microscale LA-MC-ICP-MS Re-Os isotopic analysis of molybdenite: Testing spatial restrictions for reliable Re-Os age determinations, and implications for the decoupling of Re and Os within molybdenite. *Geochim. Cosmochim. Acta* **2004**, *68*, 3897–3908.

17. Charlier, B.A.; Ginibre, C.; Morgan, D.; Nowell, G.M.; Pearson, D.G.; Davidson, J.P.; Ottley, C.J. Methods for the microsampling and high-precision analysis of strontium and rubidium isotopes at single crystal scale for petrological and geochronological applications. *Chem. Geol.* **2006**, *232*, 114–133.

18. Siivola, J.; Schmid, R. Recommendations by the IUGS Subcommission on the Systematics of Metamorphic Rocks: List of Mineral Abbreviations. IUGS Commission on the Systematics in Petrology. Available online: http://www.bgs.ac.uk/scmr/docs/papers/paper_12.pdf (accessed on 24 November 2014).

19. Sinclair, W.D. Porphyry Deposits. *Geol. Surv. Can. Spec. Publ.* **2007**, *5*, 223–243.

20. Kirkham, R.V.; Sinclair, W.D. Porphyry copper, gold, molybdenum, tungsten, tin, silver. In *Geology of Canadian Mineral Deposit Types*; Eckstrand, O.R., Sinclair, W.D., Thorpe, R.I., Eds.; Geological Survey of Canada: Ottawa, ON, Canada, 1995; pp. 421–446.

21. Singer, D.A.; Berger, V.I.; Moring, B.C. *Porphyry Copper Deposits of the World: Database and Grade and Tonnage Models*; US Geological Survey Open-File Report; US Geological Survey: Reston, VA, USA, 2008; p. 45.

22. Berzina, A.; Sotnikov, V.I.; Economou-Eliopoulos, M.; Eliopoulos, D.G. Distribution of rhenium in molybdenite from porphyry Cu-Mo and Mo-Cu deposits of Russia (Siberia) and Mongolia. *Ore Geol. Rev.* **2005**, *26*, 91–113.

23. Millensifer, T.A.; Sinclair, D.; Jonasson, I.; Lipmann, A. Chapter 14: Rhenium. In *Critical Metals Handbook*, 1st ed.; Gunn, G., Ed.; Wiley & Sons: Hoboken, NJ, USA, 2014; pp. 340–360.

24. Giles, D.L.; Shilling, J.H. Variation in rhenium content of molybdenite. In Proceedings of the 24th International Geological Congress Section, Montreal, QC, Canada, 21–29 August 1972; pp. 145–152.

25. Sinclair, W.D.; Jonasson, I.R.; Kirkham, R.V.; Soregaroli, A.E. *Rhenium and Other Platinum-Group Metals in Porphyry Deposits*; Geological Survey of Canada Open-File; Geological Survey of Canada: Ottawa, ON, Canada, 2009.

26. Bergman, S.; Kubler, L.; Martinsson, O. *Description of Regional Geological and Geophysical Maps of Northern Norrbotten County*; Sveriges Geologiska Undersökning: Uppsala, Sweden, 2001; pp. 5–100.

27. Stein, H. Low-rhenium molybdenite by metamorphism in northern Sweden: Recognition, genesis, and global implications. *Lithos* **2006**, *87*, 300–327.

28. Zweifel, H. *Aitik: Geological Documentation of a Disseminated Copper Deposit—A Preliminary Investigation*; Sveriges Geologiska Undersökning: Uppsala, Sweden, 1976.

29. Sillitoe, R. Porphyry copper systems. *Econ. Geol.* **2010**, *105*, 3–41.

30. Seo, J.; Guillong, M.; Heinrich, C. Separation of Molybdenum and Copper in Porphyry Deposits: The Roles of Sulfur, Redox, and pH in Ore Mineral Deposition at Bingham Canyon. *Econ. Geol.* **2012**, *107*, 333–356.

31. Vokes, F.M.; Spry, P.G.; Marshall, B. Ores and metamorphism: Introduction and historical perspectives. *Rev. Econ. Geol.* **2000**, *11*, 1–18.

32. Marshall, B.; Gilligan, L.B. An introduction to remobilisation: Information from orebody geometry and experimental considerations. *Ore Geol. Rev.* **1987**, *2*, 87–131.

33. Kirkham, R.V.; Pilote, P.; Sinclair, W.D.; Robert, F.; Daigneault, R. Merrill Island Cu-Au veins and Clark Lake Cu-(Mo) porphyry deposit, Doré Lake mining camp, Chibougamau. In *Geology and Metallogeny of the Chapais-Chibougamau Mining District: A New Vision of the Discovery Potential: Proceedings of the Chapais-Chibougamau 1998 Symposium*; Pilote, P., Ed.; Quebec Ministry of Communications: Quebec City, QC, Canada, 1998; pp. 85–92.

34. Aminzadeh, B.; Shahabpour, J.; Maghami, M. Variation of Rhenium Contents in Molybdenites from the Sar Cheshmeh Porphyry Cu-Mo Deposit in Iran. *Res. Geol.* **2011**, *61*, 290–295.

35. Xiong, Y.; Wood, S. Experimental determination of the hydrothermal solubility of ReS_2 and the $Re–ReO_2$ buffer assemblage and transport of rhenium under supercritical conditions. *Geochem. Trans.* **2002**, *3*, 1–10.

36. Voudouris, P.C.; Melfos, V.; Spry, P.G.; Bindi, L.; Kartal, T.; Arikas, K.; Moritz, R.; Ortelli, M. Rhenium-rich molybdenite and rheniite in the Pagoni Rachi Mo-Cu-Te-Ag-Au prospect, Northern Greece: Implications for the Re Geochemistry of porphyry style Cu-Mo and Mo mineralization. *Can. Mineral.* **2009**, *47*, 1013–1036.

37. Selby, D.; Creaser, R. Re-Os Geochronology and systematic in molybdenite from the Endako porphyry molybdenum deposit, British Columbia, Canada. *Econ. Geol.* **2001**, *96*, 197–204.

38. Ciobanu, C.L.; Cook, N.J.; Kelson, C.R.; Guerin, R.; Kalleske, N.; Danyushevsky, L. Trace element heterogeneity in molybdenite fingerprints stages of mineralization. *Chem. Geol.* **2013**, *347*, 175–189.

39. Naumov, A. Rhythms of Rhenium. *Russ. J. Non-Ferr. Met.* **2007**, *48*, 418–423.

40. Jiang, K.; Wang, Y.; Zou, X.; Zhang, L.; Liu, S. Extraction of molybdenum from molybdenite concentrates with hydrometallurgical processing. *JOM* **2012**, *64*, 1285–1289.

41. Lan, X.; Liang, S.; Song, Y. Recovery of rhenium from molybdenite calcine by a resine-in-pulp process. *Hydrometallurgy* **2006**, *82*, 133–136.

42. Fatai, I. Purification of Molybdenite Concentrates. Master's Thesis, Luleå University of Technology, Luleå, Sweden, 2008; p. 89.

43. Bulatovic, S.M. *Handbook of Flotation Reagents—Chemistry, Theory and Practice: Flotation of Sulfide Ores*; Elsevier: Amsterdam, The Netherlands, 2007.

44. Triffett, B.; Veloo, C.; Adair, B.; Badshaw, D. An investigation of the factors affecting the recovery of molybdenite in the Kennecott Utah Copper blulk flotation circuit. *Miner. Eng.* **2008**, *21*, 832–840.

45. Zanin, M.; Ametov, I.; Grano, S.; Zhou, L.; Skinner, W. A study of mechanisms affecting molybdenite recovery in a bulk copper/molybdenum flotation circuit. *Int. J. Miner. Process.* **2009**, *93*, 256–266.

7

Burial Diagenesis of Magnetic Minerals: New Insights from the Grès d'Annot Transect (SE France)

Myriam Kars [1,†,*], **Charles Aubourg** [1], **Pierre Labaume** [2], **Thelma S. Berquó** [3] **and Thibault Cavailhes** [2,‡]

[1] Laboratoire des Fluides Complexes et leurs Réservoirs, Université de Pau et des Pays de l'Adour, UMR 5150 CNRS TOTAL, Avenue de l'Université, 64013 Pau cedex, France;
E-Mail: charles.aubourg@univ-pau.fr

[2] Géosciences Montpellier, Université de Montpellier 2, UMR 5243 CNRS, Place E. Bataillon, 34095 Montpellier cedex 5, France; E-Mails: pierre.labaume@gm.univ-montp2.fr (P.L.);
thibault.cavailhes@dno.no (T.C.)

[3] Physics Department, Concordia College, 901 8th St. S., Moorhead, MN 56562, USA;
E-Mail: tberquo@cord.edu

[†] Current address: Center for Advanced Marine Core Research, Kochi University, B200 Monobe, 783-8502 Nankoku, Japan.

[‡] Current address: DNO International ASA, P.O. Box 1345 Vika, 0113 Oslo, Norway.

[*] Author to whom correspondence should be addressed; E-Mail: jm-mkars@kochi-u.ac.jp

Abstract: The diagenetic evolution of the magnetic minerals during burial in sedimentary basins has been recently proposed. In this study, we provide new data from the Grès d'Annot basin, SE France. We analyze fine-grained clastic rocks that suffered a burial temperature from ~60 to >250 °C, *i.e.*, covering oil and gas windows. Low temperature magnetic measurements (10–300 K), coupled with vitrinite reflectance data, aim at defining the magnetic mineral evolution through the burial history. Magnetite is documented throughout the entire studied transect. Goethite, probably occurring as nanoparticles, is found for a burial temperature <80 °C. Micron-sized pyrrhotite is highlighted for a burial temperature >200 °C below the Alpine nappes and the Penninic Front. A model of the evolution of the magnetic assemblage from 60 to >250 °C is proposed for clastic rocks, containing iron sulfides (pyrite) and organic matter. This work provides the grounds for a better understanding of the magnetic properties of petroleum plays.

Keywords: Grès d'Annot basin; burial diagenesis; magnetite; pyrrhotite; paleotemperature

---

1. Introduction

Deciphering the way in which chemical and mineralogical processes operate in sedimentary basins is a major issue in order to understand their thermal history. These tools have especially great importance in applied geosciences, such as coal and petroleum geology, because they allow a better understanding of the source rock and reservoirs maturities.

In this perspective, the broad contours of the diagenesis of the magnetic minerals in argillaceous rocks were proposed from early burial (subsurface) to the lower greenschist facies metamorphism [1–7]. Recently, Aubourg et al. [8] defined three magnetic windows where greigite (Fe_3S_4; from subsurface to ~8 km of depth, i.e., up to ~200 °C), magnetite (Fe_3O_4; ~2 to ~12 km of depth, i.e., ~50 to ~300 °C) and pyrrhotite (Fe_7S_8; >8 km of depth, i.e., >200 °C) formed successively from low to deep burial by considering a geothermal gradient of 25 °C/km, which is typical in a foreland context. Thus, the magnetic assemblage of argillaceous rocks can be used to assess burial conditions and, particularly, burial temperatures (e.g., [9–12]). For high temperatures, the formation of pyrrhotite at the expense of magnetite and pyrite is reported by several studies when approaching 200 °C [3,4,13,14]. More precisely, two isogrades based on the breakdown of magnetite (~250 °C) and the breakdown of pyrite (>320 °C) into monoclinic pyrrhotite were determined by Rochette [1] (see the review by [14]). For a burial temperature >300 °C, neoformed pyrrhotite completely replaces magnetite according to:

$$Fe_3O_4 + 3FeS_2 \rightarrow 6\text{"FeS"} + 2O_2 \tag{1}$$

where "FeS" is pyrrhotite [15].

The formation of pyrrhotite ($Fe_{1-x}S$ with $0 < x < 0.13$) is hence of importance, as it may inform on low-grade metamorphic conditions (>200 °C) (e.g., [1–4]).

Nevertheless, the evolution of the magnetic minerals may not be straightforward. The greigite-magnetite-pyrrhotite pattern can be disrupted by the presence of inherited magnetic minerals (e.g., [16,17]), possibly occurring as nanoparticles (e.g., [18]), and by other neoformed magnetic minerals (e.g., maghemite, goethite). The relative amount of detrital magnetic minerals with respect to the neoformed minerals is debated. This proportion is probably dependent on several factors, including the nature and concentration of eroded magnetic particles, redox conditions during deposition, the chemistry of sediments, fluid circulation or bacterial activity (e.g., [19,20]).

In any case, the concentration of the magnetic minerals is generally low (<1%). The magnetic minerals of interest in this study are sensu lato ferromagnetic, which have the capability to retain a remanent magnetization at room temperature when their size is single domain (SD) to multidomain (MD), i.e., above the blocking volume (Vb). The magnetic nanoparticles are difficult to detect, because they are in the superparamagnetic (SP) domain state with volume < Vb (<25 nm for magnetite) [21]. They do not carry a remanence at room temperature. In practical terms, the best way to detect SP grains is to use low temperature magnetic techniques (down to 10 K), where the thermal energy is considerably reduced.

The present study aims to use the magnetic assemblage in clastic rocks as an estimate of the burial conditions, particularly for assessing maximum burial temperature. In this paper, we investigate the magnetic mineral assemblage in the clay-rich rocks from the Grès d'Annot basin, southeastern French Alps. Low-temperature magnetic measurements, compared to vitrinite reflectance data, aim at defining the magnetic windows. We study a ~60 km-long transect, where rocks from the same lithostratigraphic formations suffered from moderate diagenesis (~60 °C) in the external structural domain to anchimetamorphism (~250 °C) in the internal domain.

2. Geological Background

2.1. Geological Setting

The Grès d'Annot foreland basin (SE France; Figure 1) formed in the late Eocene-early Oligocene times during the Alpine-Pyrenean orogeny [22–24]. It is characterized by a stratigraphic sequence, the so-called Trilogie Priabonnienne, that recorded the basin history [25]. The lower formations (infra-Nummulitic conglomerates, Calcaires Nummulitiques and Marnes Bleues), overlying the Mesozoic substratum, registered the initiation and deepening of the basin [26]. The Marnes Brunes Inférieures, Grès d'Annot Formation and Marnes Brunes Supérieures characterize the basin-filling phase [27]. The Grès d'Annot Formation corresponds to a Priabonian-Rupelian arkosic turbidite succession mainly sourced from the Variscan crystalline basement of the Corsica-Sardinia massif [28,29]. In the north-eastern part of the basin, these lithostratigraphic units were buried below the Embrunais-Ubaye alpine nappes, preceded by the Schistes à Blocs olistostrome, from the late Rupelian (Figure 1) [30,31]. The basin-fill constituted therefore the footwall of the nappes, the uppermost Grès d'Annot being close to the nappe sole thrust. The Grès d'Annot were then exhumed as a result of the uplift of the external basement massifs (Argentera, Pelvoux, Barrot Dôme) during the late Oligocene-early Miocene [30,32,33].

2.2. Burial History

The studied cross-section extends from Annot in the SW to Bersezio in the NE, parallel to the direction of thrusting of the Embrunais-Ubaye nappes and following their lateral erosion fringe (Figure 1). Organic matter-based, petrologic and thermochronologic studies permitted determining the burial and exhumation history of the Grès d'Annot basin along this transect [33–37].

Vitrinite reflectance (Ro) data available in the study area display an increasing trend from 0.3% in the Rouaine area (SW) up to >7% underneath the Penninic Front at Gias Vallonetto (NE) (Figure 1) [34–36,38,39]. In the southwestern part, Ro values gradually increase from 0.3% in Rouaine to 0.6%–0.7% in the Grand Coyer area. NE of Grand Coyer, the Ro values first show an abrupt increase up to 2.0%–2.5% in the Estrop and Colmars areas, then a more progressive trend in the northeastern area, with values of ~4% in La Moutière to more than 6% at Gias Vallonetto. According to the Vassoyevitch et al. [40] calibration, these Ro values correspond to temperatures increasing from ~45 °C at Rouaine to 75 °C in the Grand Coyer area, to ~160–180 °C in the Estrop and Colmars areas, to ~230–240 °C at La Moutière and ~250–270 °C at Gias Vallonetto [34–36].

Figure 1. Sampling sites and thermal data available for the study area (modified from Labaume *et al.* [33]). Dots represent the location of the samples analyzed in this study, with dark dots marking samples for which vitrinite reflectance data had been previously obtained on the same samples and white dot where vitrinite reflectance data were obtained from different samples in the same area. Vitrinite reflectance data and corresponding estimated temperatures are from Labaume *et al.* [34,35] and Cavailhes [36]. Isotherms were placed according to vitrinite data (ibid.), apatite fission tracks analyses and silica diagenesis [33,36]. 1: Embrunais-Ubaye nappes; 2: Middle Eocene to Priabonian (including the Grès d'Annot formation); 3: Mesozoic; 4: Paleozoic substratum; 5: base of allochthonous thrust units; 6: normal and strike-slip faults. A: Annot; Al: Allons; B: Braux; Co1-Co2: Colmars; GC: Grand Coyer; GV: Gias Vallonetto; MT: La Moutière; R: Rouaine; RU: Le Ruch; P: Peyresq; VC: Villars-Colmars.

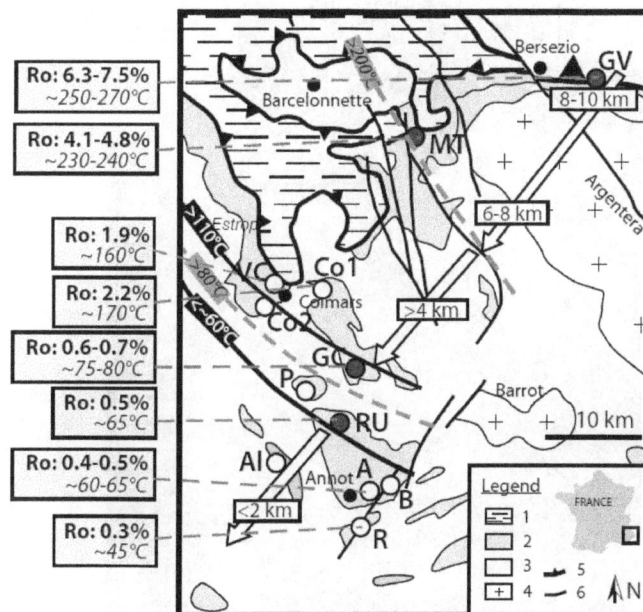

This evolution of the Ro values is consistent with the paleotemperature trends derived by Labaume *et al.* [33,34] from apatite fission tracks analysis (AFTA). These authors identified the upper boundary of the fission tracks partial annealing zone NE of Annot and the lower boundary NE of Grand Coyer, corresponding to the ~60 °C and ~110 °C isotherms, respectively (Figure 1). The trend of increasing paleotemperatures from SW to NE was also confirmed by petrological studies: the onset of silica diagenesis (marked by pressure solution of quartz), corresponding to ~80 °C, is located in the Grand Coyer area [34,37] and the abundant presence of authigenic sericite (white mica) in the La Moutière and Bersezio areas indicate temperatures >200 °C (Figure 1) [34,41,42]. In the fine-grained, clay-rich facies, the petrologic evolution is associated with a fissility that increases in intensity from the SW to the NE. This fissility is parallel to bedding in the pelitic layers of the Grès d'Annot turbidites and may correspond to an oblique cleavage in the Marnes Bleues (Figure 2).

The increase of temperature and fissility from the SW to the NE was interpreted to reflect the increasing burial, from ~2 km in the SW area to up to 8–9 km in the NE, with the abrupt increase of temperature between Grand Coyer and the Estrop-Colmars area corresponding to the front of the Embrunais-Ubaye nappes [33–36].

Figure 2. Outcrop views of the Marnes Bleues at Le Ruch (**a**), Grand Coyer (**b**) and Gias Vallonetto (**c**), located in Figure 1 as RU, GC and GV, respectively. Note the increasing fissility from rough scaly fabric in (**a**) to pencil cleavage in (**b**) to crenulation cleavage in (**c**), related to increasing compaction and burial temperature from (**a**) to (**c**) (Figure 1).

3. Methods

3.1. Sampling and Mineralogy

In this study, 36 samples from 12 different sites along the SW-NE transect, from the Rouaine area (SW) to Gias Vallonetto (NE), were analyzed (Table 1; Figure 1). They cover the entire temperature (depth) range of the Grès d'Annot basin burial history. The samples are very fine-grained argillaceous rocks corresponding to two different lithologies in the Nummulitic Trilogy: marls from the Marnes Bleues and turbiditic pelites from the Grès d'Annot and Marnes Brunes. The marls are rich in calcite (30%–50%) with a subordinate amount of quartz, whereas the turbiditic pelites are rich in quartz, with minor calcite (~10%). Labaume *et al.* [34,35] have shown that the clay fraction (<2 µm), similar in both marls and pelites from the Annot and Le Ruch areas, which suffered the weakest burial temperatures (<70 °C), comprise 40%–60% illite (+illite-smectite), 10%–20% kaolinite, 10%–20%

chlorite and, in some samples, up to 30% smectite. From Grand Coyer to NE of the study area, (1) kaolinite and smectite disappear and (2) illitisation increases, from neoformation of fiber-like illite particles at Grand Coyer ($T \sim 75$ °C) to small white mica particles at La Moutière ($T \sim 230$–240 °C).

Table 1. Location of the sampling sites (ND: no data). The lithology and name of the formation in brackets are reported. MBl: Marnes Bleues; MBr: Marnes Brunes; GA: Grès d'Annot. Vitrinite reflectance data (Ro) and total organic carbon (TOC) content are also mentioned. SD: Standard deviation.

Sampling Site	Sample	Latitude	Longitude	Lithology	Mean Ro (%)	SD Ro	TOC (%)
Allons (Al)	1A	N 43°59'05.9″	E 6°34'56.3″	Marl (MBl)			
	2A	N 44°00'09.4″	E 6°34'11.1″	Marl (MBl)			
Rouaine (R)	4A	N 43°56'01.1″	E 6°40'12.9″	Marl (MBl)			
	5A	N 43°56'06.6″	E 6°40'29.4″	Fine-grained sandstone (GA)			
Braux (B)	6A	N 43°58'16″	E 6° 42'17.2″	Marl (MBl)			
	7A	N 43°58'16″	E 6°42'17.2″	Fine-grained sandstone (GA)			
Annot (A)	A0	N 43°57'43.1″	E 6°40'34.6″	Marl (MBl)			
	11A	ND	ND	Marl (MBl)			
	12A	ND	ND	Fine-grained sandstone (GA)			
Le Ruch (RU)	RUmg	N 44°02'40.5″	E 6°40'28.1″	Marl (MBl)			
	RUmb	N 44°02'40.5″	E 6°40'28.1″	Turbiditic pelite (MBr)	0.54	0.1	0.23
Grand Coyer (GC)	CY1p	N 44°05'09.1″	E 6°41'0″	Turbiditic pelite (GA)	0.61	0.07	0.35
	CY3p	N 44°05'09.1″	E 6°41'0″	Turbiditic pelite (GA)	0.65	0.06	0.68
	CY5	N 44°05'09.1″	E 6°41'0″	Turbiditic pelite (GA)			
	CY6	N 44°05'09.1″	E 6°41'0″	Turbiditic pelite (GA)			
	CY7	N 44°05'09.1″	E 6°41'0″	Turbiditic pelite (GA)			
	CY8	N 44°05'09.1″	E 6°41'0″	Turbiditic pelite (GA)			
	CY10	N 44°05'09.1″	E 6°41'0″	Turbiditic pelite (GA)			
	CY11	N 44°05'09.1″	E 6°41'0″	Turbiditic pelite (GA)			
Peyresq (P)	20A	N 44°02'12.0″	E 6°36'25.6″	Marl (MBl)			
Colmars (Co1)	21A	N 44°09'11.1″	E 6°32'41.2″	Marl (MBl)			
	22A	N 44°09'26.2″	E 6°32'29.1″	Marl (MBl)			
	23A	N 44°09'49.3″	E 6°31'50.7″	Marl (MBl)			
Colmars (Co2)	25A	N 44°09'07.7″	E 6°40'28.7″	Marl (MBl)			
	26A	N 44°09'26.9″	E 6°39'19.0″	Marl (MBl)			
Villars-Colmars (VC)	13A	ND	ND	Turbiditic pelite (GA)			
	14A	ND	ND	Turbiditic pelite (GA)			
	15A	ND	ND	Fine-grained sandstone (GA)			
La Moutière (MT)	MT12	N 44°18'58″	E 6°47'46″	Turbiditic pelite (GA)	4.06	0.16	0.34
	MT17	N 44°18'58″	E 6°47'46″	Turbiditic pelite (GA)			
	MT29	N 44°18'58″	E 6°47'46″	Turbiditic pelite (GA)	4.13	0.17	0.28
	MT120	N 44°18'58″	E 6°47'46″	Turbiditic pelite (GA)			
	MTmg	N 44°18'58″	E 6°47'46″	Marl (MBl)			
Gias Vallonetto (GV)	GV1	N 44°21'41.7″	E 7°03'32.4″	Turbiditic pelite (GA)	6.29	0.41	0.61
	GV11	N 44°21'41.7″	E 7°03'32.4″	Turbiditic pelite (GA)	7.47	0.4	0.43
	GVmg	N 44°21'41.7″	E 7°03'32.4″	Marl (MBl)			

3.2. Analytical Methods

A series of different magnetic measurements were performed. The low-field magnetic susceptibility (χ) of these specimens was measured at the Ecole Normale Supérieure (ENS), Paris, France, with a KLY3-CS3 Kappabridge instrument. Then, the samples were crushed, and rock powders of 350–450 mg were sealed in gelatin capsules in order to perform low temperature magnetic measurement (<300 K). The evolution of a saturation isothermal remanent magnetization (SIRM) imparted at room temperature (RT-SIRM) with a 2.5 T magnetic field was monitored. Selected samples were heated to 400 K. This heating phase aims at removing the goethite contribution (if any) by heating the sample through the Néel temperature T_N of goethite (T_N goethite ~ 393 K) [43]. RT-SIRM$_{300 K}$ refers to the value of the remanence at 300 K (before the heating phase in the case where the sample is heated). Then, the samples were cooled down to 10 K in the presence of a 5-μT magnetic field inside the magnetometer in order to highlight <50 K magnetic behaviors [9,44]. The 5 μT magnetic field was either oriented downward or upward. The magnetization measured in the 10–300 K temperature range was a combination of both remanent and induced (due to the 5-μT field) magnetizations. A back curve (warming to 300 K) was measured for some samples. At 10 K, a 2.5-T magnetic field was applied to create a low temperature SIRM (LT-SIRM), and the sample was warmed up to 300 K in zero field. This remanence evolution is called hereafter zero field cooled (ZFC). Some samples were cooled down to 10 K in a 2.5-T field, before being warmed to 300 K in zero field. This remanence evolution is called FC (field cooled).

The low temperature magnetic measurements (10–300 K) were performed with two MPMS (Magnetic Properties Measurement System) cryogenic magnetometers. In addition, first-order reversal curves (FORC) and hysteresis loops were run at room temperature with a saturating field of 1 T by using a VSM (vibrating sample magnetometer). FORC diagrams were processed with FORCInel software [45]. Mössbauer spectra were measured at 300 K and 4.2 K for one selected sample in the Braux area. A conventional constant-acceleration spectrometer was used in transmission geometry with a ^{57}Co/Rh source, using an α-Fe foil at room temperature to calibrate isomer shifts and the velocity scale. Magnetic measurements (MPMS, VSM) and Mössbauer spectra were performed at the Institute for Rock Magnetism (IRM), University of Minnesota, Minneapolis, MN, USA.

Finally, some scanning electronic microscope (SEM) observations on selected specimens were made with a Zeiss SEM equipped with an X-ray energy dispersive spectrometer (EDS) at the ENS, Paris, France.

4. Results

4.1. General Trends

All of the samples have low-field magnetic susceptibility (χ < 500 μSI) and an RT-SIRM$_{300 K}$ comprised between 17 and 220 μAm^2/kg (Table 2). By plotting the data along the studied transect (Figure 3), no particular trend appears. Moreover, the evolution of these magnetic properties is not in agreement among each other, *i.e.*, a χ increase may not be associated with an RT-SIRM$_{300 K}$ increase.

Table 2. Magnetic data and low temperature magnetic features for the collected samples. χ: low field magnetic susceptibility; RT-SIRM$_{300\ K}$: room temperature saturation isothermal remanent magnetization at 300 K; LT-SIRM$_{10\ K}$: low temperature saturation isothermal remanent magnetization at 10 K.

Sampling Site	Sample	χ (μSI)	RT-SIRM$_{300\ K}$ (μAm^2/kg)	LT-SIRM$_{10\ K}$ (μAm^2/kg)
Allons (Al)	1A	74	18	387
	2A	29	29	317
Rouaine (R)	4A	54	18	314
	5A	28	29	148
Braux (B)	6A	171	42	572
	7A	123	41	399
Annot (A)	A0	101	28	378
	11A	261	111	702
	12A	170	92	905
Le Ruch (RU)	RUmg	152	116	642
	RUmb	185	59	2,894
Grand Coyer (GC)	CY1p	201	32	1,372
	CY3p	220	37	5,077
	CY5	229	37	1,334
	CY6	251	22	1,534
	CY7	231	35	1,976
	CY8	229	51	4,613
	CY10	168	24	329
	CY11	133	45	353
Peyresq (P)	20A	45	29	422
Colmars (Co1)	21A	50	20	238
	22A	60	25	205
	23A	41	28	311
Colmars (Co2)	25A	121	78	1,212
	26A	125	108	901
Villars-Colmars (VC)	13A	194	27	783
	14A	128	72	981
	15A	205	220	816
La Moutière (MT)	MT12	307	35	2,249
	MT17	259	59	7,148
	MT29	301	34	3,208
	MT120	198	39	3,359
	MTmg	130	17	467
Gias Vallonetto (GV)	GV1	239	23	10,265
	GV11	244	39	709
	GVmg	26	113	538
La Moutière (MT)	MT12	307	35	2,249
	MT17	259	59	7,148
	MT29	301	34	3,208
	MT120	198	39	3,359
	MTmg	130	17	467
Gias Vallonetto (GV)	GV1	239	23	10,265
	GV11	244	39	709
	GVmg	26	113	538

Figure 3. Magnetic susceptibility and saturation isothermal remanent magnetization at room temperature (RT-SIRM$_{300\ K}$) values along the Grès d'Annot transect. No particular trend appears.

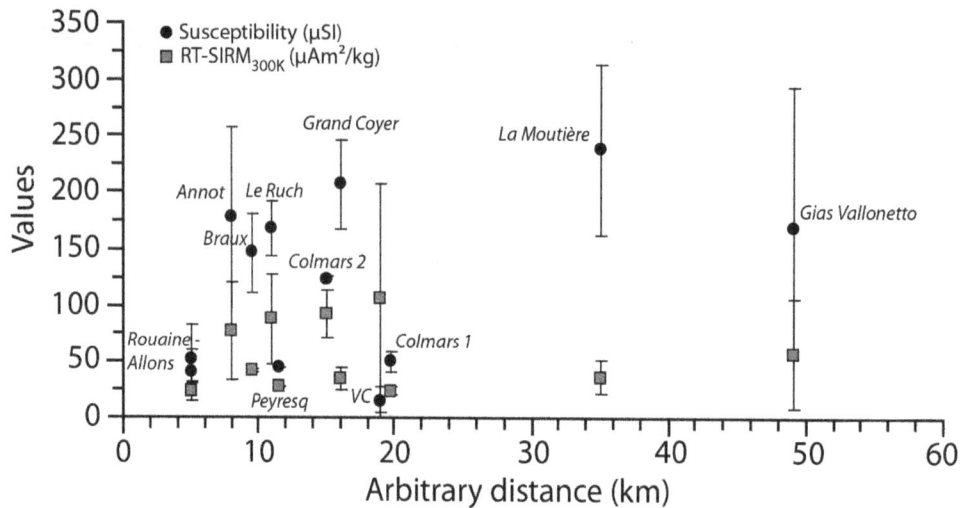

Room temperature measurements (FORC diagrams and hysteresis loops) bring general information on the mineral coercivity and grain size. In the SW part of the study area (Allons-Rouaine-Annot), the FORC diagrams show a coercivity field $Hc < 20$ mT, suggesting that a low coercivity magnetic mineral, occurring in a single domain size, is present (Figure 4a). This is supported by the hysteresis loops that saturate at 1 T (Figure 4c). This mineral is probably magnetite (e.g., [21]). In the Le Ruch-Grand Coyer area, FORCs diagrams and hysteresis loops are not interpretable, because of a very low concentration of ferromagnetic particles. At Gias Vallonetto area, for sample GVmg, the hysteresis loop is wasp-waisted, which indicates either a magnetic assemblage of at least two minerals with different coercivities or different sizes of the same mineral (Figure 4f) [46]. One is probably magnetite, because saturation is reached before 1 T. The other one presents a higher coercivity, probably pyrrhotite (identified on the low temperature measurements). The FORC diagram shows the occurrence of a low coercivity single domain mineral, probably magnetite (Figure 4d).

Additional mineralogical information is provided by the low temperature magnetic measurements that show typical magnetic features. These measurements allow the recognition of the magnetic minerals based on their characteristic magnetic behavior below room temperature. The most common observation in the studied samples is the drop of the remanence acquired at room temperature (RT-SIRM) at 110–120 K, which corresponds to the Verwey transition of magnetite [47,48]. This feature is observed throughout the entire studied transect from SW to NE. Some samples display an increase of the RT-SIRM when cooling down to 10 K. This increase is generally >30%, suggesting the occurrence of goethite [49]. Nevertheless, if the increase is lower (~10%), then it could be due to maghemite [50]. At a very low temperature (<50 K), two magnetic behaviors can be observed. The most common observation is a sudden increase (decrease) of the RT-SIRM below 50 K. This is actually a combination of both remanent and induced magnetizations as a result of the upward (downward) application of the 5-μT magnetic field inside the MPMS. This behavior is called P-behavior and characterizes the paramagnetic minerals (e.g., submicron pyrrhotite, Fe-Mn carbonates) [9,44]. The second magnetic behavior is the Besnus transition of pyrrhotite, which displays a decrease of the remanence at

~32–35 K [51–53]. This particular magnetic behavior is only observed in the NE part of the study area (Villars-Colmars, La Moutière, Gias Vallonetto). Finally, the remanence acquired at a low temperature (LT-SIRM) shows also a typical evolution. On the LT-SIRM curve, a fall of the remanence can be observed from 10 to 35 K. A parameter, called PM, was defined by Aubourg and Pozzi [9] to characterize this drop and aims at assessing the SP/SD ratio [54]. Other LT-SIRM curves display an inflection point at ~200–250 K, suggesting the occurrence of very small particles or minerals with a high Curie/Néel temperature (e.g., high-Ti titanomagnetite, hematite). Nevertheless, the occurrence of a high Curie/Néel temperature is unlikely, because the remanence is removed by ~600 °C when performing thermal demagnetization (not shown).

Figure 4. First-order reversal curves (FORC) diagrams (**a,d**) and hysteresis loops, both uncorrected (**b,e**) and corrected (**c,f**), run at room temperature for two samples. SF is the smoothing factor. Note that the corrected hysteresis loops are noisy (weak signal), as a result of the low concentration of ferromagnetic particles.

Allons area, SW (Sample 2A)

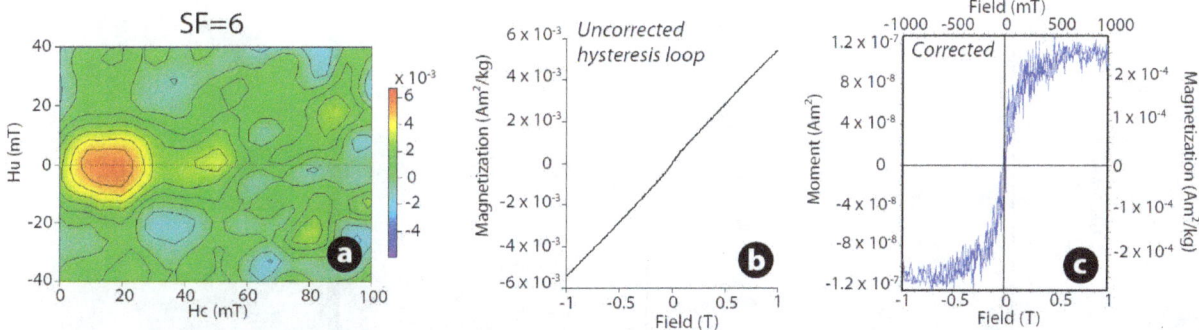

Gias Vallonetto area, NE (Sample GVmg)

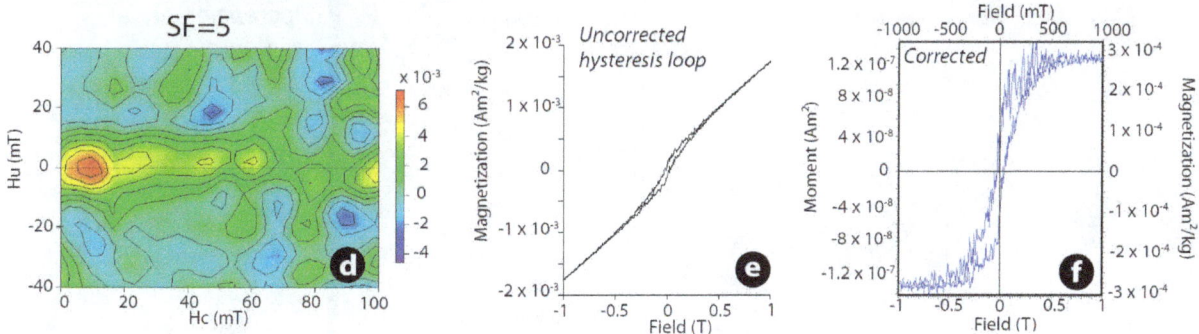

4.2. Allons, Rouaine and Annot Areas

The magnetic signal observed in those areas (Al, R and A in Figure 1, respectively) displays generally an LT-SIRM curve with a two-step pattern [9,44]. This pattern is represented by an important decrease of the LT-SIRM from 10 to 35 K and a Verwey transition of magnetite at ~120 K. For Sample 12A in the Annot area, 56% of the LT-SIRM is lost from 10 to 35 K (PM = 0.56) (Figure 5a). The Verwey transition is easily recognizable in the RT-SIRM curve, and the P-behavior is observed (Figure 5b). When warming back to 300 K, the curve is not reversible from ~70–80 K, *i.e.*, from a temperature lower than that of the Verwey transition. This is certainly due to a maghemitization effect caused by the preliminary heating at 400 K [50].

Figure 5. Evolution of the saturation isothermal remanent magnetization for four characteristic samples from the Grès d'Annot basin with increasing burial. LT-SIRM (**a,c,e,g**) and RT-SIRM (**b,d,f,h**) curves are shown. Note that the slight discontinuity on the LT-SIRM (both zero field cooled (ZFC) and FC) curves at ~50 K is due to a change in the measurement sequence. Mg, magnetite; P-b, P-behavior; Po, pyrrhotite; nG, nanogoethite.

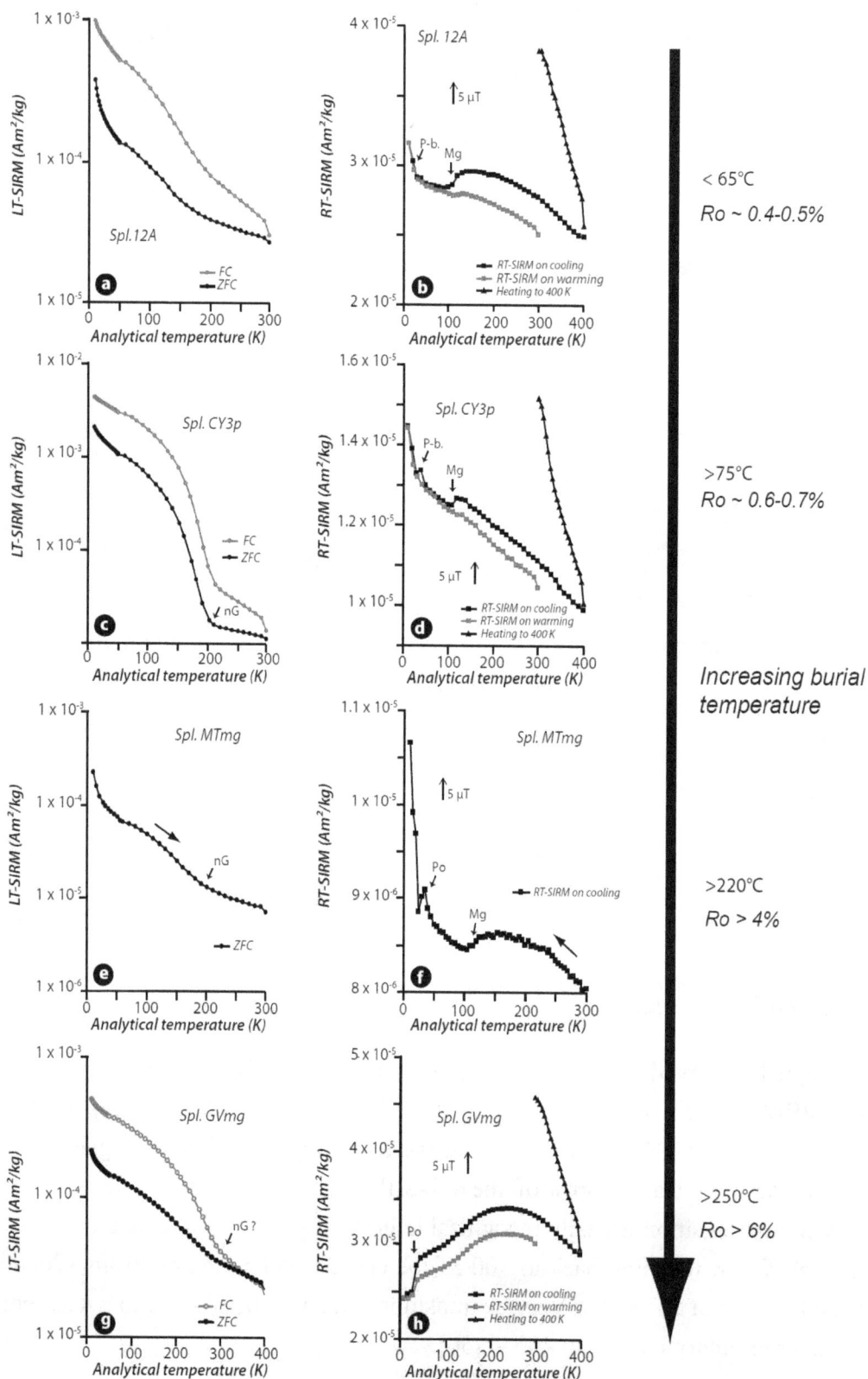

4.3. Braux Area

The LT-SIRM evolution of Sample 6A in the Braux area (B in Figure 1) appears to decrease regularly when warming to room temperature and does not display a distinct two-step pattern as identified in the Allons, Rouaine and Annot areas (Figure 6a). The RT-SIRM curve shows the Verwey transition of magnetite and an increase by 42% of the remanence from 300 to ~35 K (Figure 6a), characteristic of goethite [49]. In order to check the occurrence of goethite, Mössbauer spectra were performed at 300 and 4.2 K on the bulk sample. At room temperature, the Mössbauer spectrum is fitted with three doublets corresponding to contributions of Fe^{2+} and Fe^{3+} (not shown). These components may be associated with paramagnetic iron silicates and, in addition, the Fe^{3+} doublet of smaller QS (quadrupole splitting) may also be related to the presence of superparamagnetic iron oxide. In order to observe the presence of superparamagnetic thermal relaxation, the spectrum was taken at low temperature. At 4.2 K, a small sextet contribution (~8%) is observed and its hyperfine parameters reflect the presence of goethite (Figure 6b). The B_{HF} (magnetic hyperfine field) has a small decrease as compared with a stoichiometric goethite of B_{HF} of 50.6 T, and it could be associated with the presence of lattice defects (vacancies or isomorphic substitution) [55].

Figure 6. Sample (Spl.) 6A (Braux area). (**a**) RT-SIRM and ZFC curves in the 10–300 K range and (**b**) Mössbauer spectrum at 4.2 K. Note that the RT-SIRM increase from 300 to 35 K by 42% showing that the presence of goethite is confirmed by the occurrence of the goethite (G) sextet on the Mössbauer spectrum.

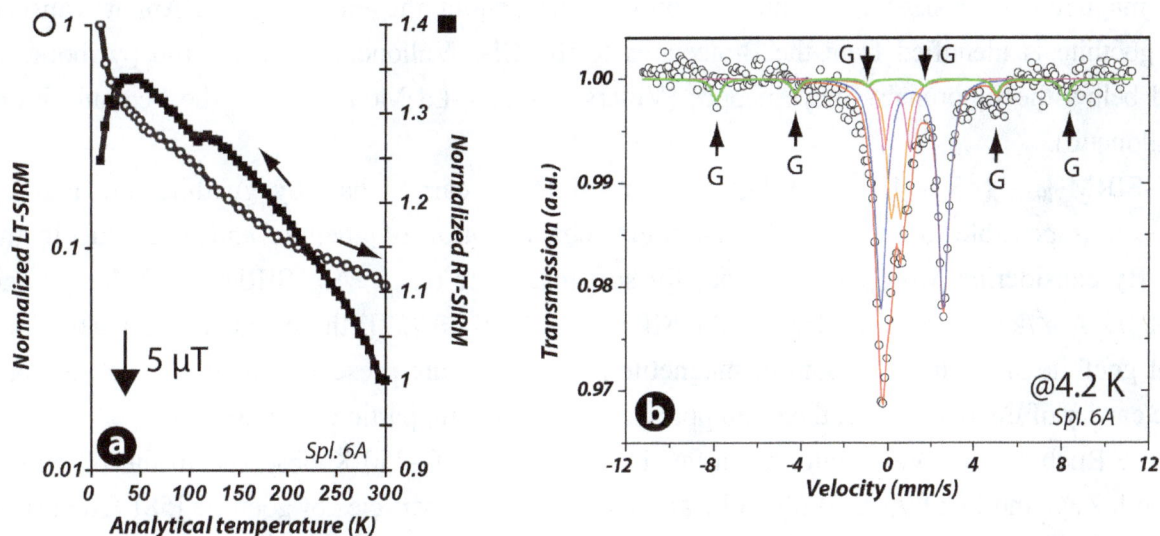

4.4. Le Ruch and Grand Coyer Areas

The RT-SIRM curves of the samples from the Le Ruch and Grand Coyer (RU and GC in Figure 1, respectively) areas show a remanence increase from 300 to 10 K, with a drop at ~120 K (Verwey transition) and a change-in-slope at ~35 K (P-behavior). For the sample, CY3p (Grand Coyer area), 47% of remanence increases from 300 to 10 K (Figure 5d), that is typical of goethite. The particular feature of that area is that the LT SIRM curves (both ZFC and FC) display an inflection point at about 200–250 K (Figure 5c), suggesting the presence of nanoparticles, likely goethite (e.g., [56]).

4.5. Villars-Colmars, La Moutière and Gias Vallonetto Areas

The LT-SIRM curves for samples from the Villars-Colmars and La Moutière areas (VC and MT in Figure 1, respectively) show an inflection point at ~150 K, whereas it occurs at ~200–250 K for Gias Vallonetto (GV in Figure 1) (Figures 5e and 5g). The RT-SIRM increase from 300 to 10 K is between 6% and 15% for the MT samples. These percentage values might be attributed to maghemite or goethite [50]. Goethite, however, usually shows a higher percentage increase (>30%), as is the case for the GV specimens. Differentiating between these two minerals in the MT area needs more data. The important magnetic observation for three samples from the VC, MT and GV areas is the Besnus transition of pyrrhotite. Sample MTmg from La Moutière (Figure 5f) displays a transition similar to the sample, 14A, from the VC area (not shown). Sample GVmg displays, however, a remanent magnetic evolution characteristic of micron-sized (>1 μm) pyrrhotite (Figure 5h) [1,51,52].

5. Discussion

Performing and combining different magnetic techniques appear to be useful and necessary to determine the magnetic assemblage of the studied samples. Low and room temperature magnetic measurements have provided complementary information on the magnetic minerals constitutive of the studied rocks.

5.1. Origin of the Magnetic Assemblage

Nano magnetite (SD size) is ubiquitous, present throughout the entire Grès d'Annot transect, whereas goethite is identified from the Braux area to the Gias Vallonetto area. Micron pyrrhotite is identified below the Embrunais-Ubaye nappes (Villars-Colmars, La Moutière) and the Penninic Front (Gias Vallonetto).

If RT-SIRM$_{300 K}$ (<5 × 10^{-4} Am^2/kg) is assumed to be carried half by goethite and half by magnetite, it is possible to calculate the maximum concentration of goethite and magnetite in the samples. By considering Mrs$_{mag}$ ~10 Am^2/kg for soft magnetite (c$_{mag}$ = RT-SIRM$_{300 K}$/2Mrs$_{mag}$) and Mrs$_{goe}$ ~0.05 Am^2/kg for goethite (c$_{goe}$ = RT-SIRM$_{300 K}$/2Mrs$_{goe}$) [57], the calculation suggests that <0.5% of goethite and a trace amount of magnetite (<25 ppmv) are present. If magnetite is the main magnetic carrier of the remanence, then <50 ppmv of ferromagnetic particles are present.

From Le Ruch to Gias Vallonetto, the inflection point at ~200–250 K observed in the LT-SIRM curves (both ZFC and FC; Figure 5) might be an indicator of nanoparticles of goethite [58]. The origin of nanogoethite is debated. It might represent an alteration product developed as a result of fluid circulation (e.g., [59]). As the tectonic contact of the nappes is not so far from the sampling sites, this possibility cannot be ruled out. A second explanation might be that nanogoethite was neoformed as a result of particular burial conditions. Indeed, based on vitrinite reflectance data, Cavailhes [36] showed that there was probably a stacking of the Embrunais-Ubaye and Parpaillon alpine nappes NE of the Grand Coyer area, leading to an abrupt increase of burial northeastwards. Nevertheless, recent weathering might also be responsible for the formation of goethite, as goethite is a common alteration by-product.

In the Villars-Colmars, La Moutière and Gias Vallonetto areas, micron pyrrhotite is identified. It has been proposed that pyrrhotite could form by the reaction of pyrite with magnetite and organic matter in metamorphic conditions [1], even at a low temperature (<200 °C) [15]. Framboidal grains, mainly composed of pyrite, were examined, because they are often associated with organic matter as a result of the bacterial activity [60]. In samples from the Villars-Colmars area, framboidal structures present a rim darker than the core (Figure 7a). An EDS transect from the core to the outer zone reveals that the Fe/S ratio varies with a decrease in S content toward the rim (Figure 7b). The oxidation rim is certainly magnetite. Replacement of pyrite by magnetite has been widely described in previous studies (e.g., [61,62]). Some alteration features are also observed in euhedral grains, which most likely represent dissolution features. In samples from La Moutière and Gias Vallonetto, EDS analyses indicate that Fe and O are the main constituents in framboidal structures: the former pyrite framboids seem totally oxidized into magnetite. The presence of pyrrhotite could not be documented in our SEM observations, most likely because of its extremely low concentration.

Figure 7. (a) Scanning Electronic Microscope (SEM) observation of a pyrite framboid with an oxidation rim (Spl. 14A) and **(b)** its associated Energy Dispersive Spectrometer (EDS) profile. Note that the Fe and S contents evolve along the profile, with S being concentrated in the inner part of the framboid.

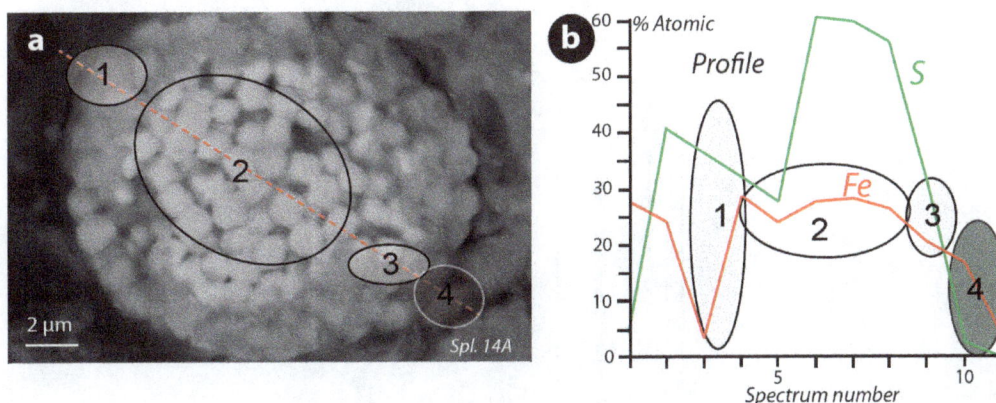

5.2. Toward a Burial Model

Based on laboratory heating experiments and the literature, Aubourg *et al.* [8] have proposed a burial model divided into three magnetic windows, as described in the Introduction. In the following, an attempt is done to refine their model based on natural observations from the Grès d'Annot transect (Figure 8). Though marls and pelites differ by their mineralogical composition, which may influence the formation of magnetic minerals, they are not distinguished in the burial model presented below. Indeed, the model is global, taking into account prior observations made from clay-rich rocks (with no distinction of the clay composition). If clay diagenesis has an influence on the magnetic mineralogy, likely on the iron supply, it is probably in terms of the temperature (±50 °C) of the first occurrence of a given magnetic mineral.

In the study area, an evolution of the magnetic assemblage from SW to NE is observed, consistent with increasing vitrinite reflectance values and maximum burial temperatures.

In the SW (Rouaine, Allons, Annot, Braux), the maximum temperature experienced by the rocks was ~60 °C (*i.e.*, a burial depth ~2–3 km assuming a 30 °C/km geothermal gradient), with vitrinite reflectance Ro < 0.6%. SD magnetite and iron sulfides are the main constituents. Neoformed magnetite is very common in sedimentary rocks. Many studies reported the presence of iron sulfides (pyrite, greigite) at a temperature < 50 °C (e.g., [6,63]). Iron sulfides are produced during early diagenesis by the destruction of the detrital magnetic minerals by bacterial activity (e.g., [19]). In the southwestern part of the study area, pyrite, greigite and submicron pyrrhotite occur.

On the other hand, Abdelmalak *et al.* [10] stated that goethite (called goethite A in Figure 8) could be used as a marker of the immature rocks (Ro < 0.5%) for claystones in volcanic margins. Blaise *et al.* [11] also identified goethite for immature rocks in the Paris Basin. The samples from the Braux area show the occurrence of goethite, suggesting that this zone experienced temperatures below 60 °C if the burial diagenesis hypothesis is favored. Even though it could be concordant with the isotherms established by Labaume *et al.* [33], alteration might also be evoked to explain the occurrence of goethite.

Figure 8. Global burial model for magnetic minerals in clay-rich rocks. The burial temperatures are deduced from the vitrinite reflectance data by using the Vassoyevitch *et al.* [40] calibration [34–36]. See the text for a discussion. References: 1: this study; 2: [10]; 3: [9]; 4: [11]; 5: [12]; 6: [64].

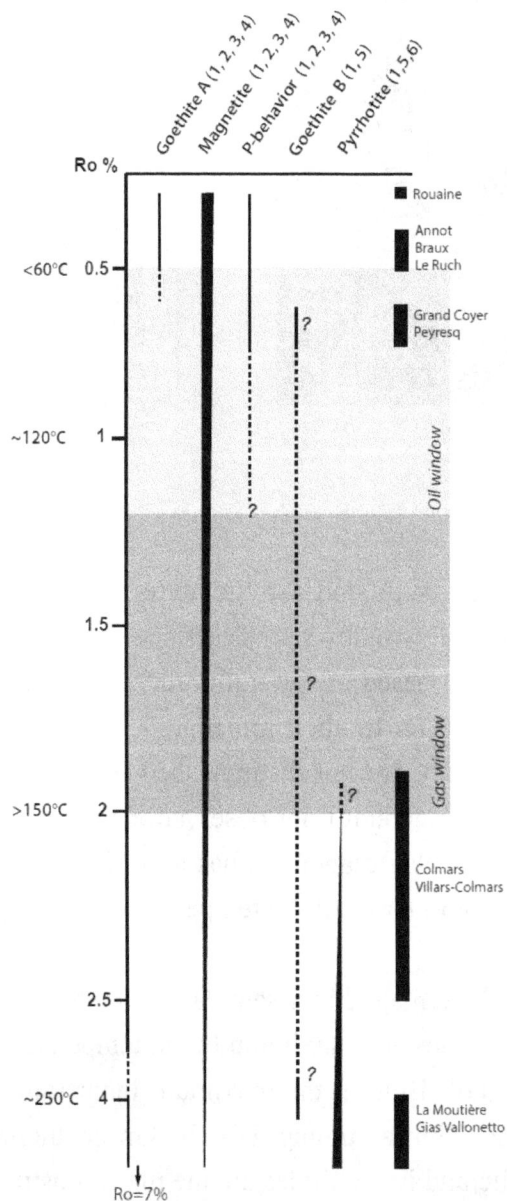

Based on heating experiments, magnetite is expected to be present in rocks experiencing a wide temperature range from ~60 °C to >200 °C, including oil and gas windows [8,9,54,65–67]. In the study area, magnetite is found throughout the entire transect in rocks that experienced a burial temperature <250 °C. Many studies reported also the occurrence of SD magnetite for such temperatures (e.g., [9–12,64]).

From Le Ruch to La Moutière, the magnetic assemblage is mainly composed of magnetite and goethite nanoparticles. The goethite origin is discussed above. Further information is needed to favor one hypothesis among others.

In the NE area (Colmars, Villars-Colmars, La Moutière, Gias Vallonetto), which is characterized by a burial temperature >200 °C (burial depth ~8–10 km), the magnetic assemblage is mainly constituted of magnetite and >1 μm pyrrhotite. Micron-sized pyrrhotite was also observed in Alpine limestone metamorphic units, northward to our study area, which experienced also such a temperature range (e.g., [1,52,68]). Micron pyrrhotite is a common finding in such metamorphic units. It is also observed in the Himalayas metamorphic limestones [4,14] and in clay-rich metamorphic rocks in Taiwan [17].

Rochette [1] was the first to map metamorphic isogrades using the ferromagnetic susceptibility in the Helvetic Jurassic black shales from the Alps. He proposed two metamorphic reactions; (1) the breakdown of magnetite in favor of pyrrhotite near 250 °C, and (2) at a higher temperature (~320–350 °C), the breakdown of pyrite in favor of pyrrhotite in lower greenschist facies. The pyrite-pyrrhotite reaction leads to an increase of ferromagnetic susceptibility by two orders of magnitude. Similarly, in Taiwan, Horng et al. [17] used hysteresis loops on magnetic extracts to map anchizone (<250 °C) from epizone (>250 °C) metamorphic grades. They observed a distinct magnetic assemblage. In anchizone, the hysteresis loops display straight lines, with subtle ferromagnetic contribution. By contrast, when entering the epizone, the hysteresis loops of the magnetic extract display a diagnostic "wide shape" that is characteristic of pyrrhotite. Similarly to Rochette [1] study, Horng et al. [17] observed a net gain of remanence by one order of magnitude for shales from the anchizone to the epizone. It can be assumed that pyrrhotite results from the pyrite breakdown according to the equilibrium proposed by Rochette [1]. Interestingly, along several sections in Taiwan, including the sampling of Horng et al. [17], Beyssac et al. [69] provided a comprehensive map of burial paleotemperatures derived from Raman spectroscopy. When comparing these burial temperatures and the map of the first occurrence of the "wide loop" of pyrrhotite [17], it is suggested that the pyrite-pyrrhotite reaction occurred for a temperature >350 ± 30 °C.

In the study area, >1 μm of pyrrhotite is detected at Gias Vallonetto, and thus, it is questionable whether the pyrrhotite results from magnetite breakdown or pyrite breakdown. Two observations suggest that the pyrite breakdown temperature is not reached. First, with respect to other southern sites, there is no significant enhancement of remanence (Figure 3 and Figure 5h). Second, magnetite is detected from the FORC diagram (Figure 4d), and the wasp-wasted shape of hysteresis loops suggests that magnetite is therefore not entirely consumed (Figure 4f). Therefore, the burial temperature at Gias Vallonetto is suggested to be lower than 350 °C, i.e., lower than the pyrite breakdown temperature. This is consistent with the burial temperature derived from vitrinite reflectance data (Ro < 7.5%) when using the Vassoyevitch et al. [40] calibration curve. It is worth reminding that this calibration curve was chosen because of its consistency with apatite fission track data [33]. The chemical kinetic model proposed by Sweeney and Burnham [70] cannot be used, because it is valid on a range of Ro between

0.3% and 4.5%. Barker and Pawlewicz [71] proposed an empirical calibration curve just as Vassoyevitch *et al.* [40]. When using this calibration, a burial temperature about 30–40 °C higher than that given by the Vassoyevitch *et al.* [40] calibration is estimated, *i.e.*, also coherent with a burial temperature < 350 °C, suggested by the observation of the magnetic assemblage (magnetite and pyrrhotite).

By summarizing and generalizing the results from the Grès d'Annot transect with other studies, a global burial model for magnetic minerals in clay-rich rocks (in a broad sense) is proposed (Figure 8). One requirement for the formation of magnetic minerals is that the rocks need to contain iron sulfides (pyrite) and organic matter as starting materials. Goethite (goethite A) for immature rocks (<50 °C) is mentioned, though its occurrence could be due to the weathering effect. Iron sulfides (and detrital iron oxides) are, however, the main magnetic minerals present in immature rocks. After early diagenesis, magnetite is the main iron oxide formed under increasing temperature (up to ~250 °C). Then, magnetite progressively disappears in favor of pyrrhotite. For a burial temperature >200–250 °C, micron-sized pyrrhotite is the main magnetic mineral present, though magnetite may still be present.

5.3. Analog for a Gas Shale System?

The magnetic mineralogy of the studied clay-rich rocks is similar to that encountered in gas shales, such as the Marcellus and Barnett Shales, USA [12,72]. First, the concentration of the ferromagnetic minerals is very low (<50 ppmv). Second, the magnetic mineralogy seems to evolve in a similar way. Micron-sized pyrrhotite occur for a high burial temperature (>150–200 °C). This finding has been reported by recent studies on the Marcellus Shales in the Appalachians [12,64]. Because they experienced a broad range of burial temperature from <60 °C to >250 °C, which covers oil and gas windows, the clay-rich rocks from the Grès d'Annot basin could be considered as an analog of a gas shale system from the immature to overmature stage.

6. Conclusions

In the Grès d'Annot basin, the magnetic assemblage evolves throughout a wide temperature range (50 to >250 °C) from SW to NE (Penninic Front). Magnetite progressively disappears in favor of pyrrhotite with increasing temperature. Nevertheless, magnetite is still present in the high temperature area (Gias Vallonetto). Pyrrhotite (>1 μm) is highlighted for rocks that experienced a temperature >200–250 °C and, thus, could be a marker of the anchimetamorphism. Based on these observations, a burial diagenesis model of the magnetic minerals, including pyrrhotite, magnetite and goethite, is attempted.

For temperature <60 °C, iron sulfides, nanogoethite and magnetite are present. For higher temperatures covering oil and gas windows (~60–150 °C), magnetite is mainly identified, though the occurrence of other magnetic minerals cannot be ruled out. For a temperature >150 °C, magnetite and micron-sized pyrrhotite are the main magnetic minerals. For a temperature >250 °C, magnetite progressively disappears in favor of pyrrhotite, indicating low-grade metamorphism. The assemblage of magnetite and pyrrhotite indicates that the burial temperature is <350 °C in the most buried site (Gias Vallonetto).

Because of their mineralogy and burial history in a foreland context, the clay-rich rocks from the Grès d'Annot basin might be considered as an analog for a gas shale system.

The comprehension of the magnetic minerals diagenesis through the oil and gas windows of source rocks has probably profound consequences for the interpretation of magnetic anomalies above petroleum plays. The next stage of this work would be to study the paleomagnetic record, assuming a continuous production and destruction of magnetic minerals.

Acknowledgments

This work was conducted as a part of Myriam Kars's Ph.D. thesis supported by Total S.A./Université de Pau et des Pays de l'Adour, France. We are grateful to the Institute for Rock Magnetism, University of Minnesota, Minneapolis, MN, USA, for providing us instrumental facilities (MPMS and VSM) and for fruitful discussions. We benefited from a grant from the University of Minnesota for our 10-day visit at the IRM in 2010. We would like to thank Alexandra Abrajevitch and an anonymous reviewer for their constructive comments on the prior versions of this manuscript.

Author Contributions

Myriam Kars performed the magnetic measurements and handled the paper, including the text and figures. Charles Aubourg, the former advisor of Myriam Kars, provided the initial idea to collect samples along this transect. He contributed to the interpretation of rock magnetic data. Pierre Labaume, who already published a study along this transect, provided additional samples and data and greatly improved the paper by providing observations and comments. Thelma S. Berquó led the Mössbauer study presented in this paper. Thibault Cavailhes provided a part of the data set.

Conflicts of Interest

The authors declare no conflict of interest.

References

1.   Rochette, P. Metamorphic control of the magnetic mineralogy of black shales in the Swiss Alps: Toward the use of "magnetic isogrades". *Earth Planet. Sci. Lett.* **1987**, *84*, 446–456.

2.   Dunlop, D.; Özdemir, O.; Clark, D.; Schmidt, P. Time-temperature relations for the remagnetization of pyrrhotite (Fe_7S_8) and their use in estimating paleotemperatures. *Earth Planet. Sci. Lett.* **2000**, *176*, 107–116.

3.   Crouzet, C.; Ménard, G.; Rochette, P. Cooling history of the Dauphinoise Zone (Western Alps, France) deduced from the thermopaleomagnetic record: Geodynamic implications. *Tectonophysics* **2001**, *340*, 79–93.

4.   Schill, E.; Appel, E.; Gautam, P. Towards pyrrhotite/magnetite geothermometry in low-grade metamorphic carbonates of the Thethyan Himalayas (Shiar Khola, Central Nepal). *J. Asian Earth Sci.* **2002**, *20*, 195–201.

5.   Roberts, A.P.; Weaver, R. Multiple mechanisms of remagnetization involving sedimentary greigite (Fe_3S_4). *Earth Planet. Sci. Lett.* **2005**, *231*, 263–277.

6.   Rowan, C.J.; Roberts, A.P.; Broadbent, T. Reductive diagenesis, magnetite dissolution, greigite growth and paleomagnetic smoothing in marine sediments: A new view. *Earth Planet. Sci. Lett.* **2009**, *277*, 223–235.

7.   Roberts, A.P.; Chang, L.; Rowan, C.J.; Horng, C.S.; Florindo, F. Magnetic properties of sedimentary greigite (Fe_3S_4): An update. *Rev. Geophys.* **2011**, *49*, doi:10.1029/2010RG000336.

8.   Aubourg, C.; Pozzi, J.-P.; Kars, M. Burial, claystones remagnetization and some consequence for magnetostratigraphy. In *Remagnetization and Chemical Alteration of Sedimentary Rocks*; Elmore, R.D., Muxworthy, A.R., Aldana, M.M., Mena, M., Eds.; Special Publications Volume 371; Geological Society London: London, UK, 2012; pp. 181–188.

9.   Aubourg, C.; Pozzi, J.-P. Toward a new <250 °C pyrrhotite-magnetite geothermometer for claystones. *Earth Planet. Sci. Lett.* **2010**, *294*, 47–57.

10. Abdelmalak, M.; Aubourg, C.; Geoffroy, L.; Laggoun-Defarge, F. A new oil window indicator? The magnetic assemblage of claystones from the Baffin Bay volcanic margin (Greenland). *AAPG Bull.* **2012**, *96*, 205–215.

11. Blaise, T.; Barbarand, J.; Kars, M.; Ploquin, F.; Aubourg, C.; Brigaud, B.; Cathelineau, M.; El Albani, A.; Gautheron, C.; Izart, A.; *et al.* Reconstruction of low temperature (<100 °C) burial in sedimentary basins: A comparison of geothermometer sensibility in the intracontinental Paris Basin. *Mar. Pet. Geol.* **2014**, *53*, 71–87.

12. Kars, M.; Aubourg, C.; Suárez-Ruiz, I. Neoformed magnetic minerals as an indicator of moderate burial: The key example of Middle Paleozoic sedimentary rocks, West Virginia, WV, USA. *AAPG Bull.* **2014**, accepted for publication.

13. Rochette, P.; Lamarche, G. Evolution des propriétés magnétiques lors des transformations minérales dans les roches: Exemple du Jurassique Dauphinois (Alpes françaises). *Bull. Mineral.* **1986**, *109*, 687–696. (In French)

14. Appel, E.; Crouzet, C.; Schill, E. Pyrrhotite Remagnetizations in the Himalaya: A Review. In *Remagnetization and Chemical Alteration of Sedimentary Rocks*; Elmore, R.D., Muxworthy, A.R., Aldana, M.M., Mena, M., Eds.; Special Publications Volume 371; Geological Society London: London, UK, 2012; pp. 163–180.

15. Gillett, S.L. Paleomagnetism of the Notch Peak contact metamorphic aureole revisited: Pyrrhotite form magnetite+pyrite under sbmetamorphic conditions. *J. Geophys. Res.* **2003**, *108*, 2446, doi:10.1029/2002JB002386.

16. Horng, C.S.; Torii, M.; Shea, K.S.; Kao, S.J. Inconsistent magnetic polarities between greigite and pyrrhotite/magnetite-bearing marine sediments from the Tsailiao-chi section. *Earth Planet. Sci. Lett.* **1998**, *164*, 467–481.

17. Horng, C.S.; Huh, C.A.; Chen, K.H.; Lin, C.H.; Shea, K.S.; Hsiung, K.H. Pyrrhotite as a tracer for denudation of the Taiwan orogen. *Geochem. Geophys. Geosyst.* **2012**, *13*, doi:10.1029/2012GC004195.

18. Roberts, A.P.; Chang, L.; Heslop, D.; Florindo, F.; Larrasoaña, J.C. Searching for single domain magnetite in the "pseudo-single-domain" sedimentary haystack: Implications of biogenic magnetite preservation for sediment magnetism and relative paleointensity determinations. *J. Geophys. Res.* **2012**, *117*, doi:10.1029/2012JB009412.

19. Canfield, D.E.; Berner, R.A. Dissolution and pyritization of magnetite in anoxic marine sediments. *Geochim. Cosmochim. Acta* **1987**, *51*, 645–659.

20. Bloemendal, J.; King, J.; Hunt, A.; DeMenocal, P.; Hayashida, A. Origin of the sedimentary magnetic record at Ocean Drilling Program sites on the Owen Ridge, western Arabian Sea. *J. Geophys. Res.* **1993**, *98*, 4199–4219.

21. Dunlop, D.; Özdemir, O. *Rock Magnetism: Fundamentals and Frontiers*; Cambridge University Press: Cambridge, UK, 1997.

22. Ford, M.; Lickorish, W.H.; Kusznir, N.J. Tertiary foreland sedimentation in the southern subalpine chains, SE France: A geodynamic appraisal. *Basin Res.* **1999**, *11*, 315–336.

23. Ford, M.; Lickorish, W.H. Foreland basin evolution around the western Alpine arc. In *Deep-water Sedimentation in the Alpine Basin of SE France: New Perspectives on the Grès d'Annot and Related Systems*; Joseph, P., Lomas, S.A., Eds.; Special Publications Volume 221; Geological Society London: London, UK, 2004; pp. 39–63.

24. Joseph, P.; Lomas, S. Deep-water sedimentation in the Alpine Foreland Basin of SE France: A new perspective on the Grès d'Annot and related systems: An introduction. In *Deep-water Sedimentation in the Alpine Basin of SE France: New Perspectives on the Grès d'Annot and Related Systems*; Joseph, P., Lomas, S.A., Eds.; Geological Society London, Special Publications: London, UK, 2004; Volume 221, pp. 1–16.

25. Ravenne, C.; Vially, R.; Riche, P.; Trémolières, P. Sédimentation et tectonique dans le bassin éocène sup-oligocène des Alpes du Sud. *Rev. Inst. Fr. Pet.* **1987**, *42*, 529–553. (In French)

26. Apps, G.; Peel, F.; Elliott, T. The structural setting and palaeogeographical evolution of the Grès d'Annot basin. In *Deep-water Sedimentation in the Alpine Basin of SE France: New Perspectives on the Grès d'Annot and Related Systems*; Joseph, P., Lomas, S.A., Eds.; Geological Society London, Special Publications: London, UK, 2004; Volume 221, pp. 65–96.

27. Du Fornel, E.; Joseph, P.; Desaubliaux, G.; Eschard, R.; Guillocheau, F.; Lerat, O.; Muller, C.; Ravenne, C.; Sztrakos, K. The southern Grès d'Annot out crops (French Alps): An attempt at regional correlation. In *Deep-water Sedimentation in the Alpine Basin of SE France: New Perspectives on the Grès d'Annot and Related Systems*; Joseph, P., Lomas, S.A., Eds.; Geological Society London, Special Publications: London, UK, 2004; Volume 221, pp. 137–160.

28. Jean, S.; Kerckhove, C.; Perriaux, J.; Ravenne, C. Un modèle Paléogène de bassin à turbidites: Les Grès d'Annot du NW du massif de l'Argentera-Mercantour. *Geol. Alp.* **1985**, *61*, 115–143. (In French)

29. Garcia, D.; Joseph, P.; Maréchal, B.; Moutte, J. Patterns of geochemical variability in relation to turbidite facies in the Grès d'Annot Formation. In *Deep-water Sedimentation in the Alpine Basin of SE France: New Perspectives on the Grès d'Annot and Related Systems*; Joseph, P., Lomas, S.A., Eds.; Geological Society London, Special Publications: London, UK, 2004; Volume 221, pp. 349–365.

30. Kerckhove, C. La "zone du flysch" dans les nappes de l'Embrunais-Ubaye (Alpes occidentales). *Geol. Alp.* **1969**, *45*, 1–202.

31. Sztrakos, K.; Du Fornel, E. Stratigraphie, paléoécologie et foraminifères du Paléogène des Alpes Maritimes et des Alpes de Haute-Provence (Sud-Est de la France). *Rev. Micropaleontol.* **2003**, *46*, 229–267. (In French)

32. Labaume, P.; Ritz, J.-F.; Philip, H. Failles normales récentes dans les Alpes sud-occidentales: Leurs relations avec la tectonique compressive. *Comptes Rendus Acad. Sci.* **1989**, *308*, 1553–1560. (In French)

33. Labaume, P.; Jolivet, M.; Souquière, F.; Chauvet, A. Tectonic control on diagenesis in a foreland basin: Combined petrologic and thermochronologic approaches in the Grès d'Annot basin (late Eocene—Early Oligocene, French-Italian external Alps). *Terra Nova* **2008**, *20*, 95–101.

34. Labaume, P.; Arnaud, N.; Buatier, M.; Charpentier, D.; Chauvet, A.; Chirouze, F.; Jolivet, M.; Monié, P.; Sizun, J.-P.; Travé, A. *Contrôle Tectonique de la Diagenèse d'une Formation Turbiditique d'Avant-Chaine, Exemple des Grés d'Annot, Alpes Externes Franco-Italiennes*; Unpublished Internal Report; TOTAL: Pau, France, 2008. (In French)

35. Labaume, P.; Sizun, J.-P.; Charpentier, D.; Travé, A.; Chirouze, F.; Buatier, M.; Chauvet, A.; Walgenwitz, F.; Jolivet, M.; Monié, P.; *et al.* Diagenesis controlled by tectonic burial in a foreland basin turbidite formation. The case example of the Grès d'Annot, French-Italian external Alps. In Proceedings of the European Geosciences Union (EGU) General Assembly, Vienna, Austria, 19–24 April 2009.

36. Cavailhes, T. Architecture et Propriétés Pétrophysiques des Zones de Failles dans une Série Gréso Pélitique Turbiditique Profondément Enfouie: Rôle de la Déformation et des Interactions Fluide-Roche. Ph.D. Thesis, Université de Montpellier 2, Montpellier, France, 2012. (In French)

37. Souquière, F. Relations Tectonique/Diagenèse dans un Bassin d'Avant-Chaîne, Exemple des Grès d'Annot: Approche Pétrologique et Thermochronologiques. Master's Thesis, Université de Montpellier 2, Montpellier, France, 2005. (In French)

38. Barlier, J.; Ragot, J.-P.; Thouray, J.-C. L'évolution des Terres Noires subalpines méridionales d'après l'analyse minéralogique des argiles et la réflectance des particules carbonées. *Bull. BRGM* **1974**, *6*, 533–548.

39. Pickering, K.; Hilton, V. *Turbidite Systems of Southern France: Application to Hydrocarbon Prospectivity*;Vallis Press: London, UK, 1998.

40. Vassoyevitch, N.B.; Korchagina, N.V.; Lopatin, N.V.; Chernyshev, V.V. Principal phase of oil formation. *Int. Geol. Rev.* **1970**, *12*, 1276–1297.

41. Leclère, H.; Buatier, M.; Charpentier, D.; Sizun, J.-P.; Labaume, P.; Cavailhes, T. Formation of phyllosilicates in fault zone affecting deeply buried arkosic sandstones. Their influence on fault zone petrophysic properties (Annot sandstones, late Eocene-early Oligocene, external Alps). *Swiss J. Geosci.* **2012**, *105*, 299–312.

42. Cavailhes, T.; Sizun, J.-P.; Labaume, P.; Chauvet, A.; Buatier, M.; Soliva, R.; Gout, C. Influence of fault rock foliation on fault zone permeability: The case of deeply buried arkosic sandstones (Grès d'Annot, SE France). *AAPG Bull.* **2013**, *97*, 1521–1543.

43. Özdemir, O.; Dunlop, D. Thermoremanence and Néel temperature of goethite. *Geophys. Res. Lett.* **1996**, *23*, 921–924.

44. Kars, M.; Aubourg, C.; Pozzi, J.-P. Low temperature magnetic behaviour near 35 K in unmetamorphosed claystones. *Geophys. J. Int.* **2011**, *186*, 1029–1035.

45. Harrison, R.; Feinberg, J. FORCinel: An improved algorithm for calculating first-order reversal curve distributions using locally weighted regression smoothing. *Geochem. Geophys. Geosyst.* **2008**, *9*, doi:10.1029/2008GC001987.

46. Tauxe, L.; Mullender, T.; Pick, T. Potbellies, wasp-waists, and superparamagnetism in magnetic hysteresis. *J. Geophys. Res.* **1996**, *101*, 571–583.

47. Muxworthy, A.R.; McClelland, E. Review of the low-temperature magnetic properties of magnetite from a rock magnetic perspective. *Geophys. J. Int.* **2000**, *140*, 101–114.

48. Özdemir, O.; Dunlop, D.; Moskowitz, B. Changes in remanence, coercivity and domain state at low-temperature in magnetite. *Earth Planet. Sci. Lett.* **2002**, *194*, 343–358.

49. Dekkers, M.J. Magnetic properties of natural goethite—II. TRM behaviour during thermal and alternating field demagnetization and low-temperature treatment. *Geophys. J. Int.* **1989**, *97*, 341–355.

50. Özdemir, O.; Dunlop, D. Hallmarks of maghemitization in low-temperature remanence cycling of partially oxidized magnetite nanoparticles. *J. Geophys. Res.* **2010**, *115*, doi:10.1029/2009JB006756.

51. Dekkers, M.J.; Mattéi, J.-L.; Fillion, G.; Rochette, P. Grain-size dependence of the magnetic behavior of pyrrhotite during its low-temperature transition at 34 K. *Geophys. Res. Lett.* **1989**, *16*, 855–858.

52. Rochette, P.; Fillion, G.; Mattéi, J.-L.; Dekkers, M.J. Magnetic transition at 30–34 Kelvin in pyrrhotite: Insight into a widespread occurrence of this mineral in rocks. *Earth Planet. Sci. Lett.* **1990**, *98*, 319–328.

53. Wolfers, P.; Fillion, G.; Ouladdiaf, B.; Ballou, R.; Rochette, P. The pyrrhotite 32 K magnetic transition. *Solid State Phenom.* **2011**, *170*, 174–179.

54. Kars, M.; Aubourg, C.; Pozzi, J.-P.; Janots, D. Continuous production of nanosized magnetite through low grade burial. *Geochem. Geophys. Geosyst.* **2012**, *13*, doi:10.1029/2012GC004104.

55. Murad, E.; Cashion, J. *Mössbauer Spectroscopy of Environmental Materials and Their Industrial Utilization*; Kluwer Academic Publishers: Boston, MA, USA, 2013.

56. Guyodo, Y.; Mostrom, A.; Penn, R.L.; Banerjee, S.K. From nanodots to nanorods: Oriented aggregation and magnetic evolution of nanocrystalline goethite. *Geophys. Res. Lett.* **2003**, *30*, doi:10.1029/2003GL017021.

57. Maher, B.A.; Thompson, R. *Quaternary Climates, Environments and Magnetism*; Cambridge University Press: Cambridge, UK, 1999.

58. Guyodo, Y.; LaPara, T.M.; Anschutz, A.J.; Penn, R.L.; Banerjee, S.K. Rock magnetic, chemical and bacterial community analysis of a modern soil from Nebraska. *Earth Planet. Sci. Lett.* **2006**, *251*, 168–178.

59. Evans, M.; Elmore, R.D. Fluid control of localized mineral domains in limestone pressure solution structures. *J. Struct. Geol.* **2006**, *28*, 284–301.

60. Wilkin, R.; Barnes, H. Formation processes of framboidal pyrite. *Geochim. Cosmochim. Acta* **1997**, *61*, 323–339.

61. Suk, D.; Peacor, D.; Van der Voo, R. Replacement of pyrite framboids by magnetite in limestones and implications for paleomagnetism. *Nature* **1990**, *345*, 611–613.

62. Suk, D.; Van der Voo, R.; Peacor, D. Origin of magnetite responsible for remagnetization of Early Paleozoic limestones of New York State. *J. Geophys. Res.* **1993**, *98*, 419–434.

63. Rowan, C.J.; Roberts, A.P. Magnetite dissolution, diachronous greigite formation, and secondary magnetizations from pyrite oxidation: Unraveling complex magnetizations in Neogene marine sediments from New Zealand. *Earth Planet. Sci. Lett.* **2006**, *241*, 119–137.

64. Manning, E.B.; Elmore, R.D. Rock Magnetism and Identification of Remanence Components in the Marcellus Shale, Pennsylvania. In *Remagnetization and Chemical Alteration of Sedimentary Rocks*; Elmore, R.D., Muxworthy, A.R., Aldana, M.M., Mena, M., Eds.; Special Publications Volume 371; Geological Society London: London, UK, 2012; pp. 271–282.

65. Cairanne, G.; Aubourg, C.; Pozzi, J.-P.; Moreau, M.-G.; Decamps, T.; Marolleau, G. Laboratory chemical remanent magnetization in natural claystones: A record of two polarities. *Geophys. J. Int.* **2004**, *159*, 909–916.

66. Moreau, M.-G.; Ader, M.; Enkin, R. The remagnetization of clay-rich rocks in sedimentary basins: Low-temperature experimental formation of magnetic carriers in natural sample. *Earth Planet. Sci. Lett.* **2005**, *230*, 193–210.

67. Aubourg, C.; Pozzi, J.-P.; Janots, D.; Sahraoui, L. Imprinting chemical remanent magnetization in claystones at 95 °C. *Earth Planet. Sci. Lett.* **2008**, *272*, 172–180.

68. Crouzet, C.; Ménard, G.; Rochette, P. High-precision three-dimensional paleothermometry derived from paleomagnetic data in Alpine metamorphic unit. *Geology* **1999**, *27*, 503–506.

69. Beyssac, O.; Simoes, M.; Avouac, J.-P.; Farley, K.A.; Chen, Y.G.; Chan, Y.C.; Goffé, B. Late Cenozoic metamorphic evolution and exhumation of Taiwan. *Tectonics* **2007**, *26*, doi:10.1029/2006TC002064.

70. Sweeney, J.J.; Burnham, A.K. Evaluation of a simple model of vitrinite reflectance based on chemical kinetics. *AAPG Bull.* **1990**, *74*, 1559–1570.

71. Barker, C.E.; Pawlewicz, M.J. Calculation of vitrinite reflectance from thermal histories and peak temperatures. In *Reevaluation of Vitrinite Reflectance*; Mukhopadhyay, P.K., Dow, W.G., Eds.; Series 570; American Chemical Society Symposium: Washington, DC, USA, 1994; pp. 216–229.

72. Bruner, K.R.; Smosna, R. A Comparative Study of the Mississippian Barnett Shale, Fort Worth Basin, and Devonian Marcellus Shale, Appalachian Basin. Available online: http://www.netl.doe.gov/File%20Library/Research/Oil-Gas/publications/brochures/DOE-NETL-2011 -1478-Marcellus-Barnett.pdf (accessed on 1 July 2014).

8

Evolution of Geochemical and Mineralogical Parameters during *In Situ* Remediation of a Marine Shore Tailings Deposit by the Implementation of a Wetland Cover

Nouhou Diaby [1,2] **and Bernhard Dold** [1,3,*]

[1] Institut de Minéralogie et Géochimie, Université de Lausanne, CH-1015 Lausanne, Switzerland

[2] Present Address: Laboratoire de Traitement des Eaux usées, Institut Fondamental d'Afrique Noire (IFAN), Université Cheikh Anta Diop, BP 206 Dakar, Senegal; E-Mail: nouhou.diaby@ucad.edu.sn

[3] Present Address: SUMIRCO (Sustainable Mining Research & Consult EIRL), Casilla 28, San Pedro de la Paz 4130000, Chile

* Author to whom correspondence should be addressed; E-Mail: bernhard.dold@gmail.com

Abstract: We present data of the time-evolution of a remediation approach on a marine shore tailings deposit by the implementation of an artificial wetland. Two remediation cells were constructed: one in the northern area at sea-level and one in the central delta area (above sea-level) of the tailings. At the beginning, the "sea-level" remediation cell had a low pH (3.1), with high concentrations of dissolved metals and sulfate and chloride ions and showed sandy grain size. After wetland implementation, the "sea-level" remediation cell was rapidly water-saturated, the acidity was consumed, and after four months the efficiency of metal removal from solution was up to 79.5%–99.4% for Fe, 94.6%–99.9% for Mn, and 96.1%–99.6% for Zn. Al and Cu concentrations decreased below detection limit. The "above sea-level" remediation cell was characterized by the same pH (3.1) and finer grain size (clayey–silty), and with some lower element concentrations than in the "sea-level" cell. Even after one year of flooding, the "above sea-level" cell was not completely flooded, showing on-going sulfide oxidation in between the wetland cover and the groundwater level; the pH increased only to 4.4 and metal concentrations decreased only by 96% for Fe, 88% for Al, 51% for Cu, 97% for Mn, and 95% for Zn. During a dry period, the water level dropped in the "sea-level" cell, resulting in a seawater ingression, which triggered the desorption of As into solution. These data show that the applied remediation approach for this tailings deposit is successful, if the system is maintained water-saturated. Metal removal from solution was possible in both systems: first, as a result of sorption on Fe(III) hydroxide/and/or clay minerals and/or co-precipitation processes after rise of pH; and then, with more reducing conditions, due to metal sulfides precipitation.

Keywords: acid mine drainage; bioremediation; sulfate-reducing bacteria; metal removal

---

1. Introduction

Sulfide oxidation in mine tailings is a major source of acid mine drainage (AMD) production [1,2]. Several treatment approaches (active and passive) have been proposed for AMD mitigation. Active treatment usually consists of the addition of lime or a chemical neutralizing agent in order to increase the pH and to precipitate metals on hydroxides and carbonates [3] or as sulfides [4] and/or adsorb them onto a sorbent. The neutralizing agents, unless they are available *in situ*, have to be quarried and transported over long distances, which can make this approach capital intensive. Contrariwise, the lower operating costs of passive treatment such as wetland construction make them attractive for the mining industry. Constructed wetlands have been successfully applied to several mining environments to remove contaminants from water, especially from metal-rich effluent [3,5]. The mechanisms of metal removal from AMD using wetland are mainly the sorption and co-precipitation processes with iron oxides and hydroxides as well as the precipitation of metal sulfides.

The *in situ* remediation at the source of AMD formation may help to mitigate AMD, but also to prevent its future formation by avoiding sulfide oxidation. Wetland cover directly on the tailings is a strategy, that was applied and in some cases, demonstrated high efficiency [6–10]. This method is based on limitation of the oxidation processes, the sorption of metals on mineral surface and metal removal by sulfide precipitation.

In addition, geochemical processes involved in tailings wetlands are mediated or accelerated by microbial activities. Wetlands promote optimal conditions for sulfate reduction by sulfate-reducing bacteria (SRB) producing hydrogen sulfide which will react and precipitate with metals. On the other hand, the geochemical reactions mediated by SRB generate alkalinity, participating to acidity neutralization. The role of this microbial group in metal removal processes has been described in various studies [11–14].

The overall study of the remediation of the Bahia de Ite tailings impoundment, Peru is published elsewhere [10]. We concentrate in the present study on the evolution of the geochemical and mineralogical changes during a real-scale *in situ* remediation experiment at the site in order to highlight the controlling parameters for the effective function of the remediation process. The aim of the study was (i) to investigate the effect on element cycling during the flooding of the oxidizing porphyry copper tailings in a marine setting with two different waters (river water and wetland water); (ii) to highlight the evolution of mineralogical and (bio)geochemical changes over time; and (iii) to evaluate the long-term effectiveness of the bioremediation process.

2. Experimental Section

2.1. Preparation of the Remediation Cell

Two remediation cells of 30 m × 30 m were constructed in September and October 2005 in the northern section (Northern Cell, NC) close to the shoreline at sea-level and in the delta area (Delta Cell, DC), about two meters above sea-level, of the tailings deposit (Figure 1). The surrounding dikes (about 70 cm high) were constructed manually using tailings material, and a piezometers nest (drive point piezometers) was installed in the center of each cell (Figure 1C). Before the flooding of the cell, the groundwater level was about 0.5 m depth in the NC and 1.5 m in the DC.

Figure 1. Overview of the Bahía de Ite (Ite bay) marine shore tailings deposit, Peru with the location of the remediation cells and the sampling points of the overall study of the system reported by Dold *et al.* [10]. (**A**) = Map of the tailings deposit; (**B**) = Profile of the northern section of the tailings deposit (the scale is exaggerated for better visibility); (**C**) = Change of the remediation cell located in the delta area from oxidizing tailings to a wetland cover.

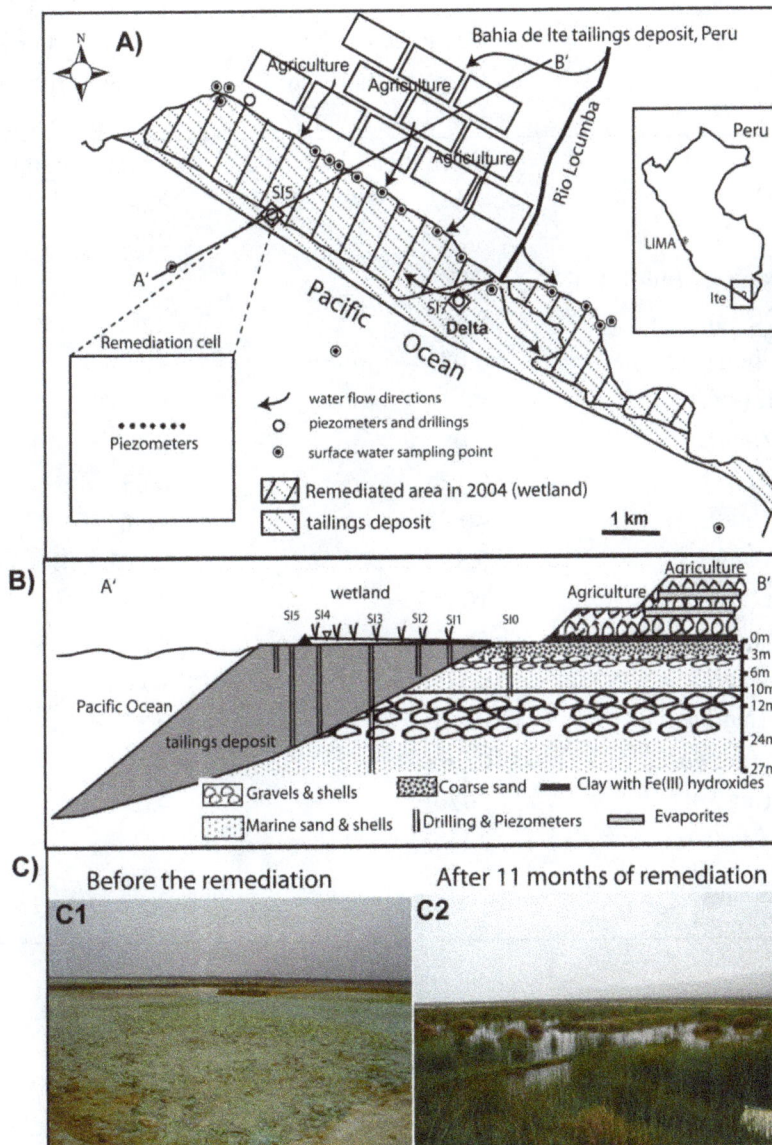

The NC cell, with mainly sandy grain size, was flooded using water from the wetland (Figure 1 and Table 1) covering the already remediated tailings areas close by. Due to the high water saturation in the wetland area, the NC was saturated after five days of flooding. The monitoring of this remediation cell was performed from October 2005 to July 2006.

The DC, which showed a more heterogenic stratigraphy with mainly clayey and fine-sandy horizons (Table 1), was flooded using water channeled from the Locumba River (Rio Locumba). This cell could not be completely saturated during all the course of the experiment, as the delta area is topographically elevated in relation to the rest of the tailings deposit (up to 2 m higher).

The monitoring of the remediation cells was performed from September 2005 to November 2007. Both cells were planted with local vegetation that developed in previously remediated sections of the tailings. Figure 1C shows the change in the delta remediation cell at the beginning and after 11 months of remediation.

Table 1. Characterization of the remediation cell at the start of the experiment and the geochemical composition of the water used for flooding. DOC = Dissolved Organic Carbon.

Characteristics		Northern Cell (NC)	Delta Cell (DC)
Grain size		Sandy	Clayey–silty
Groundwater level		Groundwater level at 1 m	Groundwater level at 1.5 m
Location		Close to shore line	Close to river inflow
Altitude		Sea water level	2 m above sea water level
Characteristics of water used for remediation	Origin	From the old wetland	From Locumba river
	pH	7.6	7.3
	Eh	341	420
	Alkalinity (mg/L CaCO$_3$)	241	133
	DOC (mg/L)	8.36	2.6
	Fe total (mg/L)	0.08	<0.005
	FeII (mg/L)	<0.1	<0.1
	Al (mg/L)	0.05	0.1
	Cu (mg/L)	0.12	0.03
	Ni (mg/L)	<0.02	<0.02
	Mn (mg/L)	0.39	<0.01
	Mg (mg/L)	100	37.4
	K (mg/L)	56	29.1
	Zn (mg/L)	<0.04	<0.04
	As (mg/L)	0.07	0.27
	Mo (mg/L)	0.06	<0.04
	Cl (mg/L)	2080	433
	Ca (mg/L)	274	141
	Na (mg/L)	1555	300
	SO$_4$ (mg/L)	1636	429

2.2. Sampling

Solid (in total 246 samples) and water (in total 376 samples) samples were taken daily during the first week from the piezometer nest and the surface waters; then weekly until the 7th week, and after that, monthly. Samples were taken for geochemical, mineralogical, and microbiological parameters (the microbiological study will be published elsewhere). Solid samples were taken using a percussion soil sampler. Paste-pH after the method described by Mine Environment Neutral Drainage (MEND) Program [15] was measured and the tailings were described by color, grain size, and texture. The water samples in the saturated zone (Solinst® Drive-point piezometers, Georgetown, ON, Canada) were taken with a peristaltic pump. In the unsaturated vadose zone, water samples were taken using squeezing technique modified after Coggans et al. [16] before the remediation process to obtain the geochemical starting conditions. After remediation started, the piezometers and the surface water in the remediation cell were sampled. All water samples were filtered (0.2 µm; regenerated cellulose) and stored at +4 °C in darkness prior to analysis. Samples for cations analyses were acidified to pH <2 with Suprapure® Merck HNO_3. Specific geochemical variables (pH, Eh, Fe^{2+}, and alkalinity) were measured immediately during sampling in a flow-cell fed by a peristaltic pump. The performance of the Eh electrodes was controlled by Light and Zobell's solutions and the values were corrected with respect to the standard hydrogen electrode (SHE). Alkalinity was measured with a Hach® digital titrator (Loveland, CO, USA) and Fe^{2+} with a Merck Reflectoquant® reflectometer (Darmstadt, Germany). The complete data set is given in the supplementary information.

2.3. Mineralogical Analysis

Polished sections and polished thin sections were prepared from selected bulk samples. Selected samples were analyzed for mineralogy by X-ray diffraction (XRD), using a Philips 3020 diffractometer (Amsterdam, The Netherlands) with CuKα ($k = 1.54056$ Å) and a monochromator. Scan settings were 3°–70° 2θ, 0.02° step size, 2-s count time per step.

2.4. Grain Size Analysis

The grain size analysis was performed using a Beckman Coulter LS 13 320 laser-particle analyzer (Brea, CA, USA) with universal liquid module (ULM).

2.5. Aquatic Geochemistry

Water samples (376 samples) were analyzed for 20 elements (Ag, Al, B, Ba, Ca, Cd, Co, Cr, Cu, Fe, K, Mg, Mn, Mo, Na, Ni, Pb, Sb, SiO_2, Zn) by inductively coupled plasma atomic emission spectroscopy (ICP-AES) (Perkin Elmer HP 3000 DV, Waltham, MA, USA), and As and Se was measured by atomic absorption spectroscopy (AAS) with hydride system (Perkin Elmer FIAS-400). SO_4 and Cl were measured by ion chromatography (IC; Dionex DX 120, Sunnyvale, CA, USA) and dissolved organic carbon (DOC) by High Temperature TOC/TNb Analyzer LiquiTOC® (Elementar, Hanau, Germany).

2.6. Sequential Extractions

The solid geochemistry was analyzed by a seven-step sequential extraction approach [17]. The analysis was done on selected samples from the oxidation zone of the two remediation cells (0–1 m). In the first step, a 1.0 g sample was mixed into 50 mL deionized H_2O and shaken for 1 h to dissolve the water-soluble secondary and tertiary sulfate (e.g., bonattite, chalcanthite, gypsum, pickeringite, magnesioauberite). The second step used NH_4–acetate at pH 4.5 shaken for 2 h to detach the exchangeable ions and carbonates (calcite, vermiculite-type-mixed-layer). The third one used 0.2 M NH_4–oxalate pH 3.0 (shaken for 1 h in darkness) to dissolves the Fe oxyhydroxides (schwertmannite, 2-line ferrihydrite, secondary jarosite, MnO_2). The fourth step used 0.2 M NH_4–oxalate pH 3.0 (heated in water bath 80 °C for 2 h) to dissolve the hydroxides and oxides (goethite, jarosite, natrojarosite, and hematite, magnetite, higher ordered ferrihydrite's (e.g., 6-line ferrihydrite). In the fifth step, 35% H_2O_2 (heat in water bath for 1 h) was used to leach the elements associated to organics, covellite, chalcocite–digenite. The sixth step combined $KClO_3$ and HCl, followed by 4 M HNO_3 boiling to dissolve the sulfides (pyrite, chalcopyrite, bornite, sphalerite, galena, molybdenite, tetrahedrite, cinnabar, orpiment, and stibnite). Finally, the seventh step used HNO_3, HF, $HClO_4$, HCl digestion to separate the residual elements attached on silicates. After extraction, multi-element (31 elements) analyses were performed in the leachates by inductively coupled plasma optical emission spectrometer (ICP-OES) at the SGS Laboratories of Toronto, Canada. The complete data set is given in the supplementary information.

3. Results and Discussion

3.1. Stratigraphy and Mineralogy before Remediation

The mineralogical and geochemical characterization from oxidizing tailings at the Bahia de Ite was discussed in detail by Dold et al. [10]. We present here the mineralogy and geochemistry of the two remediation test cells just before the remediation experiment started for the reader's convenience.

The stratigraphy of both remediation cells was characterized by a cover of efflorescent salts on the surface containing mainly water-soluble sulfates and minor chlorides: tamarugite ($NaAl(SO_4)_2 \cdot 6H_2O$, hexahydrite ($MgSO_4 \cdot 6H_2O$), pentahydrite $Mg(SO_4)(H_2O)_5$, aubertite $AlCuCl(SO_4)_2(H_2O)_{14}$, melanterite ($FeSO_4(H_2O)_7$, siderotil (Fe, Cu)$SO_4 \cdot 5H_2O$, bloedite $Na_2Mg(SO_4)_2 \cdot 4H_2O$, thenardite Na_2SO_4, dansite $Na_{21}Mg(SO_4)10C_{13}$; halite (NaCl), eriochalcite $CuCl_2 \cdot 2H_2O$, atacamite ($Cu_2Cl(OH)_3$). This enrichment of the mobile elements from the oxidation zone towards the surface was controlled by the high evaporation in the Atacama Desert, which induced a strong capillary transport to the tailings surface, a process commonly observed in arid climates [2,18]. In the northern cell, where the grain size was sandy, the efflorescent salts precipitated mainly in between the sands, whereas in the delta area, a crust of efflorescent salts up to 20 cm thick evolved above this more silty-clayey tailings surface.

The mineralogy of the underlying oxidation zone was dominated by quartz, muscovite, biotite, feldspar (albite–sanidine–anorthoclase), anhydrite, gypsum, a vermiculite-type mixed layer mineral resulting from the alteration of biotite [2] and pyrite. This low-paste pH oxidation zone (pH 2–4) showed straw-yellowish color (jarosite), with orange-ochre-brown Fe(III) hydroxides (goethite and schwertmannite), which were mainly related to dehydration cracks and grain size limits, where the Fe(III)-rich solutions percolated [10]. Below the oxidation zone, the underlying neutral primary zone was characterized by the primary mineral assemblage from the ore (quartz, muscovite, biotite, feldspar (albite–sanidine–anorthoclase), anhydrite, gypsum), with about 4% pyrite and traces of chalcopyrite, molybdenite, chalcocite-covellite.

At the northern remediation cell, the tailings consisted of sandy sediment due to grain size separation by the wave action. The primary zone had an orange-reddish color along the whole depth. This was interpreted as a consequence of the Fe(II) resulting from the tailings oxidation zone, which was pushed during the remediation towards the shoreline. The contact of the Fe(II) with more oxidizing water results in oxidation into Fe(III) hydroxide, which in turn coats the sediments. Additionally, reductive dissolution of a Fe(III) oxide-rich horizon at 10 m depth is an additional source for the iron coating [10].

In contrast, the delta area was characterized by a mixed stratigraphy of silty, clayey, and very fine-sandy layers. This cell had a clear separation between oxidation and primary zone. From 0 to 1 m were oxidizing tailings (yellow-brownish-orange). Just below and up to 1.7 m depth, a first grey primary zone was present, followed by the former oxidation zone up to 4.4 m depth and then underlain by a second primary zone. A 10 cm thick pyrite horizon was present at 3.2 m depth. This stratigraphy is explained by an ultimate fresh tailings deposition done above the old oxidation zone in 1992, due to a break of a dike in the channel transporting the tailings in the bay.

3.2. Geochemical Evolution of the Wetland

3.2.1. The Sandy Northern Remediation Cell (NC)

Starting Conditions before Remediation

At the northern remediation cell, the starting geochemical conditions in the pore water were as follows: pore water in the oxidation zone showed pH 3.1 and a redox potential of up to 358 mV (Table 2) with high concentration of heavy metal in solution of up to 12.9 mg/L Zn, 80.1 mg/L Mn, 104 mg/L Al, 176 mg/L Fe, 308 mg/L Mg, 329 mg/L Cu. Below the oxidation zone high concentrations of Fe(II) were detected (up to 242 mg/L Fe), feature typically observed in oxidizing sulfidic tailings, due to microbial iron reduction processes [18]. Arsenic concentrations averaged 0.16 mg/L, which was lower than the natural background of the inflowing water from the Locumba River (average 0.5 mg/L) [10]. This may be due to the adsorption of arsenate onto iron hydroxides at low pH conditions. Below the oxidation zone, pH increased up to 6.5 at 10 m depth and Eh decreased to 256 mV. Due to these geochemical conditions, heavy metal concentrations in the pore water were low in the primary zone. Chloride, sodium and sulfate showed concentration of 2125 and 1748, and 4167 mg/L, respectively (Table 2). Two peak concentrations of sulfate in the stratigraphy were observed. In the upper part, the oxidation zone showed concentrations of up to 4167 mg/L due to sulfide oxidation (supported by isotopic data) [10]. Below, the sulfate concentrations stabilize around 2100 mg/L due to gypsum solubility control. The increase towards 2861 mg/L SO_4^{2-} at 10 m depth was interpreted as the only sign of seawater influence detected in the deposit, which was also supported by stable isotopic data.

Table 2. Geochemistry of ground water in the Northern Cell (NC) before the start of the remediation experiment.

Samples	Depth (m)	pH	EC (ms/cm)	Eh (mV)	Alkalinity (mg/L CaCO₃)	Al (mg/L)	As (mg/L)	Ca (mg/L)	Cl (mg/L)	Co (mg/L)
SI5P1	1	3.1	7.63	358	<1.5	104	0.02	332	2125	0.89
SI5P2	2	4.0	7.20	342	<1.5	97.2	0.15	400	2085	0.74
SI5P3	3	4.4	8.00	253	<1.5	51.1	0.12	591	2245	0.53
SI5P4	4	5.4	6.15	274	30.2	0.25	<0.02	343	1830	0.04
SI5P6	6	6.5	6.25	286	105	<0.05	<0.02	223	1875	<0.02
SI5P8	8	6.4	7.55	222	116.5	<0.05	<0.02	243	2655	<0.02
SI5P10	10	6.5	10.6	256	112	<0.05	0.05	307	3860	0.06

Samples	Depth (m)	Cu (mg/L)	Fe (mg/L)	K (mg/L)	Mg (mg/L)	Mn (mg/L)	Mo (mg/L)	Na (mg/L)	SO₄ (mg/L)	Zn (mg/L)
SI5P1	1	329	176	21.0	308	80.1	0.05	1748	4167	12.9
SI5P2	2	229	154	57.5	267	64.2	0.35	1674	3807	10.4
SI5P3	3	140	242	71.2	358	23.7	0.32	1814	4581	5.98
SI5P4	4	3.57	10.6	53.3	197	1.23	0.07	1449	2124	0.22
SI5P6	6	0.17	1.13	55.9	137	0.37	<0.04	1774	2162	0.05
SI5P8	8	0.03	5.7	61.0	129	0.88	<0.04	2245	2174	0.05
SI5P10	10	0.10	54.1	75.6	157	5.57	<0.04	3167	2861	0.75

Evolution of the Aquatic Geochemistry in the Surface Waters of the Northern Remediation Cell (NC)

The water used for this remediation cell was taken directly from the wetland and had pH of 7.6, Eh of 341 mV and concentrations of heavy metals were close to or below detection limit (Figure 2). The other geochemical parameters showed the following concentrations: 100 mg/L Mg, 0.07 mg/L As, 2080 mg/L Cl, 1555 mg/L Na, 1636 mg/L SO_4^{2-}, and 8.36 mg/L DOC. Thus, the water from the wetland has about three-fold higher concentrations in DOC, nearly double in alkalinitiy and is more reducing than the Locumba river water used for the DC (Table 1), which might be an additional factor in accelerating the change to a neutral and reducing environment. After the flooding of the oxidizing tailings, the pH remained between 7 and 8. The Eh increased from 341 up to 416 mV and then was decreasing over time reaching 205 mV at the end of the experiment. Alkalinity decreased from 250 to 150 mg/L at the beginning of the experiment and stabilized around 180 mg/L. Zn and Al remained below detection limit, whereas Fe, Mn and Cu were fluctuating at low concentrations between 0 and 0.3 mg/L (Figure 2). The efflorescent salts, which were exposed at the surface of the tailings cell, were dissolved and leached downward during the five days before the water saturation of the cell and the installation of the water cover. Once the remediation cell was saturated, the height of the water cover inside the cell was about 50 cm.

Seventy-five days after the remediation started, arsenic, chloride, sodium, and sulfate showed an increase in concentration up to 0.9 mg/L, 3600 mg/L, 3200 mg/L and 3000 mg/L, respectively. These high concentrations were maintained until day 150, before dropping down again (Figure 2 and supplementary information). This increase in concentrations correlates with the dry season, where low water supply results in an increase of these elements in solution due to strong evaporation in the surface waters of the cell. Additionally, due to the lower water level in the wetland during the dry season, the hydraulic gradient decreased, so that the freshwater–seawater interface migrated towards the shore, resulting in an increase of chloride, sodium, and sulfate in the whole profile (see Figure 3G,H). This was confirmed by isotopic analyses, which showed value of $\delta^{34}S$ and $\delta^{18}O$ closer to seawater values during the period [10]. The increasing concentrations of As in the cell water seemed to be linked to the increasing concentrations of other competing anions like Cl^- and SO_4^{2-} for the sorption sites in the oxidation zone, resulting in the release of As into solution. This will be discussed in more detail in the following sections.

Figure 2. Evolution of hydrogeochemistry in the water cover of the Northern Cell during the remediation process. In the time-scale axes, the arrow (day 0) corresponds to the water sample taken at the start of the experiment before flooding.

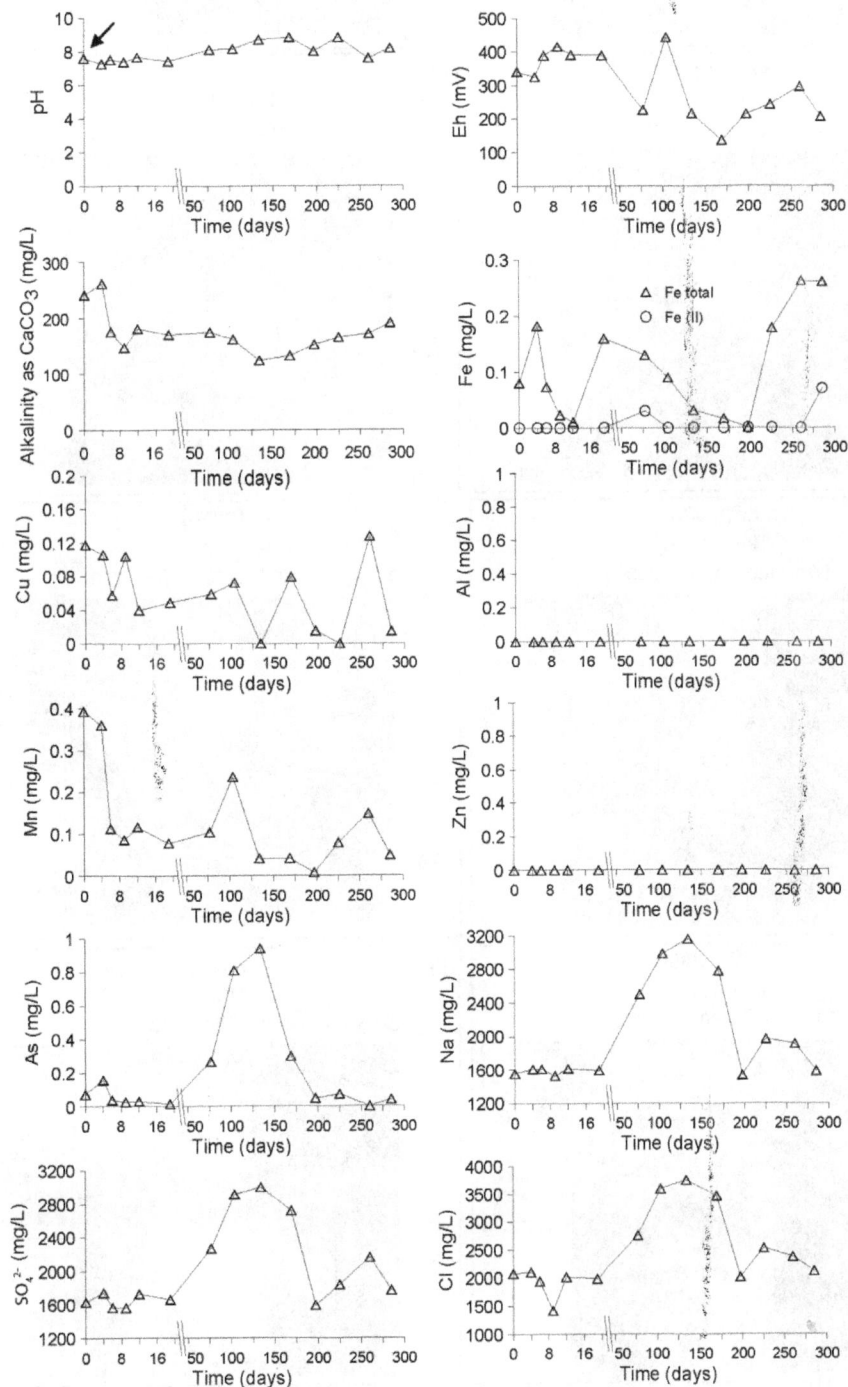

Figure 3. Evolution of hydrogeochemistry in the piezometer nest of the Northern Cell during the remediation process.

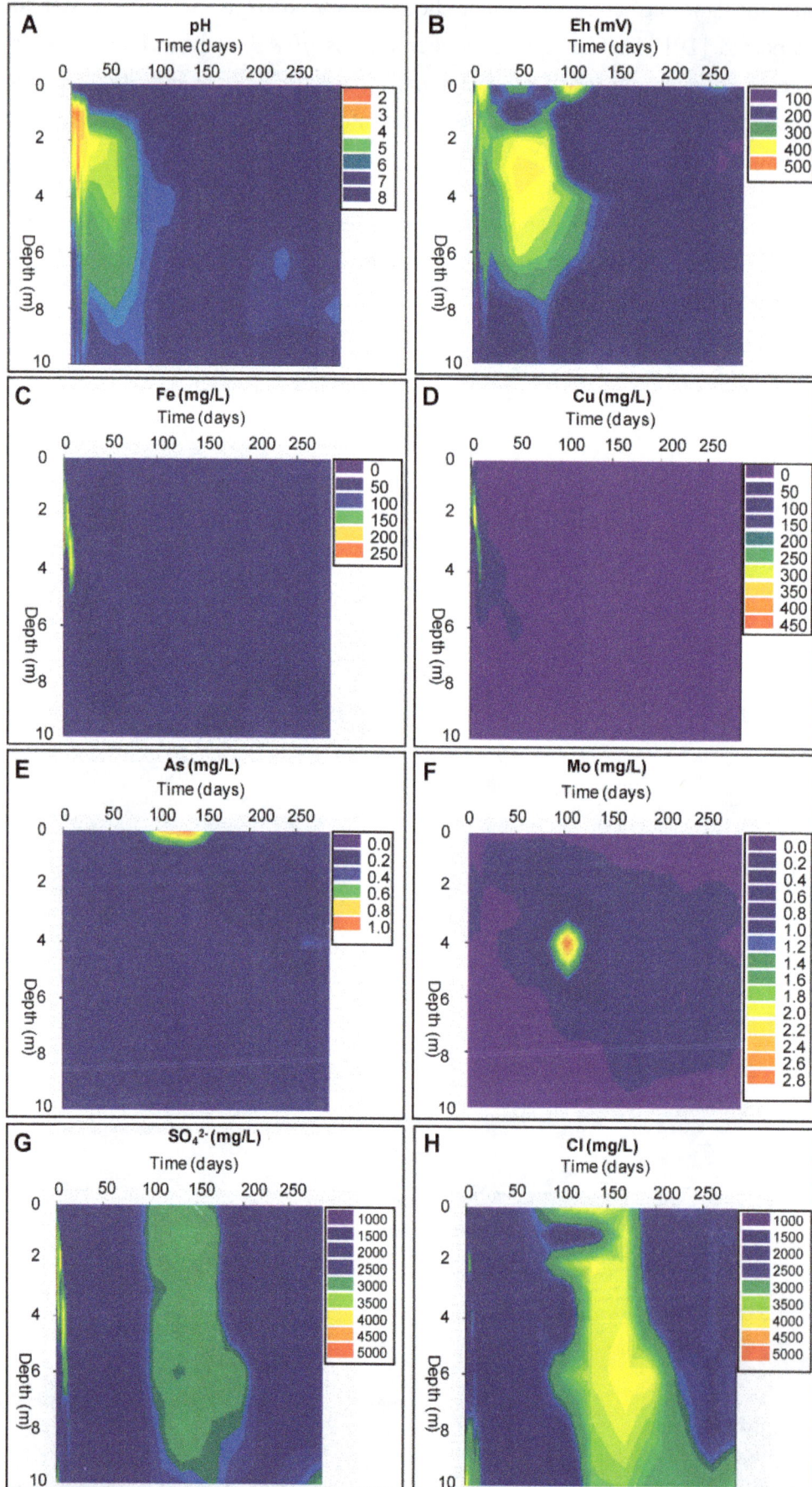

Geochemical Evolution in the NC Tailings Stratigraphy over Time

The implementation of the wetland on the former oxidation zone led immediately to a decrease of the pH down to 2.4 and an increase of the metal concentrations in solution due to the dissolution of efflorescent salts (Figure 3A). This is confirmed by sequential extraction data, which showed a decrease of the concentrations for Cu and Zn of exchangeable and/or water-soluble metals present in the tailings just after flooding (Figure 4 and supplementary information). The efflorescent salts at the top of the oxidizing tailings were composed by elements liberated during years of oxidation and transported by capillarity towards the surface. The implementation of the wetland dissolved these mainly water-soluble sulfates [19,20] and chlorides and transferred them into solution.

From day 7, the pH began to increase in the oxidation zone from 2.4 stabilizing at 6.5 after five months (Figure 3A). The increase of pH and the change to more reducing conditions (200 mV) led to a decrease of most metals in solution in both the oxidation zone and the underlying primary zone. After 100 days of remediation, the concentrations of dissolved heavy metals decreased according to the depth up to: Fe (79.5%–99.4%), Mn (94.6%–99.9%), and Zn (96.1%–99.6%). Cu was below detection limit as well as Al, and Ni. Mg had concentrations between 114 and 149 mg/L, which were close to the concentrations of the water used for flooding (100 mg/L). At the end of the 270 days remediation experiment, only Fe and Mn were mobile at high concentrations (up to 41 mg/L and 2 mg/L, respectively), most likely as a result of reductive dissolution of Fe and Mn hydroxides from the oxidation zone to the primary zone [21].

The Fe concentration in the saturated low pH oxidation zone was 176 mg/L at 1 m depth before remediation. With the flooding, the infiltration of neutral oxygenated water resulted in the precipitation of Fe(III) hydroxides through hydrolysis of Fe(III), decreasing iron concentrations in solution to 1.3 mg/L after 90 days. Additionally, due to the increase of the hydraulic gradient during flooding, the Fe(II) plume, which was underlying the oxidation front, was pushed downwards to higher pH conditions. This induced the oxidation of Fe(II) to Fe(III) and subsequent hydrolysis and precipitation as Fe(III) hydroxides.

The precipitation of iron hydroxides after flooding is supported by the saturation index calculation with the PHREEQC code [22], showing that the solution was oversaturated with respect to ferrihydrite (SI = 2.42) and goethite (SI = 5.12). In the sequential extractions, Fe concentrations increased from 0.27 wt % before remediation to 0.94 wt % after flooding in the fraction of the Fe(III) hydroxides in the top layer (Figure 4B). As the tailings were already oxidizing, the metal removal was pH controlled. With the implementation of the wetland using neutral water, the pH raised and metals were adsorbed on Fe(III) hydroxide [23,24] and/or co-precipitated [6,25,26].

Fe removal from solution is also due to formation of sulfide when reducing conditions were available after 19 days. The sulfide fraction of Fe in the sequential extractions increased from 1.38 to 2.73 wt % during the remediation process (Figure 4C).

Figure 4. Result of sequential extractions from samples of the Northern Cell during the remediation process.

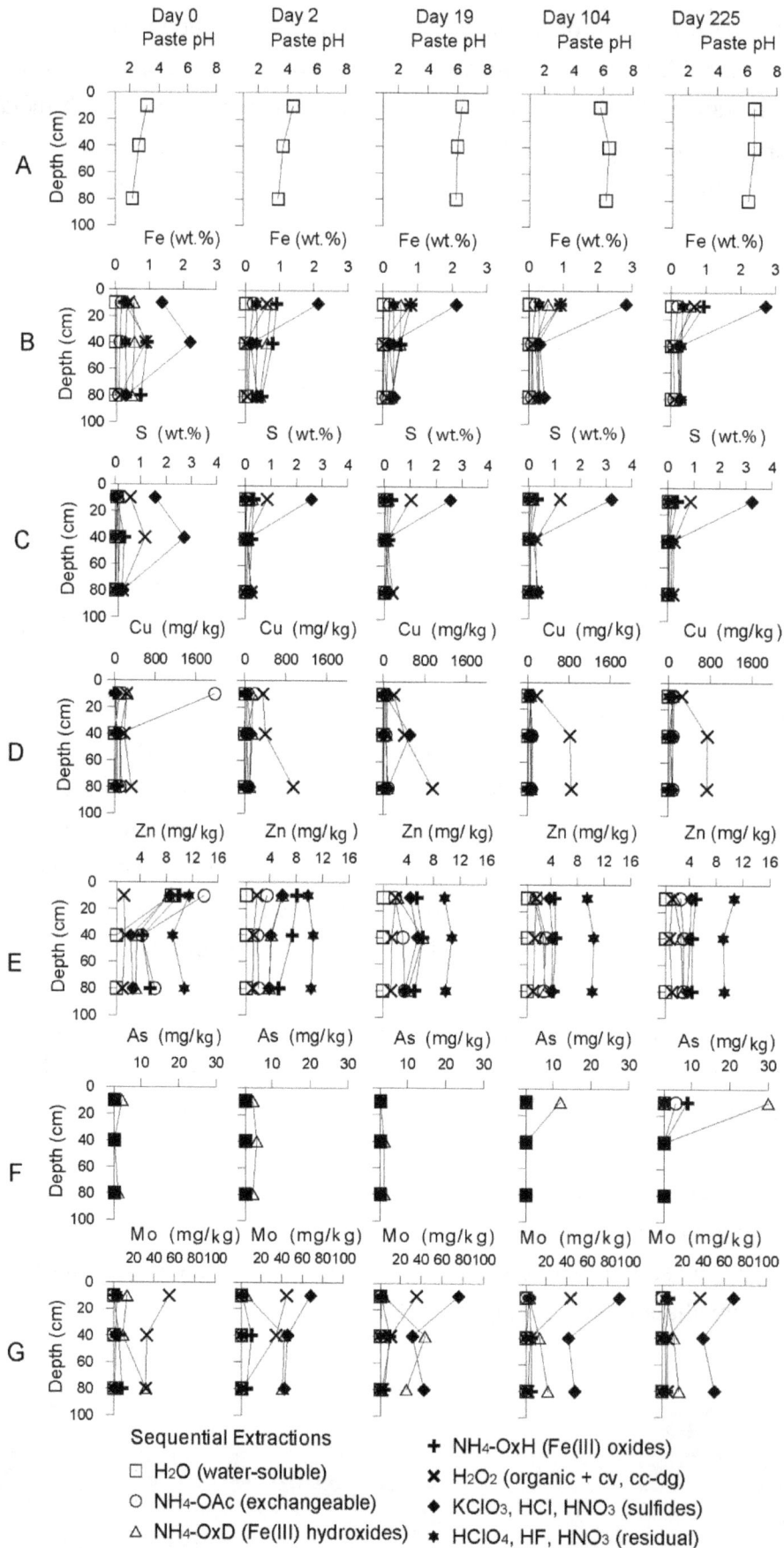

Cu concentrations were high in the oxidation zone before the remediation (up to 350 mg/L Cu) due to advanced sulfide oxidation (Figure 3D). When the remediation process was initiated the pH increased and the concentrations decreased to 200–250 mg/L. Due to the increased hydraulic gradient installed by the remediation process, a metal-rich plume was pushed downwards to higher pH and more reducing conditions (Figure 3D). After Day 19 the redox environment started to be reducing at the depth where the Cu plume was located (1–4 m depth). This change initiated the fixation of Cu by organic matter and reduction to sulphides and could be quantified by the sequential extraction data (Figure 4D): an increase in the fraction of the Cu in the H_2O_2 leach from 191 up to 824 mg/kg at 40 cm depth and from 330 up to 1240 mg/kg at 80 cm was observed. As a consequence, Cu was removed from water and its concentration decreased below the detection limit at the end of the experiment. The precipitation of iron oxides/hydroxides discussed above participated also to a minor extent to copper removal, supported by the sequential extraction data, which showed that the Cu associated to the Fe(III) hydroxide fraction increased from 4 mg/kg before remediation to 87 mg/kg after flooding (Figure 4D). Zn which concentration decreased from 12.9 mg/L to a value below detection limit in solution had the same behavior as Cu. It formed a plume downward and was precipitated as secondary sulfides [27].

The high SO_4^{2-} concentrations associated with sulfide oxidation in the upper part of the sediment column decreased from 4167 mg/L to 1829 after one month of remediation, mainly due to dilution (Figure 3G). The decreasing trend of sulfate in the late remediation experiment is interpreted as a result of sulfide precipitation subsequent to sulfate reduction by sulfate reducing bacteria (SRB). The $S_{sulfide}$ in the tailings increased from 1.57% to 3.24% during the remediation process (Figure 4C). Moreover, the microbiological study of the Bahía de Ite remediation experiment (will be published elsewhere), reported the appearance of sulfate reducing bacteria (SRB) in the late state of remediation process (after 104 days). The stable concentrations around 1800 mg/L SO_4^{2-} in the whole stratigraphy are controlled by the gypsum solubility (SI gypsum = −0.47).

Chloride had relatively constant concentrations close to the inflowing water from the wetland between 1500 and 2000 mg/L, in the whole profile, peaking at 3500 mg/L at 10 m depth (Figure 3H), where the seawater/freshwater interface is present. Na showed the same behavior as Cl.

From the third to the fifth month after remediation started, the chloride, sodium and sulfate concentrations increased to around 3000 to 4000 mg/L through the whole profile (Figure 3G,H). The water in-flow from the Locumba River can vary extremely during the year (depending on the seasons). When the water-level in the wetland was about 1 m above sea level as maximum, the freshwater/seawater interface can be assumed at 40 m depth according to the Ghyben–Herzberg approximation [10]. An average decrease to 0.5 m of the wetland water level was observed every summer (dry season) during three years of monitoring. This decrease resulted in an approximate rise of this interface to 20 m depth in the old wetland and, consequently, in a lower hydraulic gradient. Therefore, the change in salinity in the tailings pore water during the summer months can be seen as a result of the intrusion of the freshwater–seawater transition zone into the tailings deposit. This interpretation is supported by isotopic analysis. During this period of Na and Cl increase, the $\delta^{34}S$ values were at $\delta^{34}S = 14.70\%$ to 15.82% V-CDT closer to seawater value $\delta^{34}S = 20\%$ to 21% V-CDT. The $\delta^{34}S$ values on samples taken before and after this increase were lighter, at 12.04% and 13.10%, peaking at 15.15% V-CDT only for the deepest samples, which still had marine influence [10].

Arsenic showed concentrations of 0.02 mg/L in the pore water of the oxidation zone and relatively low concentration (0.15 mg/L) shortly after the remediation started, as the tailings contain very low concentrations (mean of 4 mg/kg As) of arsenic (Figure 4F). After 100 and until 150 days a strong increase in the upper part and in the water cover of the cell culminated at 0.9 mg/L arsenic. Since the tailings material is not the source of arsenic in this system, this liberation can be explained as followed: as discussed above, the remediation process initiated the precipitation of Fe(III) hydroxides in the wetland. These Fe(III) hydroxides had a high sorption capacity for the naturally incoming arsenic from the Locumba water. Therefore, it acts as a filter for arsenic, resulting in low concentration in the wetland water after eight days. This is confirmed by the sequential extraction data, which show an increase of As in the Fe(III) hydroxide fraction of up to 800% between the start and the end of the remediation experiment (Figure 4F). Then, during the dry season, increased concentration of Cl and SO_4, due to increasing seawater infiltrations into the deposit, compete with As on the sorption sites (e.g., [28]) and triggered a the release of arsenic into solution. This arsenic liberation mechanism resulted in an increase of the concentration of up to 0.3 mg/L after 6 months (Figure 3E). The liberated As migrated downwards (0.3 mg/L As), reaching 4 m depth after seven months' remediation.

Mo concentrations in the solution increased over time from 0.05 to 0.2 mg/L after three months in the oxidation zone due to pH increase and more reducing conditions, which released the formerly adsorbed Mo to the secondary Fe(III) hydroxides in the oxidation zone [2]. As a consequence, a plume (of up to 2.8 mg/L) was forming downward at 4 m depth (Figure 3F). At the end of the remediation experiment, Mo concentration in solution was close to or below the detection limit due probably to co-precipitation, as reported in other wetlands [29].

3.2.2. The Delta Area Remediation Cell (DC)

Starting Conditions before the Remediation

In the delta remediation cell, the pH of the vadose zone (0–1 m) ranged between 3.0 and 3.1 with a redox potential around 578 mV (Table 3) before the remediation started. The heavy metal concentrations in pore water were up to 2299 mg/L Fe, 4448 mg/L Al, 1035 mg/L Cu, 6138 mg/L Mg, 716 mg/L Mn, 6.77 mg/L Ni, and 213 mg/L Zn. Arsenic and sulfate concentrations were 0.15 mg/L and 33,270 mg/L, respectively (Table 3). At 1 m depth, the heavy metals concentrations were (Table 3): 703 mg/L Fe, 3.72 mg/L Al, 0.26 mg/L Cu, 166 mg/L Mg, 62.5 mg/L Mn, 0.46 mg/L Ni, 1.20 mg/L Zn, and sulfate was 3483 mg/L. Na concentrations reached 1600 mg/L in the oxidation zone, possibly due to silicate alteration and jarosite dissolution.

From 3 m depth, the pH increased up to 6.2 with reducing conditions (±200 mV), leading to a decrease of the metal concentrations in the pore water (Table 3). Fe concentrations decreased, but were still high (146 to 634 mg/L). Al, Cu, and Ni were near the detection limit, Mg decreased to around 120 mg/L, Mn to 12.8–69 mg/L, and Zn to 7.63 mg/L. Cl and Na showed concentration between 700 and 2000 mg/L.

Before the remediation, arsenic concentration was low in the 1 m depth oxidation zone (0.02 mg/L, Table 3), due to the sorption on Fe (III) hydroxides [30]. It was higher in the primary zone (up to 0.25 mg/L), but still lower than the natural background of the Locumba river water (around 0.5 mg/L) [10].

Table 3. Geochemistry of the piezometers in the delta cell before the start of the remediation experiment.

Samples	Depth (m)	pH	EC (ms/cm)	Eh (mV)	Alkalinity (mg/L CaCO$_3$)	Al (mg/L)	As (mg/L)	Ca (mg/L)	Cl (mg/L)	Co (mg/L)
SI7SI	0.2	3.0	23.50	578	<1.5	4448	0.15	980	N/A	9.80
SI7SII	0.55	3.1	5.11	566	<1.5	1653	0.10	466	N/A	5.26
SI7P1	1	3.1	2.49	533	<1.5	3.72	0.02	552	N/A	0.37
SI7P2	2	4.2	6.05	410	<1.5	0.87	0.13	551	N/A	0.15
SI7P3	3	6.2	7.01	105	165	<0.05	0.25	632	1,009	BDL
SI7P4	4	6.0	6.60	243	105.15	<0.05	0.15	539	545	0.44
SI7P6	6	5.4	5.00	305	13.8	<0.05	0.16	380	476	0.68
SI7P8	8	5.7	5.20	265	62.95	1.91	0.10	404	510	0.73
SI7P10	10	5.2	6.11	359	38.7	<0.05	<0.002	517	965	1.37

Samples	Depth (m)	Cu (mg/L)	Fe (mg/L)	K (mg/L)	Mg (mg/L)	Mn (mg/L)	Mo (mg/L)	Na (mg/L)	SO$_4$ (mg/L)	Zn (mg/L)
SI7SI	0.2	1,035	2,299	3.58	6,138	716	<0.04	1,650	33,270	213
SI7SII	0.55	335	2,001	0.91	1,631	132	<0.04	319	19,560	10.1
SI7P1	1	0.26	703	34.2	276	62.5	<0.04	692	3,483	1.20
SI7P2	2	1.37	615	87.9	120	13.2	<0.04	733	2,554	2.53
SI7P3	3	<0.005	146	78.9	256	12.8	0.05	634	2,635	0.07
SI7P4	4	<0.005	383	89.9	393	54.8	<0.04	399	3,612	10.1
SI7P6	6	0.02	598	52.5	204	28.8	<0.04	325	2,835	5.43
SI7P8	8	0.28	634	58.7	193	33.9	<0.04	377	2,873	3.16
SI7P10	10	<0.005	412	107	345	69.4	<0.04	727	3,483	7.63

Evolution of the Aquatic Geochemistry in the Surface Waters of the Remediation Cell DC

The water (from the Locumba river) used for the flooding of the remediation cell had pH of 7.3, Eh of 420 mV, 133 mg/L alkalinity as CaCO$_3$, 433 mg/L Cl, 300 mg/L Na, 429 mg/L SO$_4^{2-}$, 0.27 mg/L As and 2.6 mg/L DOC at the water inflow (Figure 5). The concentrations of heavy metals were below detection limit.

Immediately after the flooding, these values did not vary a lot, except for sulfate, which increased to 613 mg/L. Two days after flooding the pH of the water cover decreased strongly to 2.7, as well as the alkalinity and the Eh increased to 646 mV (Figure 5). Then, the concentrations of the heavy metals started to increase, reaching at day 4: 8.9 mg/L Fe, 265mg/L Al, 457 mg/L Cu, 238 mg/L Mn, 19 mg/L Zn (Figure 5). Sodium, chloride and sulfate increased up to 734 mg/L, 1621 mg/L and 5385 mg/L, respectively (Figure 5). This is explained by the dissolution of the efflorescent salts, which were abundant present at the surface of the remediation cell due to the finer grain size and subsequent stronger capillary transport as in the NC. Seven days after the wetland construction, the pH of the surface water started to increase due to the consumption of the acidity by the alkalinity (133 mg/L as CaCO$_3$) of the inflowing water, but also due to the dilution effect of the continuous flooding. The water flow into the remediation cell was controlled by an outflow pipe at 50 cm height in the dikes, discharging the excess of water to avoid breaking the dikes. As a consequence of increasing pH (7.1 at day 15) and increase of alkalinity, most of the heavy metals in the water cover were close or below the detection limit after 14 days of flooding (Figure 5). This decrease of metal concentration in the water cover with increasing pH can be explained by metal precipitation as oxide and hydroxide. Alkalinity, sodium, sulfate and chloride showed a cyclic behavior after the first peak due to dissolution of the efflorescent salts. Alkalinity showed a drop in the dry seasons, suggesting increased sulfide oxidation. Sodium, chloride, and sulfate showed an increase during the dry seasons, which correlated with an increase of the arsenic concentrations (Figure 5).

Figure 5. Evolution of hydrogeochemistry in the water cover of the delta cell during the remediation process. In the time-scale axes, the arrow (day 0) corresponds to the water sample taken at the start of the experiments before flooding.

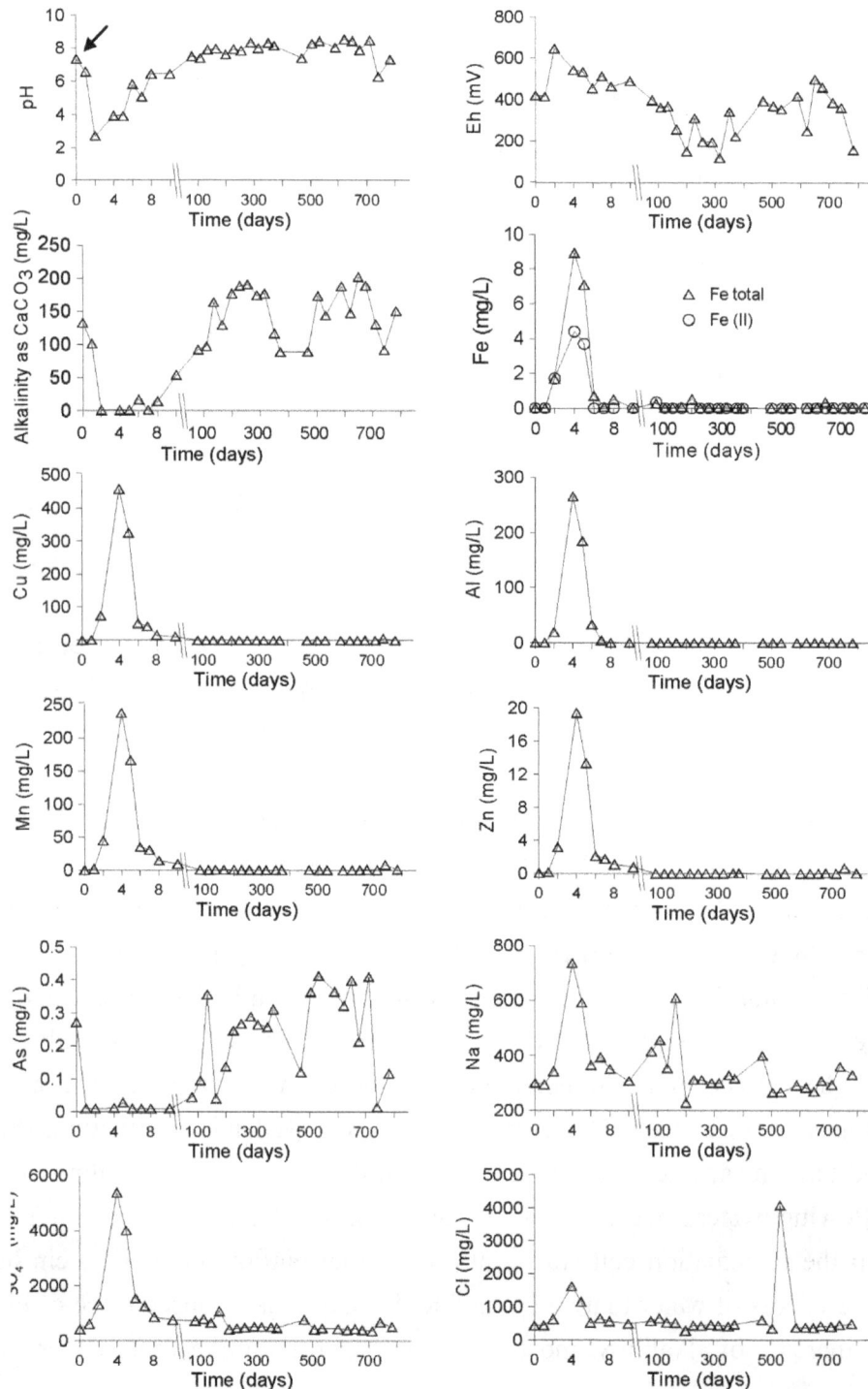

Arsenic concentration in the water cover first decreased down below the detection limit in contact with the iron hydroxide secondary minerals. After seven days, it increased up to 0.4 mg/L as a consequence of the release of As formerly adsorbed on iron hydroxides. Additionally, the increased SO_4^{2-} and Cl^- concentrations promoted the liberation of As from the Fe(III) hydroxides. When alkalinity dropped, *i.e.*, sulfide oxidation increased, the As concentration decreased, suggesting sorption to newly formed Fe(III) hydroxides through sulfide oxidation. A general trend of decrease for the Na^+, Cl^- and SO_4^{2-} peak concentrations was also observed, most likely through dilution effects.

Geochemical Evolution in the DC Tailings Stratigraphy over Time

After the implementation of the wetland, pH remained acidic (3.1) at 1 m depth in the oxidation zone (Figure 6A). Eh decreased at the beginning down to 308 mV and then was fluctuating between 300 and 500 mV (Figure 6B). The concentrations of metals in solution increased up to 46% Fe, 7900% Cu, 174% Al, and 15% Zn in the first 30 days of the remediation experiment. SO_4^{2-} and Cl^- increased up to 43% and 104%, respectively (Figure 6G,H). This can be explained by the seepage of water from the water cover (rich in metals after the flooding) into the former 1 m vadose unsaturated oxidation zone of the tailings. This infiltration dissolved also the elements liberated during years in the oxidation zone, increasing additionally their concentration in solution.

Starting 45 days after flooding, the pH showed a slight increasing trend, reaching 4.4 after 465 days of remediation (Figure 6A). During the same time period the element concentrations in solution showed a decreasing trend. Calculated with respect to the maximum value measured after flooding, these concentrations dropped down to 96% for Fe, and 51% for Cu around 15.5 months after remediation started. Al, Mn, and Zn decreased to 88%, 97%, and 95%, respectively. Sulfate and chloride were fluctuating during this time with a decreasing trend maintaining value close to the inflowing water (Figure 6G,H). The iron concentration increased up to 4432 mg/L after flooding due to dissolution of efflorescent salts (Figure 6C).

With the installation of the wetland, water seepage induced the increase of pH from 3 to 4.4, triggering Fe(III) hydroxide precipitation, removing this metal from solution down to concentration 179 mg/L after 465 days (Figure 6C). The saturation indexes showed oversaturation with respect to goethite (SI = 4.01) and ferryhydrite (SI = 1.31). The sequential extraction data showed that the Fe(III) hydroxides and Fe(III) oxide fractions were increasing over time (Figure 7B). The sulfide fraction of Fe was also increasing in the former oxidation zone during the remediation process, suggesting precipitation of a secondary sulfide, which participated in Fe removal (Figure 7B). Total iron in the tailings matrix increased from 3.58% to 5.01%. Mn showed a similar behavior as Fe and decreased from 323 to 10 mg/L after 15.5 months of remediation, due to formation of Mn oxides and hydroxides which can also act as adsorbants and co-precipitate other metals such as Zn, Co, Ni, Cu [31].

Cu was dissolving from efflorescent salts (e.g., chalcanthite, eriochalcite) and increasing to concentration in solution up to 20.8 mg/L at 1 m depth just after the flooding. Copper was present in the water-soluble and exchangeable fractions at day 3 (Figure 7D). Once the cell completely flooded, the Cu plume was pushed slowly downward, where the installation of more reducing conditions led to adsorption onto organic matter, decreasing the concentration in solution to 1.66 mg/L after one year of remediation. This was confirmed by sequential extraction data showing increase of organic related Cu fraction from 713 to 1020 mg/kg at 1 m during the remediation process (Figure 7D). Zn had a similar behavior as Cu, showing decreasing concentration in solution from 47.3 mg/L to 1.73 after 465 days, whereas its concentration in the tailings matrix was increasing in the exchangeable, organic and Fe(III) hydroxide fractions (Figure 7E). This suggested Zn removal from solution by sorption, co-precipitation, and sulfide formation [27].

Figure 6. Evolution of hydrogeochemistry in the piezometer nest of the delta cell during the remediation process.

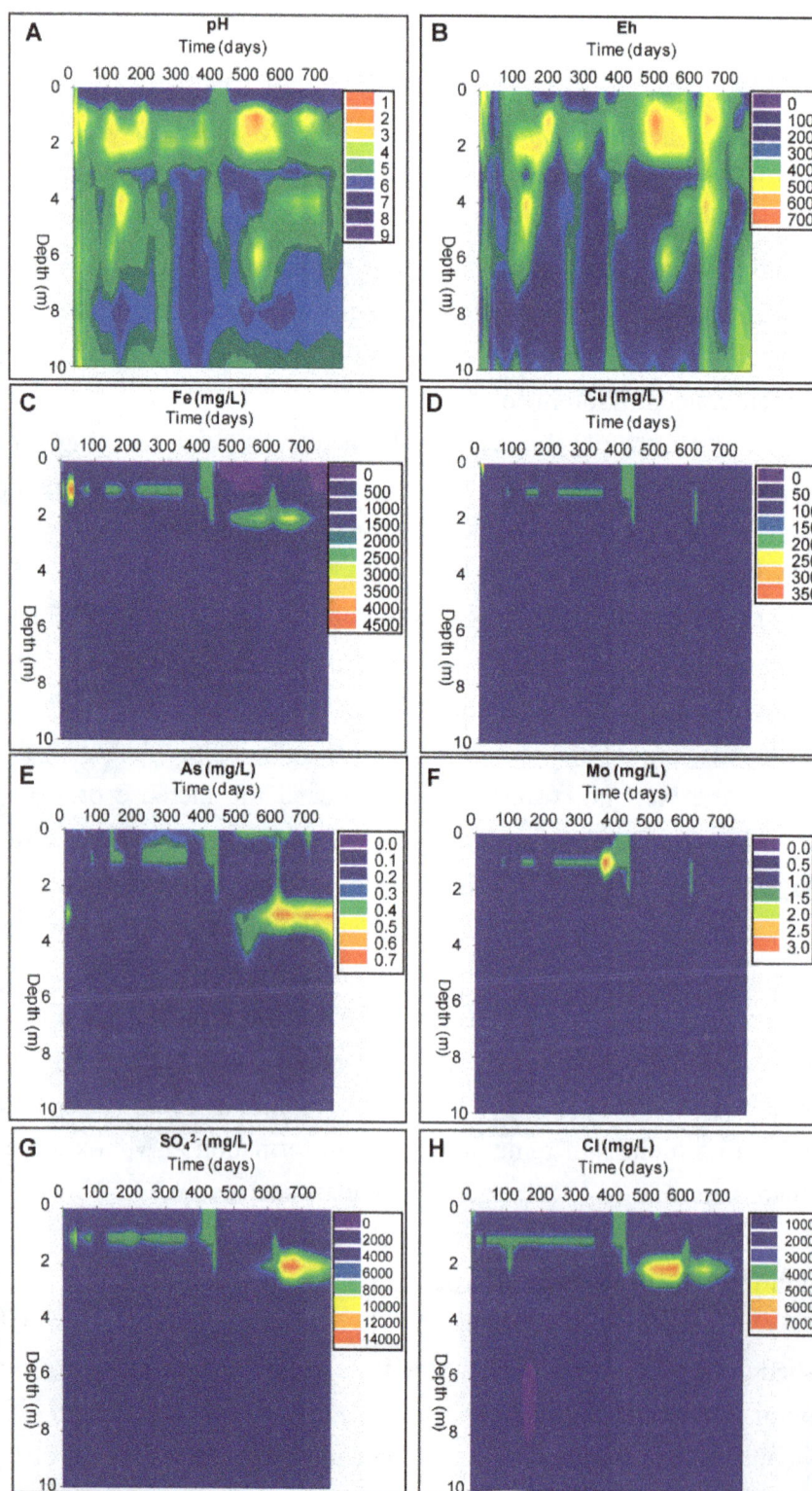

Figure 7. Result of sequential extractions from samples of the delta cell during the remediation process. For Arsenic (F), the scale at day 593 is different for better visibility.

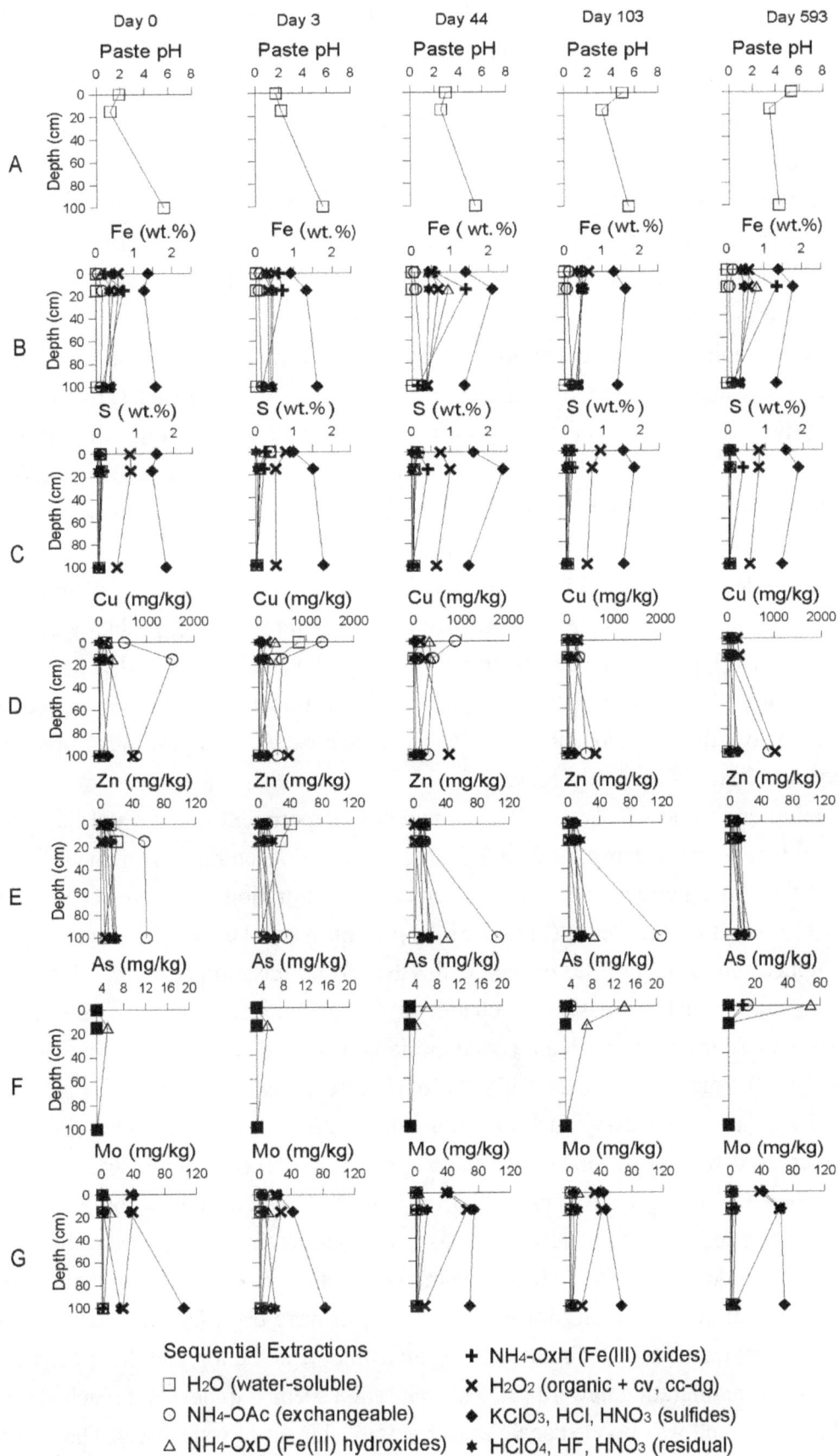

Sequential Extractions

- □ H₂O (water-soluble)
- ○ NH₄-OAc (exchangeable)
- △ NH₄-OxD (Fe(III) hydroxides)
- + NH₄-OxH (Fe(III) oxides)
- ✕ H₂O₂ (organic + cv, cc-dg)
- ◆ KClO₃, HCl, HNO₃ (sulfides)
- ✳ HClO₄, HF, HNO₃ (residual)

Before remediation, Cl^- and Na^+, showed concentrations between 600 and 2000 mg/L and the concentration of SO_4^{2-} was 3483 mg/L at 1 m (Figure 6G,H). After the implementation of the wetland Cl^-, Na^+, and SO_4^{2-} were increasing up to 3628 mg/L, 1019 mg/L, and 7815 mg/L, respectively after 180 days, before decreasing to concentrations close to the natural input from flooding water (323, 195, and 951 mg/L, respectively) after 550 days (Figure 6G,H). The increase of element concentration observed after remediation is explained by the dissolution of efflorescent salts, which were mainly composed by sulfates and chlorites.

The As concentration was 0.02 mg/L before the remediation and did not increase in the oxidation zone immediately after flooding (Figure 6E). It then increased slightly up to 0.11 mg/L at day 200 before decreasing again. This general trend of low As concentration in the tailings (compared to the natural background) is related to its affinity to iron hydroxide.

As-rich water was seeping continuously into the tailings, where As was adsorbed in contact with the Fe(III) hydroxide. The sequential extraction data showed increased As in the Fe(III) hydroxide fraction of up to 1700% (Figure 7F). At around day 660 after remediation, the As concentration in solution reached 0.63 mg/L at 3 m depth (Figure 6E). This is interpreted as the result of reductive dissolution of Fe(III) hydroxides by the installation of reducing conditions after almost two years' remediation, resulting in the release of the formerly adsorbed As [30,32].

Mo was below the detection limit before the remediation. With implementation of the wetland, its concentration in water increased up to 3 mg/L after one year of remediation, probably due to release from secondary minerals (Figure 6F). The adsorption of Mo on Fe oxide decreased with increasing pH and the implementation of the wetland favored the Mo desorption [33]. Then Mo concentration in solution decreased below the detection limit as a consequence of secondary sulfides precipitation as suggested by sequential extraction data (Figure 7G).

It was observed after around 450 days of the remediation experiment that the pH decreased again from 4.4 to 1.6 at 1 m depth and from 4.2 to 3.0 at 2 m. The concentrations of metals and the Eh increased also in the same period. The highest element concentrations located before at the 1 m oxidation zone were found at 2 m depth (Figure 6). In general it can be said that most changes were occurring in the upper 3 m or the profiles of both cells (the former unsaturated zone, which experience the main changes due to pH increase and change to reducing environment). The heavy metal concentrations were close to, or higher than, concentrations observed in the first days of remediation: up to 3200 mg/L Fe, 10.1 mg/L Cu, and 2.93 mg/L Mo, (Figure 6C,D,F). Other metals not represented in the graphs were also increasing up to 14.6 mg/L Al, 11.5 mg/L Mn, 1.8 mg/L Zn. Cl^- and SO_4^{2-} increased also up to 6639, and 6396 mg/L respectively (Figure 6G,H). The sulfate concentration reached 13,740 mg/L at months 22.5 (675 days). This change can be explained by the fact that the remediation cell was completely dry during two months (in summer) from the day 450 due to drought and a landslide in the river "Rio Locumba" flowing into the bay, and the water table dropped down below 2 m depth (the piezometers at 1 and 2 m were dry). Due to this drastic drought, and to the desert climate, oxidizing conditions were again available at the top and the liberated elements in the solution below were migrating upward [2]. This migration seemed to have not reached the top 1 m oxidation zone before the cell was flooded again after re-establishment of water flow. The re-installation of the wetland dissolved again the elements increasing their concentration at 2 m. The mobilization of heavy metals downward is typical for humid raining climate so that it can be assumed that the implementation of the wetland is acting like a rainfall, dissolving and transporting the metals down to the water table [2]. These data showed that the water saturation is a key factor for oxidizing tailings wetland remediation and controls the kinetics of the remediation effectiveness together with the geochemical composition of the water used for remediation.

These concentrations started again to decrease three months after the incident, and this trend was maintained until the end of the experiment—780 days after the remediation started (Figure 6). The decrease of metal concentrations is probably due to adsorption on Fe oxides/hydroxides and co-precipitation, as well as to sulfide formation [34], as described before.

4. Conclusions

Two systems were studied in this temporal *in situ* bioremediation process, the Northern Cell at sea-level (NC) and the delta cell (DC) about 2 m above sea-level.

At the NC, the tailings consisted of sandy sediment due to grain size separation by the wave action, and the cell was close to the already remediated wetland-covered tailings. The water used for the remediation had 241 mg/L alkalinity and 8.36 m/L DOC. After flooding, this water infiltrated rapidly, saturated the system (Figure 8A), consumed the acidity, oxidized the Fe(II) and hydrolyzed the subsequent Fe(III). As a consequence, iron was precipitating and co-precipitating with other metals, removing them from solution. The DOC contained in the flooding water provided additional carbon source for the heterotrophic bacteria, which degraded the organic matter and consumed the oxygen, leading to the creation of more reducing conditions. The upcoming of reducing conditions and the availability of sulfate promoted bacterial sulfate reduction and metal sulfide precipitation as well as alkalinity generation. Moreover, the organic matter binds some metals such as Cu and Zn, removing them from water.

Figure 8. Characteristic of the remediation cells after the implementation of the wetland above the tailings.

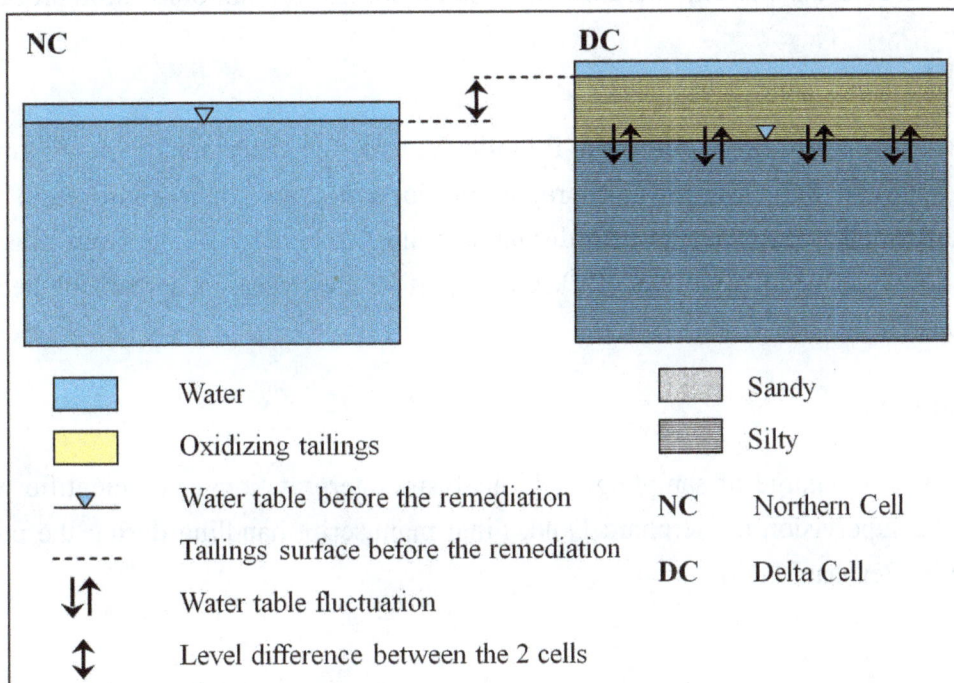

The arsenic coming with the inflowing water was adsorbed to the Fe(III) hydroxide. The water monitoring of this Northern Cell carried out during 15 months after the end of the experiment showed neutral pH and heavy metals below their detection limit, suggesting a good stability of the system.

In contrast, the delta cell was characterized by a mixed stratigraphy of silty, clayey, and very fine-sandy layers from 0 to 1 m depth. It is topographically situated 1 to 2 m higher than the northern cell and the old remediated tailings wetland due to additional tailings deposition at this area, following a break of a dike in 1992. The water used for the wetland implementation was neutral with 133 mg/L alkalinity, and 2.6 mg/L DOC. After flooding the water did not infiltrate rapidly due to grain size; the cell was not saturated during the entire course of the experiment. The cell had a water cover on the surface, followed by a vadose zone; and the groundwater table was fluctuating between 0.8 and 1.2 m (Figure 8B). However, slow infiltration of water was noted over time and observed during sampling. The same mechanisms of metal removal as described for the NC were also occurring in the DC, but they were slower and more progressive. The depletion of heavy metals in solution started after one year, and was much slower than in the northern remediation cell where the total metal removal took four months. The installation of reducing conditions also took a longer time, probably due to the non-saturation of the cell and to less DOC than in the NC.

Our data additionally has shown that changes in water supply (e.g., dry periods) and subsequent changes in the hydrogeological system might compromise the water saturation of the system and thus influence the efficiency of the approach. Therefore, to assure a sustainable remediation solution, the long-term water supply has to be ensured. Temporal changes might induce re-oxidation and/or seawater ingression, which might lead to liberation of contaminants into solution.

This study showed that the wetland implementation on oxidizing acidic tailings using alkalinity-rich and DOC-rich water is effective and rapid when the system is water saturated, *i.e.*, water saturation is the key factor that triggers the geochemical condition necessary for the effective remediation. It showed also that the water chemistry is an important factor for the kinetics and effectiveness of the bioremediation. The pH and the redox are key parameters, which influence the dissolution and the sorption/precipitation mechanisms, and therefore the metal mobilization/immobilization processes.

Acknowledgments

We would like to thank Ezio Buselli and Rodolfo Vicetti, the directors of the Environmental Services of the Southern Peru Copper Corporation in Ilo, Peru, and their entire team from the laboratory for their excellent support and help during sampling and analysis. The study was financed by the Southern Peru Copper Corporation (SPCC) as part of the research agreement between the University of Lausanne and SPCC.

Author Contributions

Nouhou Diaby was in charge of sampling, field analysis, interpretations and scientific writing of this study under the supervision of Bernhard Dold. Final manuscript handling during the publication process was done by Bernhard Dold.

Conflicts of Interest

The authors declare no conflict of interest.

References

1. Nordstrom, D.K.; Alpers, C.N. Geochemistry of Acid Mine Waste. In *The Environmental Geochemistry of Ore Deposits. Part A: Processes, Techniques, and Health Issues*; Plumlee, G.S., Logsdon, M.J., Eds.; Reviews in Economic Geology Volume 6A; Society of Economic Geologist: Littleton, CO, USA, 1999; pp. 133–160.

2. Dold, B.; Fontboté, L. Element cycling and secondary mineralogy in porphyry copper tailings as a function of climate, primary mineralogy, and mineral processing. *J. Geochem. Explor.* **2001**, *74*, 3–55.

3. Johnson, D.B.; Hallberg, K.B. Acid mine drainage remediation options: A review. *Sci. Total Environ.* **2005**, *338*, 3–14.

4. Machemer, S.D.; Reynolds, J.S.; Laudon, L.S.; Wildeman, T.R. Balance of S in a constructed wetland built to treat acid mine drainage, Idaho Springs, Colorado, USA. *Appl. Geochem.* **1993**, *8*, 587–603.

5. Sheoran, A.S.; Bhandari, S. Treatment of mine water by a microbial mat: Bench-scale experiments. *Mine Water Environ.* **2005**, *24*, 38–42.

6. Catalan, L.J.J.; Yanful, E.K.; St-Arnaud, L. Field assessment of metal and sulfate fluxes during flooding of pre-oxidized mine tailings. *Adv. Environ. Res.* **2000**, *4*, 295–306.

7. Das, M.; Maiti, S.K. Comparison between availability of heavy metals in dry and wetland tailing of an abandoned copper tailing pond. *Environ. Monit. Assess.* **2008**, *137*, 343–350.

8. Moreno, L.; Neretnieks, I. Long-term environmental impact of tailings deposits. *Hydrometallurgy* **2006**, *83*, 176–183.

9. Widerlund, A.; Ebenå, G.; Landin, J. Potential biogeochemical and ecological development of a flooded tailings impoundment at the Kristineberg Zn-Cu mine, northern Sweden. *Sci. Total Environ.* **2004**, *333*, 249–266.

10. Dold, B.; Diaby, N.; Spangenberg, J.E. Remediation of a marine shore tailings deposit and the importance of water-rock interaction on element cycling in the coastal aquifer. *Environ. Sci. Technol.* **2011**, *45*, 4876–4883.

11. Elliott, P.; Ragusa, S.; Catcheside, D. Growth of sulfate-reducing bacteria under acidic conditions in an upflow anaerobic bioreactor as a treatment system for acid mine drainage. *Water Res.* **1998**, *32*, 3724–3730.

12. Fortin, D.; Roy, M.; Rioux, J.P.; Thibault, P.J. Occurrence of sulfate-reducing bacteria under a wide range of physico-chemical conditions in Au and Cu-Zn mine tailings. *FEMS Microbiol. Ecol.* **2000**, *33*, 197–208.

13. Luptakova, A.; Kusnierova, M. Bioremediation of acid mine drainage contaminated by SRB. *Hydrometallurgy* **2005**, *77*, 97–102.

14. Praharaj, T.; Fortin, D. Indicators of microbial sulfate reduction in acidic sulfide-rich mine tailings. *Geomicrobiol. J.* **2004**, *21*, 457–467.

15. Canmet and Coastech Research Inc. *Acid Rock Prediction Manual*; Canada Centre for Mineral and Energy Technology: Ottawa, ON, Canada, 1991.

16. Coggans, C.J.; Blowes, D.W.; Robertson, W.D.; Jambor, J.L. The Hydrogeochemistry of a Nickel-Mine Tailings Impoundment—Copper Cliff, Ontario. In *The Environmental Geochemistry of Ore Deposits. Part A: Processes, Techniques, and Health Issues*; Plumlee, G.S., Logsdon, M.J., Eds.; Reviews in Economic Geology Volume 6A; Society of Economic Geologist: Littleton, CO, USA, 1999; pp. 447–478.

17. Dold, B. Speciation of the most soluble phases in a sequential extraction procedure adapted for geochemical studies of copper sulfide mine waste. *J. Geochem. Explor.* **2003**, *80*, 55–68.

18. Dold, B. Element flows associated with marine shore mine tailings deposits. *Environ. Sci. Technol.* **2006**, *40*, 752–758.

19. Hammarstrom, J.M.; Seal, R.R., II; Meier, A.L.; Kornfeld, J.M. Secondary sulfate minerals associated with acid drainage in the eastern US: Recycling of metals and acidity in surficial environments. *Chem. Geol.* **2005**, *215*, 407–431.

20. Smuda, J.; Dold, B.; Friese, K.; Morgenstern, P.; Glaesser, W. Mineralogical and geochemical study of element mobility at the sulfide-rich Excelsior waste rock dump from the polymetallic Zn-Pb-(Ag-Bi-Cu) deposit, Cerro de Pasco, Peru. *J. Geochem. Explor.* **2007**, *92*, 97–110.

21. Nealson, K.H.; Saffarini, D. Iron and Manganese in Anaerobic Respiration: Environmental Significance, Physiology, and Regulation. *Annu. Rev. Microbiol.* **1994**, *48*, 311–343.

22. Parkhurst, D.L.; Appelo, C.A.J. *User's Guide to PHREEQC (Version 2)—A Computer Program for Speciation, Batchreaction, One-Dimensional Transport, and Inverse Geochemical Calculations*; Water-Resources Investigations Report 99-4259; U.S. Geological Survey: Reston, VA, USA, 1999.

23. Woulds, C.; Ngwenya, B.T. Geochemical processes governing the performance of a constructed wetland treating acid mine drainage, Central Scotland. *Appl. Geochem.* **2004**, *19*, 1773–1783.

24. Webster, J.G.; Swedlund, P.J.; Webster, K.S. Trace metal adsorption onto an acid mine drainage iron(III)oxyhydry sulfate. *Environ. Sci. Technol.* **1998**, *32*, 1362–1368.

25. Carlsson, E.; Öhlander, B.; Holmström, H. Geochemistry of the infiltrating water in the vadose zone of a remediated tailings impoundment, Kristineberg mine, northern Sweden. *Appl. Geochem.* **2003**, *18*, 659–674.

26. Holmstrom, H.; Salmon, U.J.; Carlsson, E.; Petrov, P.; Ohlander, B. Geochemical investigations of sulfide-bearing tailings at Kristineberg, northern Sweden, a few years after remediation. *Sci. Total Environ.* **2001**, *273*, 111–133.

27. Peltier, E.F.; Webb, S.M.; Gaillard, J.F. Zinc and lead sequestration in an impacted wetland system. *Adv. Environ. Res.* **2003**, *8*, 103–112.

28. Wilkie, J.A.; Hering, J.G. Adsorption of arsenic onto hydrous ferric oxide: Effects of adsorbate/adsorbent ratios and co-occurring solutes. *Colloids Surf. A Physicochem. Eng. Asp.* **1996**, *107*, 97–110.

29. Fox, P.M.; Doner, H.E. Accumulation, release, and solubility of arsenic, molybdenum, and vanadium in wetland sediments. *J. Environ. Qual.* **2003**, *32*, 2428–2435.

30. Bone, S.E.; Eagle, M.; Charette, M.A. Geochemical cycling of arsenic in a coastal aquifer. *Environ. Sci. Technol.* **2006**, *40*, 3273–3278.

31. Miyata, N.; Tani, Y.; Sakata, M.; Iwahori, K. Microbial manganese oxide formation and interaction with toxic metal ions. *J. Biosci. Bioeng.* **2007**, *104*, 1–8.

32. O'Day, P.A.; Vlassopoulos, D.; Root, R.; Rivera, N. The influence of sulfur and iron on dissolved arsenic concentrations in the shallow subsurface under changing redox conditions. *Proc. Natl. Acad. Sci. USA* **2004**, *101*, 13703–13708.

33. Goldberg, S.; Forster, H.S.; Godfrey, C.L. Molybdenum adsorption on oxides, clay minerals, and soils. *Soil Sci. Soc. Am. J.* **1996**, *60*, 425–432.

34. O'Sullivan, A.D.; Moran, B.M.; Otte, M.L. Accumulation and fate of contaminants (Zn, Pb, Fe and S) in substrates of wetlands constructed for treating mine wastewater. *Water Air Soil Pollut.* **2004**, *157*, 345–364.

9

Approaches to the Low Grade Metamorphic History of the Karakaya Complex by Chlorite Mineralogy and Geochemistry

Sema Tetiker [1], Hüseyin Yalçın [2,*] and Ömer Bozkaya [3]

[1] Department of Geological Engineering, Batman University, 72100 Batman, Turkey;
    E-Mail: sema.tetiker@batman.edu.tr
[2] Department of Geological Engineering, Cumhuriyet University, 58140 Sivas, Turkey
[3] Department of Geological Engineering, Pamukkale University, 20070 Denizli, Turkey;
    E-Mail: obozkaya@pau.edu.tr

* Author to whom correspondence should be addressed; E-Mail: yalcin@cumhuriyet.edu.tr

Academic Editor: Antonio Simonetti

Abstract: In this study, chlorite is used to investigate the diagenetic-metamorphic evolution and accurate geological history of the different units belonging to the Karakaya complex, Turkey. Primary and secondary chlorite minerals in the very low-grade metamorphic rocks display interference colors of blue and brown and an appearance of optical isotropy. Chlorites are present in the matrix, pores, and/or rocks units as platy/flaky and partly radial forms. X-ray diffraction (XRD) data indicate that Mg-Fe chlorites with entirely IIb polytype (trioctahedral) exhibit a variety of compositions, such as brunsvigite-diabantite-chamosite. The major element contents and structural formulas of chlorite also suggest these were derived from both felsic and metabasic source rocks. Trace and rare earth element (REE) concentrations of chlorites increase with increasing grade of metamorphism, and these geochemical changes can be related to the tectonic structures, formational mechanics, and environments present during their generation.

Keywords: petrography; XRD; major and trace elements; geological evolution

1. Introduction

Chlorite is a typical rock-forming mineral present in very low grade facies of metamorphic rocks. Chlorite may originate from aggradation (positive transformation) following the evolution of neomineralization of trioctahedral smectite → interlayered smectite-chlorite (C-S) → chlorite, which implies prograde metamorphism due to burial and subsequent diagenesis. Moreover, it represents degradation (negative transformation) in igneous rocks from trioctahedral mica and ferromagnesian minerals, and residual and authigenic minerals in sedimentary rocks [1]. The most significant mineralogical and chemical changes/transformations in chlorite occur during burial diagenesis/metamorphism (e.g., [2–13]). The fact that the half-height width of the 7-Å chlorite peak [14] decreases with the increasing diagenetic/metamorphic grade has been used as a mineralogical parameter for the determination of the diagenetic/metamorphic grade; in particular for metabasic (metavolcanic) rocks, despite these not being as widespread as illites (e.g., [15–18]).

Chlorite occurs pervasively in the various units of the Karakaya Complex, one of the most debated tectonic settings within Turkey [19–21]. In this study, the utility of chlorite is demonstrated relative to the interpretation of the diagenetic-metamorphic evolution of the different units within the Karakaya complex.

2. Geotectonic Setting and Lithology

The Karakaya Complex [22] within the Sakarya Tectonic Unit [23] consists of units representing the remnants of the subduction accretionary prism formed by the closure of the Pre-Jurassic Paleotethys Ocean [24,25]. It is divided into two areas [26] (Figure 1) with the Lower Karakaya Complex (LKC), identified as the Nilüfer unit in northwestern Anatolia and Turhal Metamorphites in the central NE Anatolia [27,28]; this unit is structurally and stratigraphically located at the bottom, and consists of rocks subjected to metamorphism at the end of the Paleozoic or during the Triassic [24]. The LKC units are divided into the upper and lower (LKC-LP and LKC-UP) parts in accordance with their petrographic properties and the mineralogy of phyllosilicates, which correspond to blueschist and greenschist facies, respectively [21]. These are represented by metapsammite (metasandstone), metapelites (slate, phyllite, schist), metacarbonates (metalimestone, metadolomite, marble), and metabasic rocks (metavolcanic, metatuff) in NW Anatolia, and metapelitic (phyllite, slate, metasiltstone), metasandstone, metamagmatic (metabasalt, metatuff, metagabbro, metadiabase), and metalimestone in the central NE Anatolia region.

The Upper Karakaya Complex (UKC) contains many exotic limestone blocks within the Permo-Carboniferous-aged units [29–31]. It consists mainly of three units (Hodul, Orhanlar and Çal) of Permian or Triassic age that is strongly deformed in the NW Anatolia [24] and Devecidağ Melange [27] in the central NE Anatolia. Among these, the Hodul unit comprises arkosic sandstones and intercalated (meta-) pelitic rocks (mudstone, shale, slate, siltstone); the Orhanlar unit contains shale intercalated sandstones; the Çal unit consists of spilitic basaltic lava-agglomerates and clastic rocks (mudstone, shale, sandstone); and the Devecidağ Melange unit contains metapelitic rocks (metashale, slate, metasiltstone), metasandstone, spilitic metabasalt, and metalimestones.

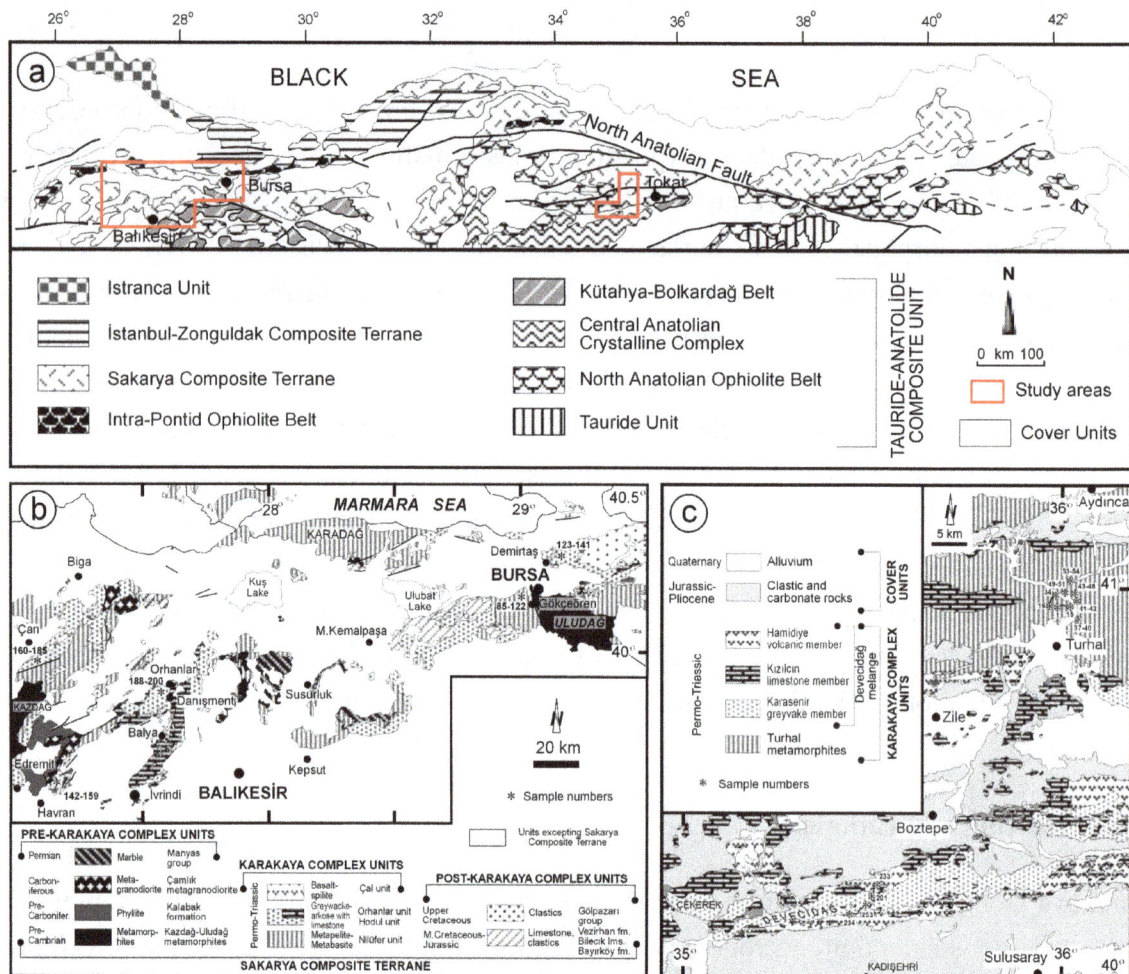

Figure 1. Simplified geologic-geotectonic maps of the Karakaya Complex and related units (unit boundaries [32] and unit determinations [23], (**a**) The Alpine tectonic units of northern Turkey (unit boundaries [32]; (**b**) NW Anatolia (modified after MTA [32]); (**c**) Central NE Anatolia [27].

3. Materials and Methods

In total, 253 rock samples were taken from various units within the area of investigation. The weathered surface and soil on the samples were removed by crushing and then washing with ultrapure water. Subsequently, samples underwent various processes such as thin sectioning, crushing–grinding–screening, clay separation, X-ray diffraction (XRD), and optical microscopy (OM) at Cumhuriyet University (CU), Department of Geology and Mineralogy-Petrography and Geochemistry Research Laboratories (MİPJAL). Scanning electron microscopy (SEM) was conducted at the Turkish Petroleum Corporation Research Center (Ankara).

The samples investigated by XRD were first crushed in pieces of 3–5 cm in diameter with a hammer and then into fine particles (<5 mm) in a Fritisch rock crusher. Depending on their hardness, samples were then ground for 10–30 min in a silicon carbide bowl grinder and/or agate mortar. The powder material was then placed in plastic bags and labeled and ready for analysis. The XRD analyses were conducted with a DMAX III X-ray diffractometer (Rigaku Corporation, Tokyo, Japan) (Anode = Cu (Cu Kα = 1.541871 Å), Filter = Ni, Voltage = 35 kV, Current = 15 mA, Gonimeter speed = 2°/min,

Paper speed = 2 cm/min, Time constant = 1 s, Slits = 1°–0.15 mm–1°–0.30 mm, Paper interval $2\theta = 5°–35°$).

As a result of the XRD analyses, all rock and clay-sized components (<2 µm) of the samples were defined and their semi-quantitative percentages were calculated using the external standard method [33]. Mineral intensity factors were used in all rock and clay fraction calculations and the reflections were measured in mm. In this method, kaolinite was taken as a reference from the glycolylated preparations for clay fraction and dolomite for whole rock [34]. Quartz was used as the internal standard in the measurement of d-spaces. The definitions of clay minerals were based mainly according to their (001) basal reflections.

The separation of phyllosilicate/clay minerals was conducted at the Clay Separation Laboratory, Department of Geological Engineering, Cumhuriyet University. The clay separation process necessary for the XRD-CF analyses essentially consists of chemical solving (removal of non-clay fraction), centrifugation–decantation/resting and washing, suspending–sedimentation–siphoning–centrifugation, and bottling. In the absence of suspended particles, the separation process was accelerated by the addition of Calgon. Centrifugation was performed using a Sepatech Varifuge 3.2S (Heraeus Group, Frankfurt, Germany) model centrifuger at 5600 cycles/min speed and 200 cc capacities with stainless steel and/or plastic bowls. Three oriented slide preparations from each clay mud were prepared as plastering, or in suspension and these were dried at room temperature. Clay fraction diffractograms were obtained by putting through the processes of normal-AD (air-dried), glycolylated-G (leaving in ethylene glycol vapour in a dessicator for 16 h at 60 °C) and heated-H (heating in the oven for 4 h at 490 °C). The goniometer speed at the intervals was set at 1°/min and the recording interval was set as $2\theta = 2–30°$ ($\pm 0.04°$).

In the measurement of the illite crystallinity, the width at the half-height of the 10-Å illite peak $\Delta°2\theta$ (Kübler index—[35,36]) was used. The peak widths were calibrated according to Kisch [37] and Warr and Rice [38] standards. WINFIT [39] programme (http://xray.geol.uni-erlangen.de/html/software/soft.html) was used to distinguish the single and two illite phases with the asymmetric and symmetric peaks as well as crystalline illite-WCI and poor crystalline illite-PCI. For the lower and upper limits of the anchizone, 0.21 and 0.37 $\Delta°2\theta$, and 0.25 and 0.42 $\Delta°2\theta$ were taken respectively for Kisch [37,40] and CIS [38] standards.

In the measurement of chlorite "crystallinity", the half-height width ($\Delta°2\theta$; Árkai index—AI: [15,36]) of 7-Å chlorite peak was used. The diagenesis-anchizone and anchizone-epizone limits for AI were suggested as 0.37 and 0.26 ($\Delta°2\theta$) respectively [15–17]; and according to the Kübler index (KI; [35]) limits (0.42 and 0.25 $\Delta°2\theta$), the anchizone-epizone limit is similar, while the diagenesis-anchizone limit is lower. WINFIT [39] programme was used for precise determination of peak widths. Polytype examinations in chlorite minerals were performed between the recording interval of $2\theta = 31°–52°$ as from the nonoriented plaques. Diagnostic peaks suggested by Bailey [41] were used in the determination of the polytypes.

The following method was used in order to determine the chemical components of the chlorites: $d_{(001)}$ reflection values were found as from $d_{(005)}$ peaks and the amount of tetrahedral Al was established according to $d_{(001)} = 14.55 \text{ Å} - 0.29\text{Al}^{IV}$ formula [42]. The amount of octahedral Fe^{2+} was obtained from the diagram using the $[I_{(002)} + I_{(004)}]/[I_{(001)} + I_{(003)}]$ [43] and $I_{(002)}/I_{(001)}$ and $I_{(004)}/I_{(003)}$ [44] ratios. In this method, the Fe contents in the talc and brucite layers can be determined with the help of the $I_{(004)}/I_{(003)}$ ve $I_{(002)}/I_{(001)}$ basal peak ratios of the chlorite minerals. Mg contents were determined in accordance with the $Fe + Al^{VI} + Mg = 6$ equation by presuming that $Al^{IV} = Al^{VI}$.

The major, trace, and rare earth element (REE) analyses of four pure (or nearly pure) chlorite fractions were performed at Activation Laboratories Ltd. (Actlabs, Ancaster, ON, Canada). Fusion inductively coupled plasma (ICP) was used for major element analysis, while inductively coupled plasma-mass spectrometry (ICP-MS) was employed for the analyses of trace and REEs. Details of the analysis method and the instrument detection limits are provided on Actlabs' website (http://www.actlabs.com/).

4. Results

4.1. Petrography

4.1.1. Optical Microscope Examinations

Plagioclase microliths are present in microlithic porphyritic metavolcanic rocks representing the Nilüfer Unit within the LKC-LP in NW Anatolia. The groundmass material consists mainly of volcanic glass, phyllosilicates, and amphiboles. The volcanic glass within the matrix has been subjected to mainly chloritization, silicification and partly Fe-oxidation, and carbonization. Chlorite minerals show blue and brown interference colors, but are mainly green in plane polarized light. Chlorite plates oriented parallel to the crystallographic c-axis and quartz-rich zones that are sequenced in the direction of c-crystallographic axis in the blue schists {001} of the Nilüfer Unit (LKC-LP), and are typical textures indicative of metamorphic differentiation (Figure 2a). Isotropiclike chlorite minerals in the matrix are observed in porphyroblasts that form in tremolite-rich metabasalts of the LKC-UP, which formed under greenschist facies conditions (Figure 2b).

Platy chlorite minerals (displaying blue interference color) result from the neoformation within greenschist facies metasandstones defined as Turhal Metamorphites of LKC-UP in the central NE Anatolia. Muscovite, biotite, and chlorite stack (pod) structures showing micro-orientation and bending in certain metasandstones are typical. The stacks look platy/flaky, and partly radial, and these have three different mineralogical compositions that are chlorite-muscovite (CMS), chlorite-biotite (CBS), and biotite-muscovite (BMS) stacks (Figure 2c). In metavolcanic rocks, green chlorite minerals are characterized by blue and brown interference colours. The chlorites in the matrix and pores may also show optical isotropic-like features as in corresponding rocks in NW Anatolia (Figure 2d).

The chlorites in the pores of sandstones in the Orhanlar Unit among UKC units in NW Anatolia display a blue interference colour (Figure 3a,b). Chlorite has a distinctive reaction rim that indicates retrograde metamorphism. Furthermore, calcite and rare epidote minerals are other components in the pores. Volcanic rocks constituting the main lithology of the Çal Unit have hypocrystaline porphyritic, vitrophyric-microlitic porphyritic, amygdaloidal, and partly glomeroporfiric textures. The groundmass consists of plagioclase microlites and mostly chloritized volcanic glass (Figure 3c).

The groundmass for the metasandstones from the Devecidağ Melange among UKC units in the central NE Anatolia consists of Fe oxidation and phyllosilicates matrix in addition to calcite and/or dolomite and silica cement. Chlorite minerals are typically observed with blue interference color developed within pores and matrix (Figure 3d).

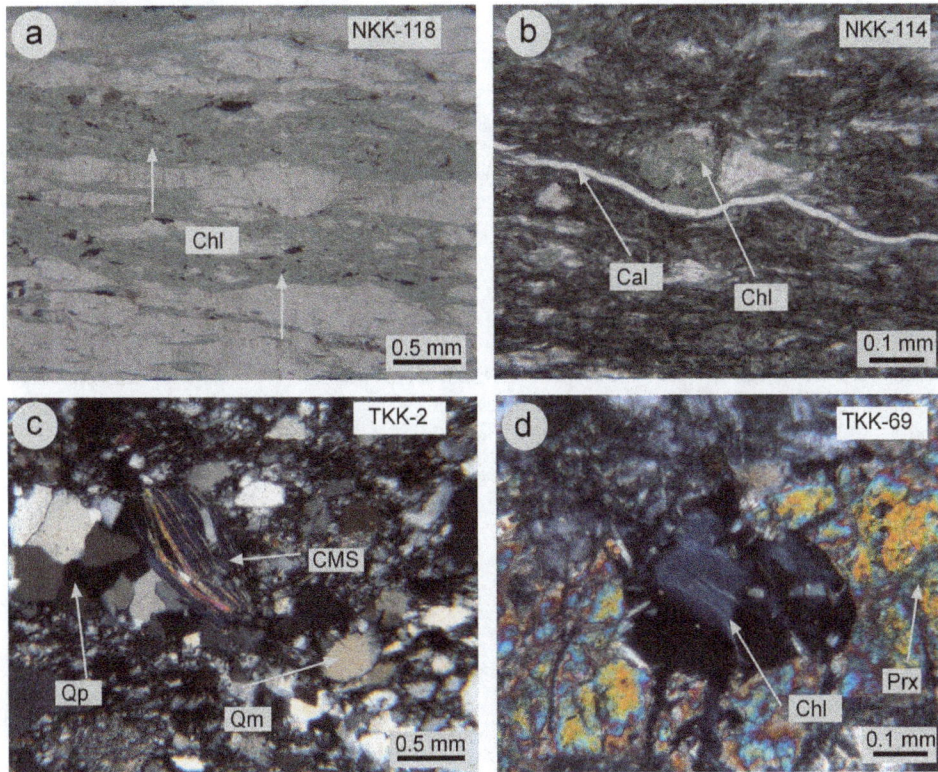

Figure 2. Optical microphotograps of chlorite and associated minerals in the Lower Karakaya Complex units (Chl = Chlorite, Qp = Polycrystalline quartz, Qm = Monocrystalline quartz, CMS = Chlorite-muscovite stack, Cal = Calcite, Prx = Pyroxene, ppl = plane polarized light, cn = crossed polars); (**a**) Typical metamorphic differentiation in the schists of the Nilüfer Unit (ppl); (**b**) Isotropic-view chlorite porphyblasts and pore filling calcite occurrences within the matrix with oriented texture of tremolitized metabasalts from Nilüfer Unit (ppl); (**c**) Monocrystalline and polycrystalline quartzs, and chlorite-muscovite stack in the metasandstones of Turhal Metamorphites (cp); (**d**) Blue interference colored chlorites within the metabasalts of Turhal Metamorphites (cp).

4.1.2. Scanning Electron Microscopy

The greenschist facies sample of shale (NKK-93) from the Nilüfer Unit, LKC-UP in NW Anatolia consists of chlorite, illite, quartz, feldspar, and calcite, with chlorite minerals forming clusters in the form of plates (Figure 4a). The illites are mainly in the form of parallel, partly radially sequenced long and thin filaments (2–30 μm) (Figure 4b).

In the greenschist facies metasandstone sample (NKK-93) of the Turhal Metamorphites, LKC-UP in central NE Anatolia comprises illite, chlorite, quartz, and feldspar, with the chamosite Fe-chlorites occurring as thick platy forms (Figure 4c). The illites range between 10–50 μm in length and are observed in part as thin filamentous balls (Figure 4d).

Figure 3. Optical microphotographs of chlorite and associated minerals from the Upper Karakaya Complex units (Chl = Chlorite, ppl = plane polarized light, cn = crossed polars); (**a**) Blue interference colored chlorite within the metasandstones of Orhanlar Unit (cp); (**b**) Green colored chlorite within the metasandstones of Orhanlar Unit (ppl); (**c**) Green colored chlorites within the metabasalts of Çal Unit (ppl); (**d**) Green colored chlorites within the metabasalts of Devecidağ Melange (ppl).

4.2. X-Ray Diffraction Mineralogy

4.2.1. Whole Rock and Clay Fraction

The mineralogical composition of rocks within the KC units is given in Table 1. Volcanogenic (feldspar, augite), metamorphic-metasomatic (glaucophane, tremolite/actinolite, epidote, stilpnomelane, phyllosilicates), and secondary minerals (calcite, dolomite, quartz, hematite, goethite) were recorded in metapelitic and metavolcanogenic rocks constituting the LKC-LP Nilüfer Unit in NW Anatolia. Phyllosilicate minerals consist of chlorite + illite unity and stilpnomelane, smectite and mixed-layer (C-S and C-V) minerals are involved in this unity to a lesser amount. In slate, chlorite wholly represents the clay fraction (Figure 5a).

Volcanogenic (feldspar), metamorphic-metasomatic (glaucophane, tremolite/actinolite, epidote, stilpnomelane, paragonite, phyllosilicates) and secondary chemical minerals (calcite, dolomite, quartz, hematite, goethite) were noted in metamagmatic and metapelitic rocks of LKC-LP Turhal Metamorphites in the central NE Anatolia. Pure chlorite (Figure 5b) represents the phyllosilicate mineral in metatuffites and the chlorite + illite association in the remaining rocks.

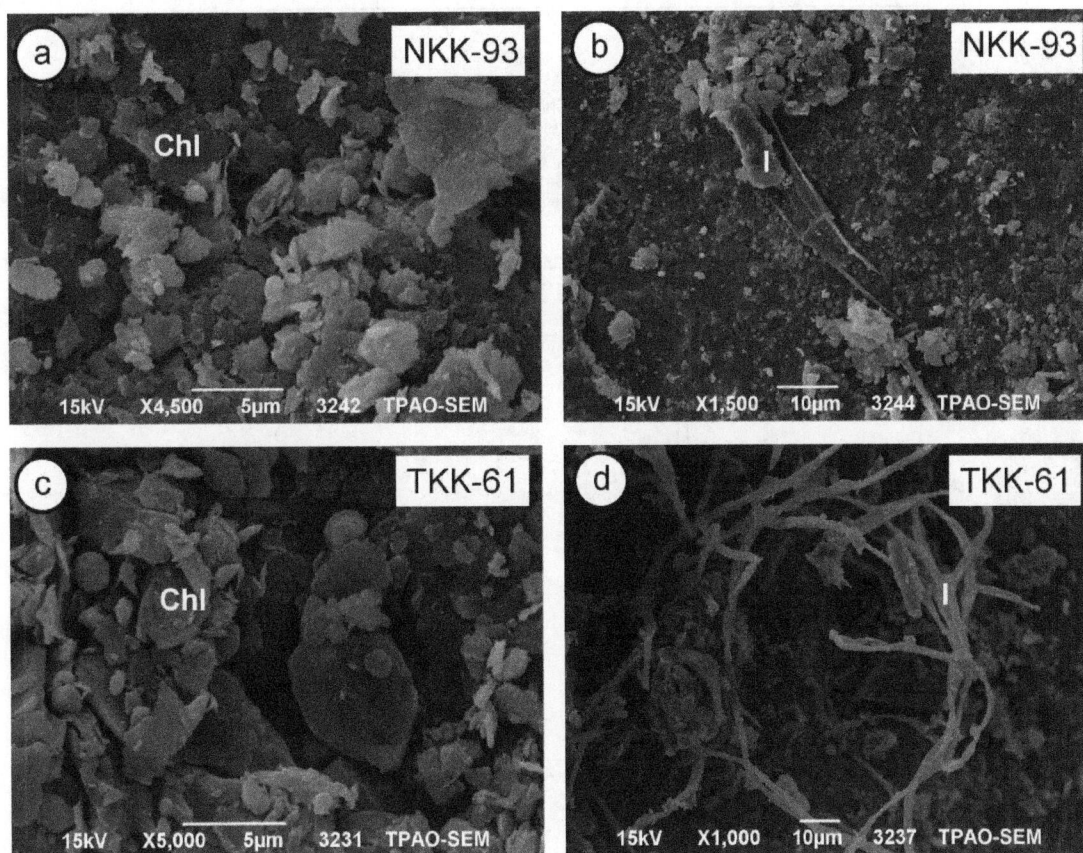

Figure 4. Scanning electron microscopy (SEM) photos of chlorite and associated minerals in the Lower Karakaya Complex units (Chl = Chlorite, I = Illite), (**a**) Platy chlorites in the shale sample of the Nilüfer Unit; (**b**) Thin-long filaments of illites in the shale sample of the Nilüfer Unit; (**c**) Radial needle-like chlorites and surrounding illites within the pores of metasandstones in the Turhal Metamorphites; (**d**) Fibrous illites in the metasandstones of the Turhal Metamorphites.

Table 1. The dominant mineralogy of samples from the Karakaya Complex (KC) units.

Unit	Area	KC	Facies	Minerals
Çal				chlorite, illite, C-V, C-S
Orhanlar		UKC		chlorite, illite, C-S, C-V, I-S, epidote
Hodul			Greenschist	illite, chlorite, I-S, C-V
	NW Anatolia	LKC-UP		augite, hornblende, epidote, stilpnomelane, illite, chlorite, C-S, C-V, I-S
Nilüfer		LKC-LP	Blueschist	glaucophane, tremolite/actinolite, augite, epidote, stilpnomelane, chlorite, illite, C-V, C-S
Devecidağ Melange		UKC		augite, epidote, prehnite, illite, chlorite, C-S, C-V, I-C, I-S
	Central NE Anatolia	LKC-UP	Greenschist	augite, epidote, illite, chlorite, paragonite, stilpnomelane, I-C, I-S, C-S, C-V
Turhal Metamorphites		LKC-LP	Blueschist	glaucophane, tremolite/actinolite, epidote, stilpnomelane, paragonite, illite, chlorite

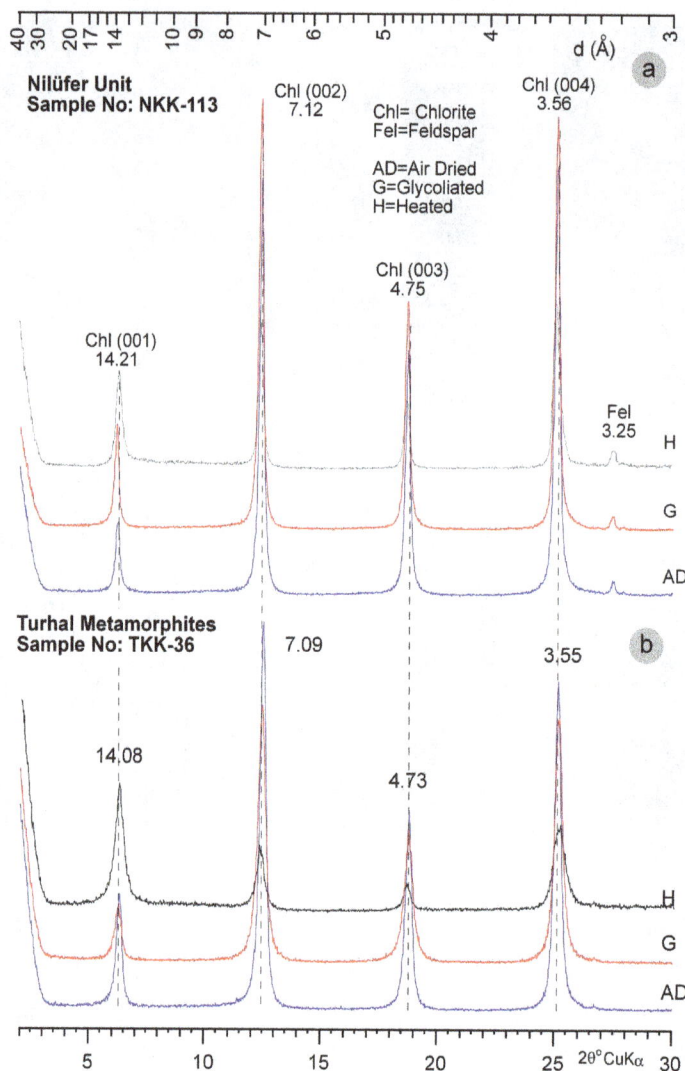

Figure 5. Oriented X-ray diffraction (XRD) patterns of chlorite minerals in the units of Lower Karakaya Complex-Lower Part, (**a**) Slate sample of blueschist facies in the Nilüfer Unit; (**b**) Metatuffite sample of greenschist facies in the Turhal Metamorphites.

Volcanogenic (pyroxene, feldspar), metamorphic-metasomatic (amphibole, epidote, stilpnomelane), and secondary minerals (calcite, dolomite, quartz, hematite, goethite) were detected in metapsammite, metapelite, metacarbonate, and metabasic rocks representing LKC-UP in NW Anatolia. While phyllosilicates are mostly represented by illite + chlorite paragenesis, in some instances this assemblage is accompanied by stilpnomelane, smectite, kaolinite, and mixed-layer (C-S, C-V and I-S) minerals.

Metapelitic, metapsammitic, metamagmatic, and metalimestones in LKC-UP in the central NE Anatolia contain feldspar, phyllosilicate, quartz, pyroxene, epidote, stilpnomelane, paragonite, calcite, dolomite, hematite, and goethite minerals. In these rocks, illite, chlorite, kaolinite, smectite, paragonite, stilpnomelane, and mixed layer (I-C, I-S, C-S and C-V) minerals represent phyllosilicates in the order of abundance. In these samples, illite + chlorite association constitutes the widespread phyllosilicate paragenesis. While the assemblage of illite + chlorite is observed in certain metasedimentary rocks, the association of chlorite + stilpnomelane was detected in metavolcanic rocks. Illite + kaolinite + smectite, illite + chlorite + paragonite + I-S and illite + chlorite + I-C or C-V assemblages constitute other parageneses.

Clastic rocks and (meta-) limestones of the Hodul Unit within the UKC in NW Anatolia contain quartz, phyllosilicate (illite, kaolinite, I-S, C-V, smectite and chlorite), calcite, and feldspar. Clastic rocks and (meta-) limestones of the Orhanlar Unit have quartz, feldspar, phillosilicate (chlorite, illite, C-S, C-V and I-S), calcite, and epidote minerals. Calcite, phyllosilicate (chlorite, C-S, vermiculite, illite and C-V), quartz, feldspar, hematite, and dolomite minerals were recorded in metavolcanic, shale and limestones of the Çal Unit.

Metapelitic, metapsammitic, metavolcanic, and metalimestones from the Devecidağ Melange within the UKC units in central NE Anatolia consist of quartz, feldspar, phyllosilicate (illite, chlorite, C-S, C-V, smectite, kaolinite, I-C and I-S), pyroxene, epidote, hematite, dolomite, and prehnite minerals.

4.2.2. Polytype

The results of polytype examinations of pure chlorite fractions for the Nilüfer unit in NW Anatolia and Turhal Metamorphites in the central NE Anatolia of the LKC are given in Table 2. It was detected that the chlorite associations of the LKC-LP and LKC-UP, which correspond to blueschist and greenschist facies in metavolcanic and metapelitic rocks of the Nilüfer Unit and Turhal Metamorphites are entirely of IIb polytype (Figure 6).

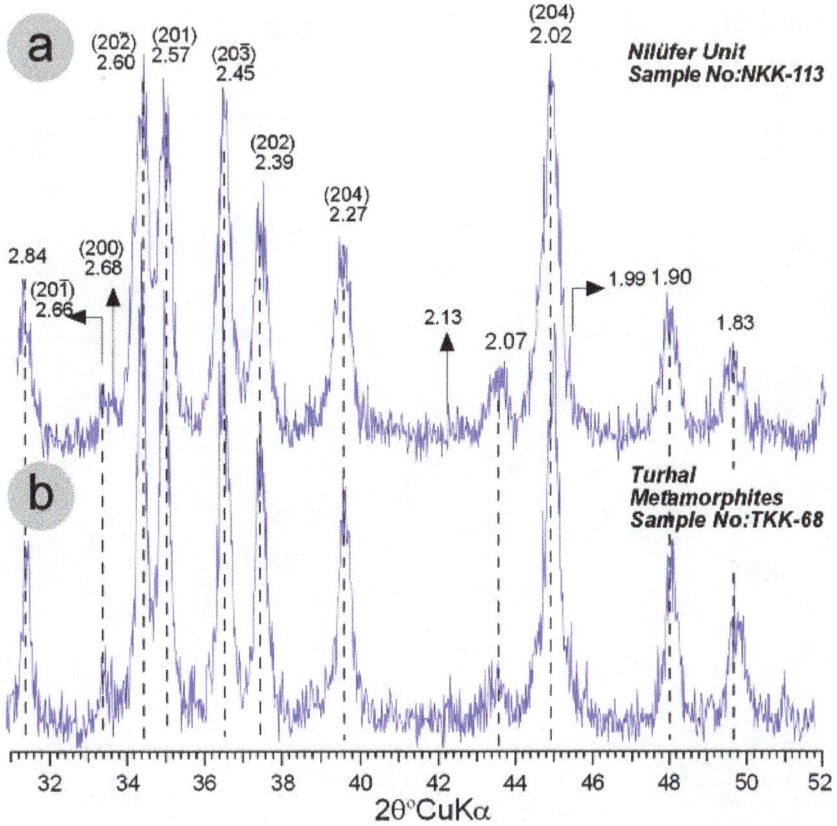

Figure 6. Unoriented XRD patterns of chlorite minerals in the units of Lower Karakaya Complex-Lower Part, (a) Slate sample of blueschist facies in the Nilüfer Unit; (b) Metagabbro sample of greenschist facies in the Turhal Metamorphites.

Table 2. The results of polytype examination of chlorite minerals in the Karakaya Complex units.

Unit	Area	KC	Facies	Sample No	Polytype	Lithology
Nilüfer	NW Anatolia	LKC-UP	Greenschist	NKK-103	IIb	Calcslate
		LKC-LP	Blueschist	-113	IIb	Slate
Turhal Metamorphites	Central NE Anatolia	LKC-UP	Greenschist	TKK-36	IIb	Metatuffite
				-58	IIb	Metagabbro
				-63	IIb	Metavolcanic
				-64	IIb	Metavolcanic
				-68	IIb	Metagabbro
				-73	IIb	Metabasalt
				-82	IIb	Phyllite

4.2.3. Composition

The chemical compositions of the chlorites for the LKC units determined in accordance with basal peak values and intensity ratios are listed in Tables 3 and 4 and shown in Figure 7. The reflection values of the (001) surface in the chlorites in NW Anatolia vary between 14.189–14.338 Å (14.249 Å on average) for LKC-LP and 14.172–14.392 Å (14.262 Å on average) for LKC-UP; and 14.202–14.366 Å (14.275 Å on average) for the Orhanlar Unit. For the Çal Unit, the average reflection value is 14.269 Å.

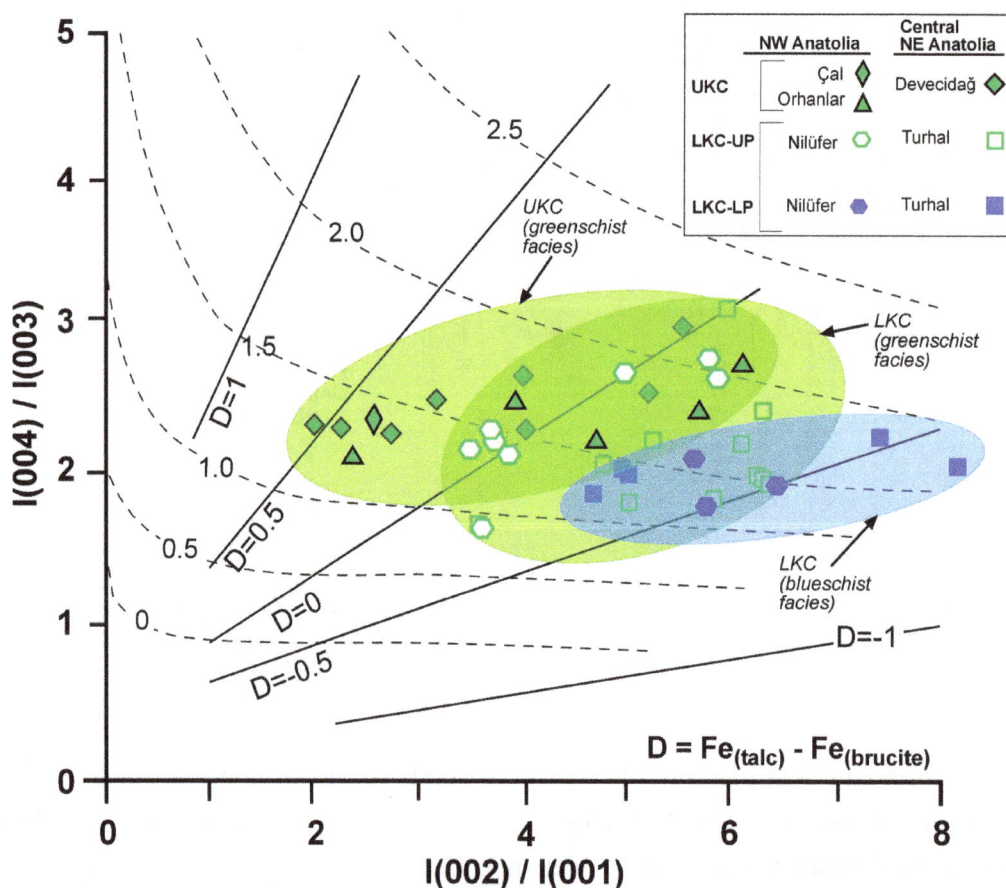

Figure 7. Fe content within the talc layer and distribution of talc and brucite layers of chlorite minerals according to basal peak ratios [44] in the Karakaya Complex units.

Table 3. X-ray diffraction (XRD) basal reflections and chemical compositions of chlorite minerals in the Lower Karakaya Complex units. B: Brindley [42], B&B: Brindley and Brown [43], C&D: Chagnon and Desjardins [44]), $Al^{IV} = Al^{VI}$ and $Mg = 6 - (Fe + Al^{IV})$.

Sample No.	$d_{(001)}$ (Å)	$I(003)/I(001)$	$I(002+004)/I(001+003)$	$I(002)/I(001)$	$I(004)/I(003)$	Talc Fe^{+2} B&B	Talc Fe^{+2} C&D	Brucite Fe^{+2} B&B	Brucite Fe^{+2} C&D	ΣFe^{+2} B&B	ΣFe^{+2} C&D	Al^{IV} B&B	Mg^{VI} C&D	Fe/(Fe + Mg) C&D
NW Anatolia														
LKC-UP (Greenschist facies)														
NKK-95	14.231	1.26	4.05	5.89	2.59	1.87	1.98	2.58	2.13	4.45	4.11	1.10	0.79	0.84
-96	14.392	2.16	3.39	5.01	2.64	1.07	1.90	2.60	1.90	3.67	3.80	0.54	1.66	0.70
-97	14.172	1.36	4.02	5.79	2.71	1.79	2.10	2.61	2.20	4.40	4.30	1.30	0.40	0.92
-98	14.311	1.29	2.88	3.89	2.10	1.13	1.43	1.88	1.44	3.01	2.87	0.82	2.31	0.55
-103	14.280	1.98	2.72	3.77	2.19	0.69	1.48	2.09	1.40	2.78	2.88	0.93	2.19	0.57
-106	14.200	1.18	2.81	3.51	2.22	1.15	1.50	1.76	1.30	2.91	2.80	1.21	1.99	0.58
-108	14.287	1.13	2.98	3.70	2.34	1.30	1.60	1.84	1.40	3.14	3.00	0.91	2.09	0.59
-111	14.220	1.95	3.33	5.65	2.15	1.11	1.60	2.49	1.95	3.60	3.55	1.14	1.31	0.73
LKC-LP (Blueschist facies)														
-113	14.220	3.44	2.67	5.77	1.77	0.24	1.20	2.47	1.70	2.71	2.90	1.14	1.96	0.60
-115	14.189	2.53	3.23	6.43	1.97	0.85	1.50	2.62	2.00	3.47	3.50	1.24	1.26	0.74
-117	14.338	1.72	2.36	3.63	1.62	0.54	0.90	1.72	1.20	2.26	2.10	0.73	3.17	0.40
Central NE Anatolia														
LKC-UP (Greenschist facies)														
TKK-4	14.372	1.51	4.31	5.97	3.22	1.81	2.30	2.79	2.30	4.60	4.60	0.61	0.79	0.85
-10	14.192	3.56	3.81	9.27	2.28	1.11	1.90	3.39	2.60	4.50	4.50	1.23	0.27	0.94
-28	14.269	1.88	2.38	3.59	1.74	0.34	0.85	1.66	1.15	2.00	2.00	0.97	3.03	0.40
-49	14.231	1.94	2.97	5.03	1.90	0.79	1.27	2.15	1.67	2.94	2.94	1.10	1.96	0.60
-50	14.251	1.54	3.48	5.27	2.32	1.27	1.65	2.28	1.90	3.55	3.55	1.03	1.42	0.71
-58	14.251	2.85	3.28	6.11	2.29	0.83	1.65	2.77	1.95	3.60	3.60	1.03	1.37	0.72
-63	14.220	2.85	3.15	6.31	2.04	0.79	1.55	2.74	1.98	3.53	3.53	1.14	1.33	0.73
-64	14.182	2.22	3.37	6.25	2.07	0.97	1.55	2.53	1.95	3.52	3.50	1.27	1.23	0.74
-73	14.311	2.23	3.70	6.30	2.53	1.24	1.90	2.81	2.15	4.05	4.05	0.82	1.13	0.78
-82	14.259	1.76	3.11	4.78	2.16	1.02	1.50	2.23	1.75	3.25	3.25	1.00	1.75	0.65
-84	14.190	2.20	3.13	5.85	1.90	0.86	1.40	2.42	1.88	3.28	3.28	1.24	1.48	0.69
LKC-LP (Blueschist facies)														
-42	14.200	1.69	2.72	4.30	1.79	0.72	1.10	1.88	1.50	2.60	2.60	1.21	2.19	0.54
-36	14.200	1.67	2.32	3.45	1.64	0.53	1.00	1.67	1.20	2.20	2.20	1.21	2.59	0.46
-38	14.311	1.45	2.06	2.81	1.55	0.38	0.74	1.29	0.93	1.67	1.67	0.82	3.51	0.32
-54	14.220	2.09	2.45	4.29	1.57	0.30	0.80	1.78	1.28	2.08	2.08	1.14	2.78	0.43
-68	14.291	1.79	2.86	4.23	2.09	0.85	1.40	2.10	1.55	2.95	2.95	0.89	2.16	0.58
-79	14.239	1.71	3.43	5.43	2.27	0.75	1.65	1.92	2.02	2.67	2.67	1.07	2.26	0.54

Table 4. XRD basal reflections and chemical compositions of chlorite minerals in the Upper Karakaya Complex units (subgreenschist facies). B: Brindley [42], B&B: Brindley and Brown [43], C&D: Chagnon and Desjardins [44]), $Al^{IV} = Al^{VI}$ and $Mg = 6 - (Fe + Al^{IV})$.

Sample No.	$d_{(001)}$ (Å)	$I(003)/I(001)$	$I(002+004)/I(001+003)$	$I(002)/I(001)$	$I(004)/I(003)$	Talc Fe^{2+} B&B	Talc Fe^{2+} C&D	Brucite Fe^{2+} B&B	Brucite Fe^{2+} C&D	Σ Fe^{2+} B&B	Σ Fe^{2+} C&D	Al^{IV} B&B	Mg^{VI} C&D	Fe/(Fe+Mg) C&D
NW Anatolia														
CKK-163	14.269	0.71	2.49	2.58	2.35	1.30	1.42	1.15	1.02	2.45	2.44	0.98	2.58	0.48
BKK-188	14.366	1.38	3.11	3.96	2.49	1.23	1.65	2.08	1.50	3.31	3.25	0.63	2.12	0.61
-189	14.281	1.29	3.84	5.70	2.39	1.73	1.90	2.47	2.20	4.20	4.10	0.93	0.97	0.81
-190	14.220	1.50	3.22	4.72	2.22	1.24	1.55	2.22	1.70	3.46	3.25	1.14	2.61	0.46
-191	14.220	0.64	2.27	2.38	2.10	1.22	1.23	0.91	0.88	2.13	2.11	1.14	2.75	0.43
-193	14.287	1.40	4.12	6.11	2.69	1.82	2.15	2.69	2.35	4.51	4.50	0.91	0.59	0.88
Central NE Anatolia														
DKK-222	14.364	0.92	3.35	3.99	2.66	1.69	1.80	1.93	1.80	3.62	3.60	0.64	1.76	0.67
-223	14.130	1.23	3.09	4.04	2.31	1.30	1.65	1.98	1.60	3.28	3.25	1.45	1.30	0.71
-224	14.368	0.91	3.34	3.94	2.69	1.69	1.85	1.92	1.70	3.51	3.55	0.63	1.82	0.66
-225	14.269	0.55	2.28	2.27	2.29	1.35	1.35	0.79	0.90	2.24	2.25	0.97	2.78	0.45
-226	14.226	1.59	3.59	5.23	2.55	1.43	1.95	2.49	2.00	3.92	3.95	1.12	0.93	0.81
-234	14.216	0.95	2.84	3.18	2.48	1.33	1.60	1.62	1.30	2.95	2.90	1.15	1.95	0.60
-235	14.267	0.63	2.12	2.00	2.32	1.12	1.27	0.79	0.73	1.91	2.00	0.98	3.02	0.40
-243	14.262	1.21	4.15	5.55	2.99	1.95	2.25	2.60	2.20	4.45	4.45	0.99	0.56	0.89
-250	14.351	0.72	2.54	2.74	2.27	1.34	1.40	1.20	1.15	2.54	2.55	0.69	2.76	0.48

According to the methods outlined by Brindley and Brown [43] and Chagnon and Desjardins [44], the Fe content in talc and brucite layers of the chlorites are similar, and the Chagnon and Desjardins [44] method was taken as a basis for determining octahedral Fe (Figure 7). According to the diagram, the Fe contents in the talc layer of the chlorites range between 1.0 and 2.4. Therefore, 2.10–3.50 (2.83 on average) was established for the octahedral Fe contents of the chlorites for LKC-UP in NW Anatolia, 2.80–4.11 (3.41 on average) for LKC-UP, and 2.11–4.50 (3.44 on average) and 2.44 for UKC Orhanlar and Çal units, respectively. Octahedral Al contents are 0.73–1.24 (1.04 on average) for LKC-LP, 0.54–1.21 (0.99 on average) for LKC-UP, and 0.63–1.14 (0.94 on average) and 0.98 for UKC (Orhanlar and Çal units), respectively. Octahedral Mg contents are 1.26–3.17 (2.13 on average) for LKC-LP, 0.40–2.31 (1.59 on average) for LKC-UP, and 0.56–2.75 (1.62 on average) and 2.58 for UKC (Orhanlar and Çal units), respectively. Total amounts of octahedral cations for LKC-L, LKC-U and UKC (Orhanlar and Çal units) is 6.00 and it was noted that the chlorites are of trioctahedral component and the average structural formulae are provided below:

$$(Al_{0.94}Fe_{3.44}Mg_{1.62})[(Si_{3.06}Al_{0.94})O_{10}](OH)_8 \quad (\text{UKC Orhanlar})$$

$$(Al_{0.98}Fe_{2.44}Mg_{2.58})[(Si_{3.02}Al_{0.98})O_{10}](OH)_8 \quad (\text{UKC Çal})$$

$$(Al_{0.98}Fe_{3.47}Mg_{1.55})[(Si_{3.02}Al_{0.98})O_{10}](OH)_8 \quad (\text{LKC-UP})$$

$$(Al_{1.00}Fe_{3.26}Mg_{1.74})[(Si_{3.00}Al_{1.00})O_{10}](OH)_8 \quad (\text{LKC-LP})$$

$d_{(001)}$ values of the chlorites of Turhal Metamorphites LKC-LP, LKC-UP and UKC-Devecidağ Melange in the central NE Anatolia are 14.200–14.311 Å (14.244 Å on average), 14.190–14.372 (14.248 on average) and 14.130–14.368 Å (14.273 Å on average). Octahedral Fe contents are 1.67–2.95 (2.36 on average), 2.00–4.60 (3.53) and 2.00–4.45 (3.17 on average); octahedral Al contents are 0.82–1.21 (1.06 on average), 0.61–1.27 (1.04 on average) and 0.63–1.45 (0.96 on average). Octahedral Mg contents are 2.16–3.51 (2.58 on average), 0.27–1.96 (1.43 on average) and 0.56–3.02 (1.88 on average), respectively. Total octahedral cation amounts are 6.00 and it was determined that they are closer to trioctahedral composition. The overall average structural formulas of the chlorites are as follow:

$$(Al_{0.96}Fe_{3.16}Mg_{1.88})[(Si_{3.04}Al_{0.96})O_{10}](OH)_8 \text{ (UKC Devecidağ Melange)}$$

$$(Al_{1.04}Fe_{3.53}Mg_{1.43})[(Si_{2.96}Al_{1.04})O_{10}](OH)_8 \text{ (LKC-UP Turhal Metamorphites)}$$

$$(Al_{1.06}Fe_{2.36}Mg_{2.58})[(Si_{2.94}Al_{1.06})O_{10}](OH)_8 \text{ (LKC-LP Turhal Metamorphites)}$$

The substitution of Al in the place of tetrahedral Si leads to a reduction in layer thickness and thus a reduction in $d_{(001)}$ value as a result of the strengthening of the layer and the bond between the layers [45].

On the basis of the diagram using $Al^{IV} - Fe/(Fe + Mg)$ ratios [46] the, LKC-LP chlorites examined here fall in brunsvigit-diabantite zone, whereas LKC-UP chlorites cluster in brunsvigit zone; UKC chlorites plot within the diabandite-chamosite zone and show variable compositions. According to the AIPEA classification [47], the chlorites investigated here classify mainly as chamosite and partly clinochlore compositions (Figure 8). In contrast, it is observed that only one (NKK-113) of the two samples defined as IIb polytype falls within the area defined by Foster [46].

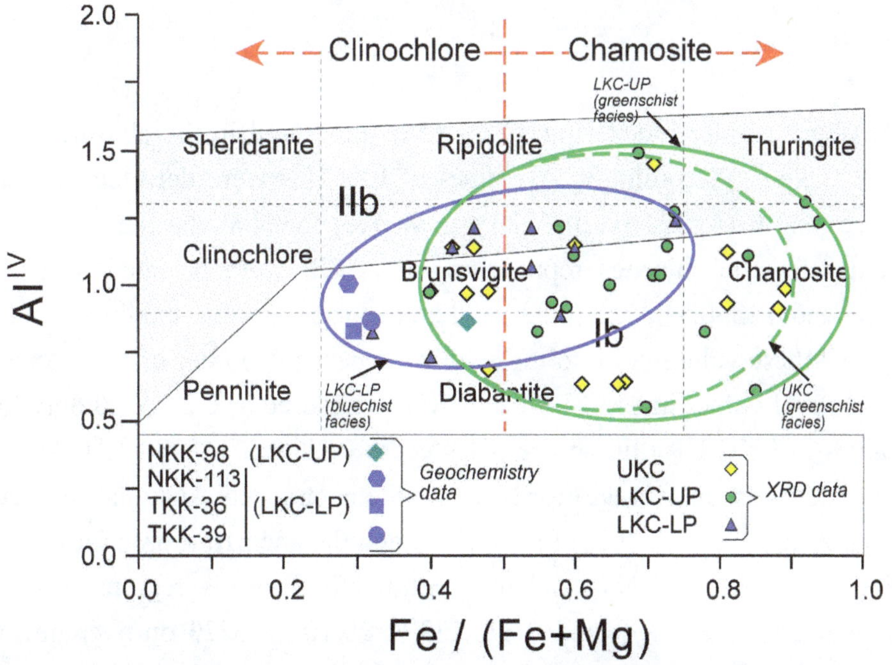

Figure 8. The setting of chlorite minerals in the Karakaya Complex units in tetrahedral Si-octahedral Fe/(Fe + Mg) diagram (Classification and IIb polytype area: Foster [46]; Ib polytype area: Curtis *et al.* [5]; chamosite-clinochlore boundary: Bailey [47]).

Distributions of the first three basal reflections (I_{001}, I_{002} and I_{003}) of the chlorites [16] indicate that the LKC and UKC units are distinct (Figure 9). The intensities of the chlorites shift in a way such that they get closer to the I_{002}-I_{003} line as their Fe/(Fe + Mg) contents increase. The difference in intensities for the LKC-UP and LKC-LP units relative to their I_{002} and I_{003} can be attributed to their relatively higher iron and magnesium contents, respectively. Moreover, the high intensities related to 003 peaks in LKC-LP are the result of higher Fe contents in brucite layers rather than talc layers (Figure 7).

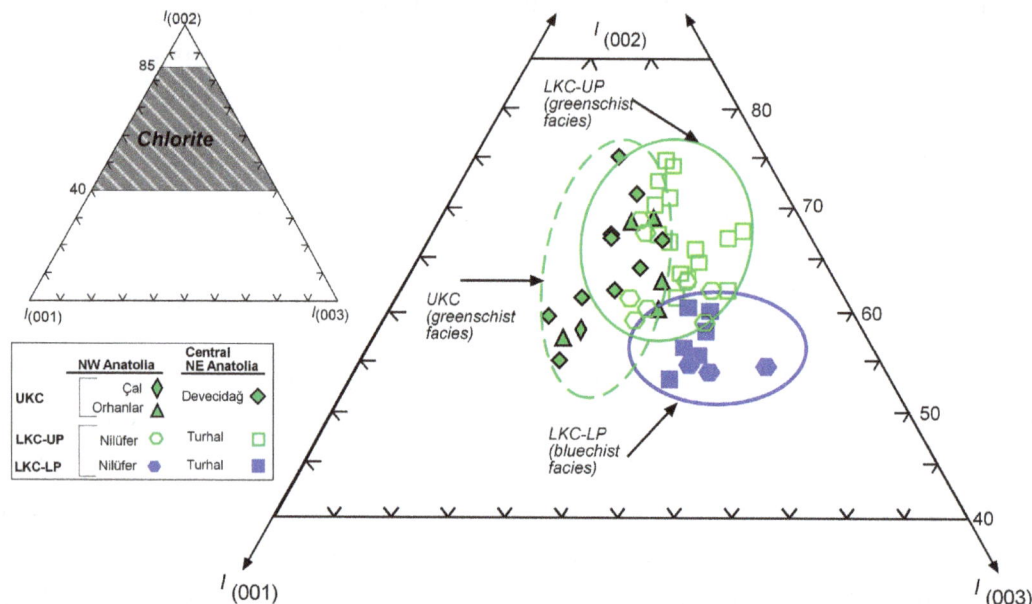

Figure 9. The distribution of chlorite minerals in the Karakaya Complex units in the triangular diagram according to the first three basal reflections (Chlorite area: Árkai and Ghabrial [16]).

4.2.4. Crystallinity

Arkai (AI: [15,36]) and Kübler Index values (KI: [35]) measured in the chlorite minerals of KC units are provided in Table 5. Accordingly, AI values of LKC-LP were detected as 0.23–0.28 Δ°2θ (0.26 Δ°2θ on average) and 0.14–0.31 (0.20 Δ°2θ on average) for NW and the central NE Anatolia, respectively. AI values for the chlorites representing LKC-UP were detected as 0.22–0.35 Δ°2θ (0.28 Δ°2θ on average) and 0.13–0.34 Δ°2θ (0.22 Δ°2θ on average) for NW and the central NE Anatolia regions respectively. All these values point to the anchizone near the border of anchizone-epizone and epizone area (AI < 0.26). KI values in LKC-LP illites were measured as 0.21 Δ°2θ only for the central NE Anatolia. KI values in LKC-UP illites were detected as 0.22–0.37 Δ°2θ (0.31 Δ°2θ on average) and 0.23–0.28 Δ°2θ (0.25 Δ°2θ on average) for NW and the central NE Anatolia, respectively. These values represent the epizone area (KI < 0.25) for LKC-LP and the anchizone area for LKC-UP.

As for UKC units, AI values for NW and the central NE Anatolia regions were measured as 0.37–0.56 Δ°2θ (0.46 Δ°2θ on average) and 0.27–0.39 Δ°2θ (0.31 Δ°2θ on average), respectively. Average values are distributed in anchizone (AI = 0.26–0.33 Δ°2θ) and diagenesis (AI > 0.33 Δ°2θ) areas. KI values in UKC units were found as 0.31–0.49 Δ°2θ (0.41 Δ°2θ on average) and 0.21–0.38 Δ°2θ (0.29 Δ°2θ on average) for NW and the central NE Anatolia regions respectively. Average values indicate the anchizone area.

Combined AI and KI data indicate that crystallinity decreases for samples from LKC units compared to the UKC units and the units diverge (Figure 10). There is a weak correlation relation between KI and AI values and it is different than (epizone: 0.21 $\Delta°2\theta$, anchizone: 0.53 $\Delta°2\theta$) the AI values corresponding to the epizone (AI < 0.26 $\Delta°2\theta$) and anchizone (AI = 0.26–0.37 $\Delta°2\theta$) as suggested by Árkai [15].

Table 5. Crystallinity index values of chlorite (AI) and illite (KI) in the Karakaya Complex units.

KC	Area	Unit	Sample No.	AI (001)	KI (001)	Area	Unit	Sample No.	AI (001)	KI (001)	Sample No.	AI (001)	KI (001)
UKC	NW Anatolia	Çal	CKK-163	0.36	0.49	Central NE Anatolia	Devecidağ	DKK-222	0.32	0.21	-234	0.30	0.25
		Orhanlar	BKK-188	0.37	0.41			-223	0.30	0.29	-235	0.30	0.26
			-197	0.56	0.31			-224	0.33	0.35	-243	0.31	0.38
			-199	0.40	0.41			-225	0.31	0.31	-245	0.29	0.31
			-	-	-			-226	0.27	0.29	-250	0.39	-
LKC-UP		Nilüfer	NKK-95	0.22	0.22		Turhal Metamorphites	TKK-4	0.13	0.26	-51	0.26	0.27
			-96	0.25	-			-10	0.15	-	-58	0.14	-
			-97	0.28	0.29			-11	0.22	0.23	-63	0.34	-
			-98	0.24	-			-25	0.22	-	-64	0.26	-
			-103	0.31	-			-28	0.15	-	-65	0.25	0.30
			-106	0.33	0.34			-30	0.24	0.26	-73	0.24	-
			-108	0.35	0.37			-31	0.19	0.15	-82	0.27	-
			-111	0.27	-			-49	0.20	0.23	-84	0.30	-
			-	-	-			-50	0.21	0.27	-	-	-
LKC-LP			-113	0.23	-			-36	0.16	-	-54	0.15	-
			-115	0.26	-			-38	0.31	-	-68	0.24	-
			-117	0.28	-			-42	0.22	0.21	-79	0.14	-

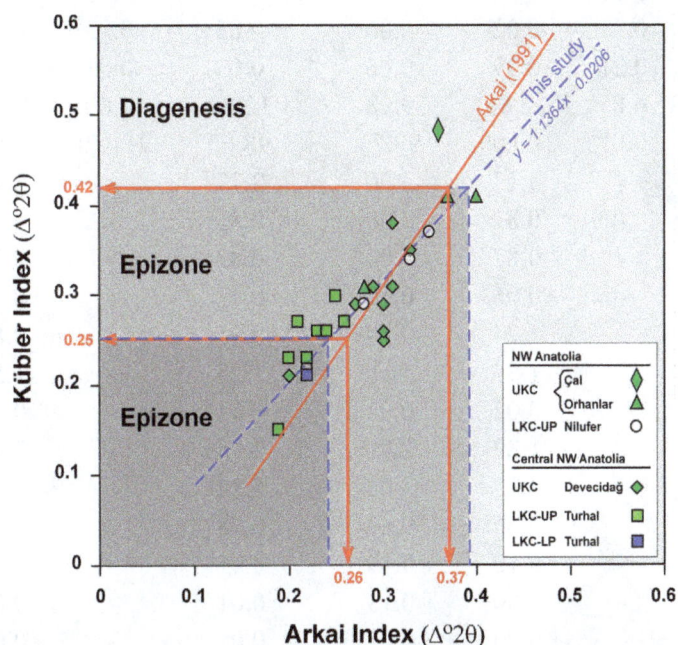

Figure 10. Relationships between Kübler Index ($\Delta°2\theta$)-Arkai Index ($\Delta°2\theta$) in the Karakaya Complex units (red line and values: Arkai [15]).

4.3. Geochemistry

Structural formulas [48] of the chlorite minerals for the KC units were calculated based on the major and trace element contents and 14 oxygen atoms (per formula unit) and are listed in Tables 6 and 7. The structural formulas of two samples were recalculated by removing Al_2O_3 and SiO_2 percentages depending on K_2O, Na_2O and CaO with 1:1:6 and 1:1:2 ratios of theoretical feldspar composition (K-feldspar $K_2O \cdot Al_2O_3 \cdot 6SiO_2$, albite $Na_2O \cdot Al_2O_3 \cdot 6SiO_2$, anortite $CaO \cdot Al_2O_3 \cdot 2SiO_2$) in the clay fraction. The charge deficiency resulting from the introduction of Al from feldspar within chlorite is balanced by K, Na, and Ca from the feldspars.

Table 6. Major element composition and structural formulas of chlorite minerals in the Karakaya Complex units (*: The values show that theoretical feldspar composition is removed, $\sum Fe_2O_3$: Total iron as Fe_2O_3, LOI: Loss on ignition at 1000 °C, TC: Tetrahedral charge, TOC: Total octahedral charge, OC: Octahedral charge, ILC: Interlayer charge, TLC: Total layer charge).

Area	NW Anatolia			Central NE Anatolia		
Unit	LKC-UP		LKC-LP	LKC-LP	LKC-LP	
Rock	Metatuffite		Slate	Metatuffite	Schist	
% Weigth	NKK-98*		NKK-113	TKK-36	TKK-39*	
SiO_2	51.73	32.07	31.45	32.98	44.03	32.43
TiO_2	0.470	0.882	2.355	0.717	1.111	1.54
Al_2O_3	17.74	18.32	15.95	15.71	11.35	11.44
$\sum Fe_2O_3$	11.22	21.06	16.49	17.21	15.05	20.80
MnO	0.157	0.295	0.377	0.160	0.192	0.27
MgO	6.90	12.95	20.49	20.76	16.31	22.54
CaO	0.60	1.13	1.84	0.10	2.27	0.01
Na_2O	3.78	0.02	0.70	0.04	1.93	0.01
K_2O	0.77	0.02	0.06	0.05	0.12	0.03
P_2O_5	0.21	0.39	0.06	0.03	0.03	0.04
LOI	6.85	12.86	9.98	10.89	7.89	10.90
Total	100.73	100.00	99.75	98.65	100.28	100.00
Si	-	3.13	3.00	3.17	-	3.13
Al	-	0.87	1.00	0.83	-	0.87
TC	-	0.87	1.00	0.83	-	0.87
Ti	-	0.06	0.17	0.05	-	0.11
Al	-	1.24	0.79	0.95	-	0.87
Fe	-	1.55	1.18	1.25	-	1.51
Mn	-	0.02	0.03	0.01	-	0.02
Mg	-	1.89	2.91	2.98	-	3.25
TOC	-	4.77	5.08	5.24	-	5.33
OC	-	0.46	0.47	0.78	-	0.84
Ca	-	0.12	0.19	0.01	-	0.00
Na	-	0.00	0.13	0.01	-	0.00
K	-	0.00	0.01	0.01	-	0.00
P	-	0.03	0.00	0.00	-	0.00
ILC	-	0.40	0.54	0.05	-	0.00
TLC	-	0.40	0.54	0.05	-	0.03

Tetrahedral Si-Al (1.00–0.83) and octahedral Mg-Fe-Al (4.64–4.79) substitutions yield total calculated octahedral cations in excess of 4 (4.77–5.33) in samples of chlorite from slate of the Nilüfer unit and the metatuffite of Turhal Metamorphites, which indicate their di-trioctahedral composition and is verified by the XRD data.

Chlorites plot within the Type I zone represented by Mg-Fe chlorites (trioctahedral) in Al + □ − Mg − Fe diagram [49]. According to the Zane and Weiss [49] classification, the source for the LKC-LP chlorites generally correspond to metabasic rocks, whereas those for the LKC-UP chlorites are felsic rocks, and those for the UKC chlorites correspond to both felsic and metabasic sources, with felsic being predominant (Figure 11).

The trace element concentrations of chlorite minerals are compared in Figure 12, and these are extremely variable. The substitution of elements with similar geochemical behaviour is mostly observed in the metatuff sample of Nilüfer unit. The highest trace element enrichment for the chlorites is observed in the LKC-LP unit (TKK-36) and the lowest in the LKC-UP unit (NKK-98). Cr, Ni, V, and Cu (transition elements), Ga (low field strength elements-LFSEs), and Nb, Zr and U (high field strength elements-HFSEs) record the highest enrichment in LKC-LP chlorites. As for LKC-UP chlorites, Zn (transition elements), W (lithophile elements); and Rb, Ba and Sr (LFSEs) are present at high concentrations. Hence, trace element concentrations are a function of the protoliths, *i.e.*, igneous *versus* metamorphic parent materials.

Chondrite-normalized trace element patterns for chlorite are shown in Figure 13. Trace element concentrations for the North American Shale Composite (NASC) are from Gromet *et al.* [50], with the exception of Nb and Y (from Condie, [51]). The chlorite minerals exhibit marked variation when compared to chondrite values, and generally have lower concentrations relative to NASC, except for Ti and U. The latter shows an enrichment of up to 5600 times in the volcanogenic chlorite for sample LKC-LP (TKK-36) and therefore exhibits a strong positive anomaly (Figure 13). In all chlorites, K, Sr, and P display significant negative anomalies. The chlorite from sample TKK-30 within the LKC-LP exhibits the lowest chondrite normalized values.

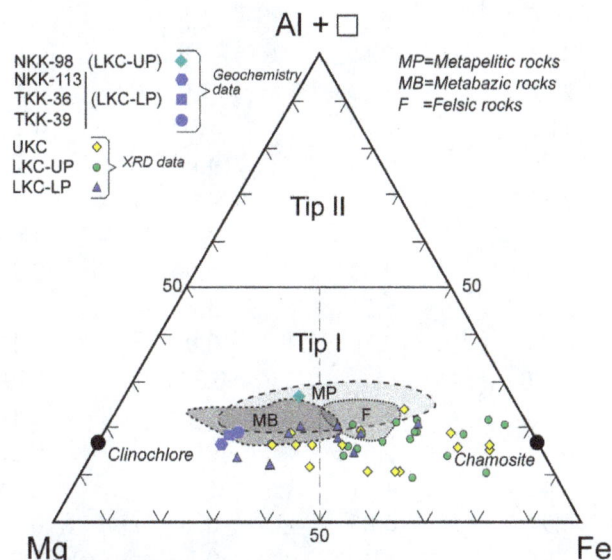

Figure 11. The distribution of chlorites in the Karakaya Complex units in the Al+ □ − Mg − Fe triangular diagram according to octahedral compositions (□: Octahedral vacancy).

Table 7. Trace element composition of chlorite minerals in the Karakaya Complex units (<: Concentrations are below the detection limit).

Area	NW Aantolia		Central NE Anatolia	
Unit	LKC-UP	LKC-LP	LKC-LP	LKC-LP
Rock	Metatuffite	Slate	Metatuffite	Schist
ppm	NKK-98	NKK-113	TKK-36	TKK-39
Cr	270	440	1330	850
Ni	130	230	890	400
Co	23	73	102	62
Sc	29	10	24	24
V	214	231	327	226
Cu	40	280	160	90
Pb	11	<5	34	<5
Zn	290	230	160	130
Bi	<0.1	1.0	0.7	<0.1
In	<0.1	<0.1	<0.1	<0.1
Sn	2	3	<1	<1
W	72.6	3.4	20.8	16.4
Mo	<2	<2	<2	3
As	21	<5	19	5
Sb	11.1	0.7	7.1	0.3
Ge	2.3	0.5	2.0	2.1
Be	2	2	3	<1
Ag	1.0	<0.5	<0.5	0.8
Rb	32	1	6	3
Cs	2.3	<0.1	0.4	2.1
Ba	174	17	14	7
Sr	172	8	6	14
Tl	0.24	<0.05	<0.05	<0.05
Ga	26	33	26	20
Ta	0.63	1.24	1.02	0.56
Nb	3.1	21.8	10.9	6.3
Hf	3.7	5.3	1.5	2.6
Zr	139	204	71	104
Y	27.4	22.9	7.0	12.3
Th	11.2	1.19	1.13	1.22
U	3.70	0.55	44.8	0.34
La	27.6	2.04	4.42	8.81
Ce	60.0	6.2	9.0	18.9
Pr	6.56	0.83	0.96	2.20
Nd	24.70	5.55	3.85	9.28
Sm	4.84	2.30	0.93	2.19
Eu	1.190	0.863	0.331	0.796
Gd	4.00	3.38	1.12	2.18
Tb	0.71	0.67	0.20	0.38
Dy	4.42	4.31	1.27	2.34
Ho	0.94	0.85	0.24	0.46
Er	2.96	2.39	0.66	1.23
Tm	0.462	0.345	0.087	0.175
Yb	3.11	2.09	0.49	1.12
Lu	0.510	0.314	0.066	0.186

The chlorite chondrite normalized REE abundances are shown in Figure 14. NASC values (with the exception of Ho and Tm; Haskin *et al.* [52]) are taken from Gromet *et al.* [50]. REE contents of all chlorite minerals are lower than NASC and exhibit varying degrees of enrichment compared to chondrite [53]. The highest chondrite normalized values are displayed by the volcanogenic chlorite (NKK-98) from the LKC-UP unit, whereas the lowest values are recorded in the volcanogenic chlorite (TKK-36) from the LKC-LP unit. Furthermore, light REE (LREEs; La-Gd) concentrations of chlorite minerals show a slight decrease when compared to those for the heavy REEs (HREEs; Tb-Lu). In the chlorite for the LKC-LP unit within the NW-Anatolia, Pr defines a negative anomaly, while Sm yields a slight positive anomaly.

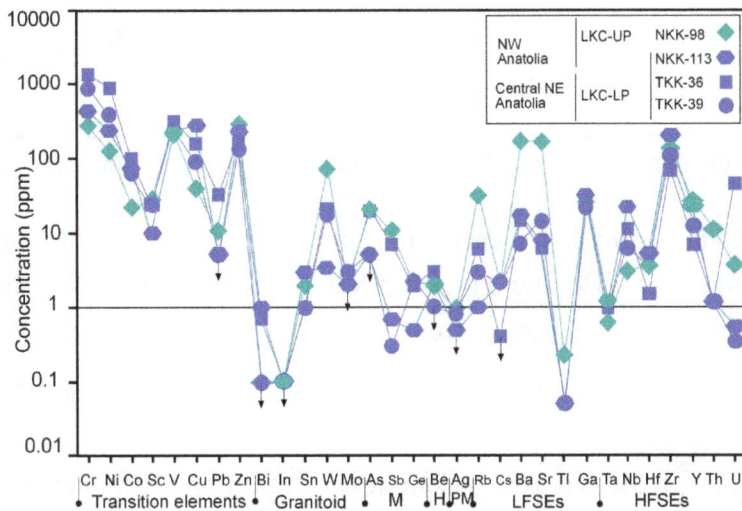

Figure 12. The correlation of chlorite minerals in the Karakaya Complex units according to the contents of trace elements (Arrows show values below the detection limits, M = Miscellaneous elements, H = Halogen elements, PM = Precious metals, LFSEs = Low field strength elements, HFSEs = High field strength elements).

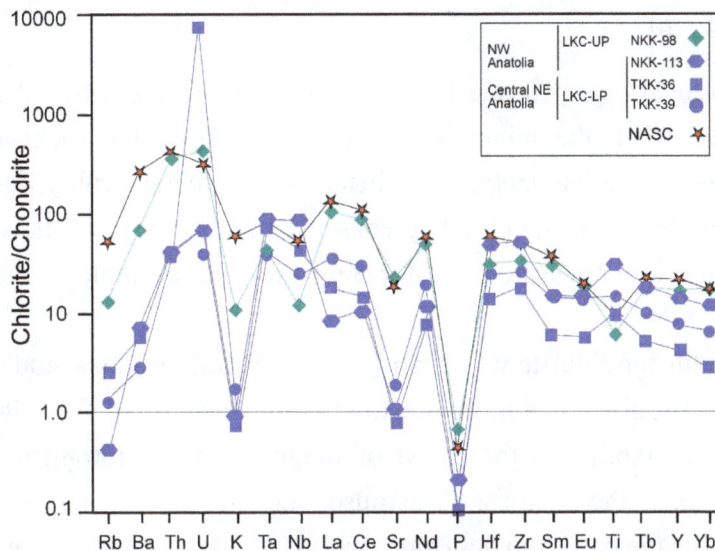

Figure 13. Chondrite-normalized trace element pattern of chlorite minerals in the Karakaya Complex units (Chondrite: Sun and Mcdonough [53]; Nb and Y for North American Shale Composite (NASC): [51]; other elements: Gromet *et al.* [50]).

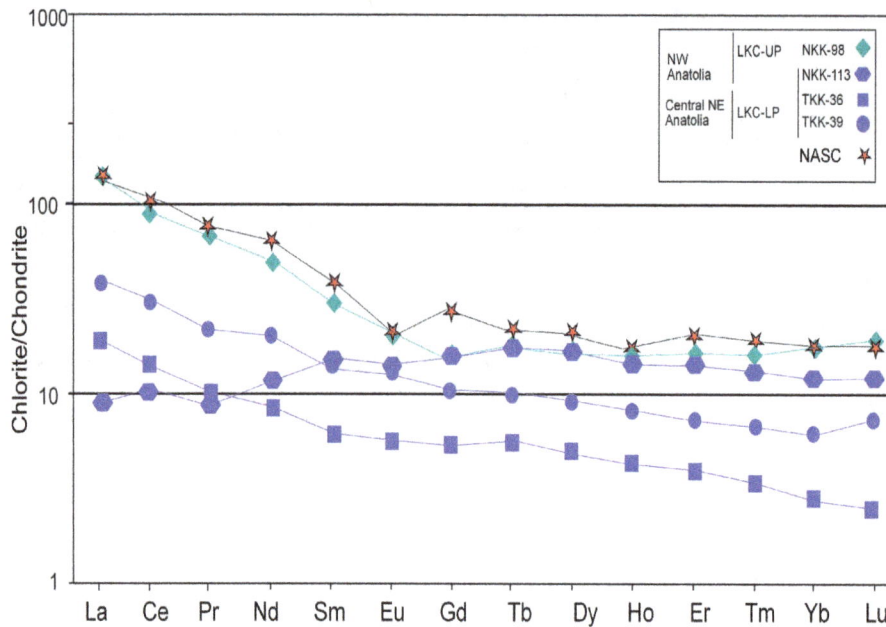

Figure 14. Chondrite-normalized rare earth element (REE) pattern of chlorite minerals in the Karakaya Complex units (Ho and Tm for NASC: Haskin *et al.* [52], other elements: Gromet *et al.* [50]; Chondrite: Sun and McDonough [53]).

5. Conclusions

Chlorite within the units of the Karakaya Complex representing NW and NE Anatolia are of neoformed origin in the groundmass and/or matrix and pores, and display different optical properties and morphologies. It is proposed that volcanic glass and/or matrix-chlorite transformation occurs via a permeating aqueous MgFeAl-silicate gel intermediate phase (Volcanic glass and/or matrix + Ions → Aqueous MgFeAl-silicate gel → Chlorite + Water: Yalçın *et al.* [54]):

$$7SiO_2 \cdot Al_2O_3 \cdot 2MgO \cdot 2FeO + 24OH^- + 10H^+ \rightarrow 9MgO \cdot 2FeO \cdot 2Al(OH)_3 \cdot 7Si(OH)_4 \rightarrow Mg_9Fe_2Al[AlSi_7O_{20}](OH)_{16} + 9H_2O$$

The Mg content of chlorite is higher in the LKC-LP blueschist facies, while the Fe content in the LKC-UP greenschist is related to the mineralogical composition of the rocks as a result of the increasing heat/pressure due to burial depth. The difference in metamorphic degree, in particular between the LKC-LP (blueschist) and LKC-UP (greenschist) units, seems to have also affected the chemical compositions of the chlorites since the latter can be related to lithological differences and the nature of the protolith.

The 7 Å (Δ°2θ) peak width for chlorite was examined by several previous studies [15,55–62], and these indicate that chlorite crystallinity is more apparent compared to that for illite, in particular for basic-metabasic rocks [15,58,60,62]. As the grade of diagenesis/metamorphism increases, chlorite crystallinity also increases as is the case for illite/muscovite. In the samples investigated here, AI values correspond to those for the epizone-anchizone for LKC units, whereas those for the UKC units are consistent with low-degree anchimetamorphic-diagenetic stage. For units within NW and the central NE Anatolia, illite crystallinity (Kübler Index: Kübler [35]) seems to be compatible with AI values [19–21].

The features of transition to semi-regular or regular stacking are observed in chlorite layers [47,63,64], and these structural (polytype) changes may be partly attributed to heating, which then serves as a qualitative method for geothermometric assessment [65]. It has been established that the transformation from Ib $\rightarrow$ IIb type chlorites occurs at temperatures <200 °C [10]; the chlorites investigated here are IIb and those for Turhal Metamorphites within the UKC units are of $2M_1$ polytype [20], both of which indicate a similar degree of metamorphism.

The combined optical, mineralogical, morphological, and geochemical evidence reported here for chlorite are shown to be related to the structures, formation mechanisms, and tectonic environments present during their formation. Consequently, chlorite plays a key role in distinguishing units with distinct geological evolutions within the Karakaya Complex.

Acknowledgments

The authors extend their thanks for their help to Cumhuriyet University Directorate of Scientific Research Projects Commission (Project No: M 301) that gave financial support for the realization of this study, the deceased Mehmet DURU (General Directorate of Mineral Research and Exploration) and Ahmet GÖKÇE (Cumhuriyet University) who ensured the introduction of the units in NW Anatolia and the central NE Anatolia in the field respectively, Cumhuriyet University Department of Geology Engineering laboratory workers in XRD works and the preparation of thin sections and to TPAO workers for SEM examinations. In addition, M. Cemal GÖNCÜOĞLU has contributed to the shaping of the article with his suggestions and criticisms on the understanding of the geological evolution of the Karakaya Complex. We are also grateful to the academic editor and anonymous referees for their constructive comments leading to important improvements in the manuscript.

Author Contributions

Sema Tetiker was in charge of sampling, field analysis, interpretations, and scientific writing of this study under the supervision of Hüseyin Yalçın and Ömer Bozkaya.

Conflicts of Interest

The authors declare no conflict of interest.

References

1. Millot, G. *Geology of Clays*; Farrand, W.R., Paquet, H., Eds.; Springer Verlag: Berlin, Germany, 1970; p. 429.
2. Weaver, C.E.; Highsmith, P.B.; Wampler, J.M. Chlorite. In *Shale-Slate Metamorphism in the Southern Appalachians*; Weaver, C.E., Ed.; Elsevier: Amsterdam, The Netherlands, 1984; Volume 10, pp. 99–139.
3. Ahn, J.; Peacor, D.R. Transmission electron microscopic study of diagenetic chlorite in Gulf Coast argillaceous sediments. *Clays Clay Miner.* **1985**, *33*, 228–236.
4. Cathelineau, M.; Nieva, D. A chlorite solid solution geothermometer, the Los Azufres geothermal system (Mexico). *Contrib. Mineral. Petrol.* **1985**, *91*, 235–244.

5. Curtis, C.D.; Hughes, C.R.; Whiteman, J.A.; Whittle, C.K. Compositional variations within some sedimentary chlorites and some comments on their origin. *Mineral. Mag.* **1985**, *49*, 375–386.

6. Cathelineau, M. Cation site occupancy in chlorites and illites as a function of temperature. *Clay Miner.* **1988**, *23*, 471–485.

7. Velde, B.; Medhioub, M. Approach to chemical equilibrium in diagenetic chlorites. *Contrib. Mineral. Petrol.* **1988**, *98*, 122–127.

8. Jahren, J.S.; Aagard, P. Diagenetic illite-chlorite assemblages in arenites. I. Chemical evolution. *Clays Clay Miner.* **1992**, *40*, 540–546.

9. Hillier, S.; Velde, B. Octahedral occupancy and the chemical composition of diagenetic (low temperature) chlorites. *Clay Miner.* **1991**, *26*, 149–168.

10. Walker, J.R. Chlorite polytype geothermometry. *Clays Clay Miner.* **1993**, *41*, 260–267.

11. Xie, X.G.; Byerly, G.R.; Ferrell, R.E. IIb trioctahedral chlorite from the Barberton greenstone belt: Crystal structure and rock composition constraints with implications to geothermometry. *Contrib. Mineral. Petrol.* **1997**, *126*, 275–291.

12. Bozkaya, Ö.; Yalçın, H. Relationships between degree of diagenesis-metamorphism and chemistry of phillosilicates from eastern Taurus Aotochtonous. In Proceedings of the IXth Turkish National Clay Symposium, İstanbul, Turkey, 15–18 September 1999; pp. 21–30. (In Turkish).

13. Bozkaya, Ö.; Yalçın, H.; Schroeder, P.A.; Crowe, D. New insights in the definition of phyllosilicate stacks in diagenetic-metamorphic environments-examples from clastic to metaclastic rocks in Turkey. In Proceedings of the MECC14 7th Mid-European Clay Conference, Dresden, Germany, 16–19 September 2014; p. 117.

14. Árkai, P. Chlorite crystallinity: An empirical approach and correlation with illite crystallinity, coal rank and mineral facies as exemplified by Palaeozoic and Mesozoic rocks of northeast Hungary. *J. Metamorph. Geol.* **1991**, *9*, 723–734.

15. Árkai, P.; Tóth, M. Illite and chlorite "crystallinity" indices, I: An attempted mineralogical interpretation. In Proceedings of the IGCP Conference on Phyllosilicates as indicators of very low-grade metamorphism and diagenesis (IGCP 294), Manchester, UK, 4–6 July 1990.

16. Árkai, P.; Ghabrial, D.S. Chlorite crstallinity as an indicator of metamorphic grade of low-temperature meta-igneous rocks: A case study form the Bükk Mountains, Northeast Hungary. *Clay Miner.* **1997**, *32*, 205–222.

17. Árkai, P.; Sassi, F.P.; Sassi, R. Simultaneous measurements of chlorite and illite crystallinity: A more reliable geothermometric tool for monitoring low- to very low-grade metamorphisms in metapelites. A case study from the Southern Alps (NE Italy). *Eur. J. Mineral.* **1995**, *7*, 1115–1128.

18. Potel, S.; Ferreiro Mählmann, R.; Stern, W.B.; Mullis, J.; Frey, M. Very low-grade metamorphic evolution of pelitic rocks under high-pressure/low-temperature condition, NW New Caledonia (SW Pacific). *J. Petrol.* **2006**, *47*, 991–1015.

19. Tetiker, S.; Yalçın, H.; Bozkaya, Ö. Diagenesis-low grade metamorphism of Karakaya Complex units in the NW Anatolia. Hacettepe University. *Earth Sci. J.* **2009**, *30*, 193–212. (In Turkish)

20. Tetiker, S.; Yalçın, H.; Bozkaya, Ö. Low grade metamorphism of the units from Karakaya Complex (Tokat region). In Proceedings of the 14th Turkish National Clay Symposium, Trabzon, Turkey, 1–3 October 2009; pp. 155–173. (In Turkish)

21. Tetiker, S.; Yalçın, H.; Bozkaya, Ö.; Göncüoğlu, M.C. Diagenetic to low-grade metamorphic evolution of the Karakaya Complex in northern Turkey based on phyllosilicate mineralogy. *Mineral. Petrol.* **2015**, *109*, 201–215.

22. Şengör, A.M.C.; Yılmaz, Y.; Sungurlu, O. Tectonics of the Mediterranean Cimmerides: Nature and evolution of the western termination of Paleo-Tethys. In *The Geological Evolution of the Eastern Mediterranean*; Dixon, J.E., Robertson, A.H.F., Eds.; Special Publications Volume 17; Geological Society London: London, UK, 1984; pp. 77–112.

23. Göncüoğlu, M.C.; Dirik, K.; Kozlu, H. General chracteristics of pre-Alpine and Alpine Terranes in Turkey: Explanatory notes to the terrane map of Turkey. *Ann. Geol. Pays Hell.* **1997**, *37*, 515–536.

24. Okay, A.İ.; Göncüoğlu, M.C. The Karakaya Complex: A review of data and concepts. *Turk. J. Earth Sci.* **2004**, *13*, 77–95.

25. Robertson, A.H.F.; Ustaömer, T. Testing alternative tectono-stratigraphic interpretations of the Late Palaeozoic-Early Mesozoic Karakaya Complex in NW Turkey: Support for an accretionary origin related to northward subduction of Palaeotethys. *Turk. J. Earth Sci.* **2012**, *21*, 961–1007.

26. Tekeli, O. Subduction complex of pre-Jurassic age, Northern Anatolia, Turkey. *Geology* **1981**, *9*, 68–72.

27. Özcan, A.; Erkan, A.; Keskin, A.; Keskin, E.; Oral, A.; Özer, S.; Sümergen, M.; Tekeli, O. *Kuzey Anadolu Fayı Ile Kırşehir Masifi Arasının Temel Jeolojisi*; Appendix: Geological Map and Sections; Unpublished Report No: 6722; Institute of Mineral Research and Exploration: Ankara, Turkey, 1980; p. 136. (In Turkish)

28. Gökçe, A. Turhal Antimon Yataklarının Maden Jeolojisi. In *Doktora Tezi, Hacettepe Üniversitesi Jeoloji Mühendisliği Bölümü*; Fen Bilimleri: Ankara, Turkey, 1983; p. 150. (In Turkish)

29. Bailey, S.W.; McCallien, W.J. The Ankara melange and the Anatolian Thrust. *Nature* **1950**, *166*, 938–941.

30. Bailey, S.W.; McCallien, W.J. Serpentinite lavas, the Ankara melange and the Anatolian Thrust. *Trans. R. Soc. Edinb.* **1953**, *62*, 403–442.

31. Erol, O. Ankara Güneydoğusundaki Elma Dağı ve Çevresinin Jeolojisi ve Jeomorfolojisi Üzerine bir Araştırma. In *Bulletin of the Mineral Research and Exploration*; Institute of Mineral Research and Exploration: Ankara, Turkey, 1956; p. 99. (In Turkish)

32. Maden Tetkik ve Arama. Geologic Map of Turkey with the 1:500 000 scale, Istanbul Section. *General Directorate of Mineral Research and Exploration*; Institute of Mineral Research and Exploration: Ankara, Turkey, 2002. (In Turkish)

33. Brindley, G.W. Quantitative X-ray mineral analysis of clays. In *Crystal Structures of Clay Minerals and Their X-Ray Identification*; Brindley, G.W., Brown, G., Eds.; Mineralogical Society: London, UK, 1980; pp. 125–195.

34. Moore, D.M.; Reynolds, R.C., Jr. *X-Ray Diffraction and the Identification and Analysis of Clay Minerals*; Oxford University Press: New York, NY, USA, **1990**; p. 332.

35. Kübler, B. Evaluation quantitative du métamorphisme par la cristallinité de l'illite. *Bull. Cent. Rech. Pau-SNPA* **1968**, *2*, 385–397. (In French)

36. Guggenheim, S.; Bain, D.C.; Bergaya, F.; Brigatti, M.F.; Drits, A.; Eberl, D.D.; Formoso, M.L.L.; Galan, E.; Merriman, R.J.; Peacor, D.R.; *et al.* Report of the AIPEA nomenclature committee for 2001: Order, disorder and crystallinity in phyllosilicates and the use of the "Crystallinity Index". *Clay Miner.* **2002**, *37*, 389–393.

37. Kisch, H.J. Illite crystallinity and coal rank associated with lowest-grade metamorphism of the Taveyanne greywacke in the Helvetic zone of the Swiss Alps. *Eclogae Geol. Helv.* **1980**, *73*, 753–777.

38. Warr, L.N.; Rice, A.H.N. Interlaboratory standartization and calibration of clay mineral crystallinity and crystallite size data. *J. Metamorph.Geol.* **1994**, *12*, 141–152.

39. Krumm, S. WINFIT 1.2: Version of November 1996 (The Erlangen Geological and Mineralogical Software Collection) of WINFIT 1.0: A Public Domain Program for Interactive Profile-Analysis under Windows. In Proceedings of the XIII Conference on Clay Mineralogy and Petrology, Praha, Czech Republic, 29 August–2 September 1994; Volume 38, pp. 253–261.

40. Kisch, H.J. Calibration of the anchizone: A critical comparison of illite "crystallinity" scales used for definition. *J. Metamorph. Geol.* **1990**, *8*, 31–46.

41. Bailey, S.W. X-ray diffraction identification of the polytypes of mica, serpentine, and chlorite. *Clays Clay Miner.* **1988**, *36*, 193–213.

42. Brindley, G.W. Chlorite Minerals. In *The X-Ray Identification and Crystal Structures of Clay Minerals*; Brown, G., Ed.; Mineralogical Society: London, UK, 1961; pp. 242–296.

43. Brindley, G.W.; Brown, G. X-ray Diffraction Procedures for Clay Mineral Identification. In *Crystal Structures of Clay Minerals and Their X-ray Identification*; Brindley, G.W., Brown, G., Eds.; Mineralogical Society: London, UK, 1980; pp. 305–360.

44. Chagnon, A.; Desjardins, M. Determination de la composition de la chlorite par difraction et microanalyse aux rayons X. *Can. Mineral.* **1991**, *29*, 245–254. (In French)

45. Shirozu, H. X-ray powder patterns and cell dimensions of some chlorites in Japan with a note on their interference colors. *Mineral. J.* **1958**, *2*, 209–223.

46. Foster, M.D. Interpretation of the composition and a classification of the chlorites. *US Geol. Surv. Prof. Pap.* **1962**, *414*, 1–33.

47. Bailey, S.W. Summary of recommendations of AIPEA nomenclature committee on clay minerals. *Am. Mineral.* **1980**, *65*, 1–7.

48. Weaver, C.E.; Pollard, L.D. The Chemistry of Clay Minerals. In *Developments in Sedimentology*; Elsevier: Amsterdam, The Netherlands, 1973; Volume 15, p. 272.

49. Zane, A.; Weiss, Z. A procedure for classification of rock-forming chlorites based on microprobe data. *Rend. Fis. Accad. Lincei* **1998**, *9*, 51–56.

50. Gromet, L.P.; Dymek, R.F.; Haskin, L.A.; Korotev, R.L. The "North American shale composite": Its compilation, major and trace element characteristics. *Geochim. Cosmochim. Acta* **1984**, *48*, 2469–2482.

51. Condie, K.C. Chemical composition and evolution of the upper continental crust: Contrasting results from surface samples and shales. *Chem.Geol.* **1993**, *104*, 1–37.

52. Haskin, L.A.; Haskin, M.A.; Frey, F.A.; Wildeman, T.R. Relative and absolute terrestrial abundances of the rare earths. In *Origin and Distribution of the Elements*; Ahrens, L.H., Ed.; Pergamon Press: Oxford, UK, 1968; pp. 889–912.

53. Sun, S.S.; McDonough, W.F. Chemical and Isotopic Systematics of Oceanic Basalts: Implications for Mantle Composition and Processes. In *Magmatism in the Ocean Basins*; Saunders, A.D., Norry, M.J., Eds.; Geological Society London: London, UK, 1989; Special Publications Volume 42; pp. 313–345.

54. Yalçın, H.; Bozkaya, Ö.; Tetiker, S. Clay mineralogy and geochemistry of Kangal coal deposit. In Proceedings of the XIIth Turkish National Clay Symposium, Yüzüncü Yıl University, Van, Turkey, 5–9 September 2005; pp. 16–31. (In Turkish)

55. Ludwig, V. Zum übergang eines tonschiefers in die metamorphose: "Grieffelschiefer" des Ordoviziums in NE-Bayren (mit einem beitrag zum problem der illit-kristallinität). *Neues Jahrb. Geol. Paläontol. Abh.* **1973**, *144*, 50–103. (In German)

56. Schamel, S. Eocene Subduction in Central Liguria, Italy. Unpublished Ph.D. Thesis, Yale University, New Haven, CT, USA, 1973.

57. Le Corre, C. Analyse comparée de la cristallinité des micas dans le Briovérien et le Paléozoique Centre-Armoricains: Zonéographie et structure d'un domaine épizonal. *Bull. Soc. Géol. Fr.* **1975**, *17*, 547–553. (In French)

58. Schaer, J.P.; Persoz, F. Aspects strutcruaox et petrographtqoes du Hant Atlas calcaire de Midelt (Maroc). *Bull. Soc. Géol. Fr.* **1976**, *18*, 1239–1250. (In French)

59. Deutloff, O.; Teichmüller, M.; Teichmüller, R.; Wolf, M. Inkohlungs-untersuchungen im Mesozoikum des Massivs von Vlotho (Niedersädhsisches Tektogen). *Neues Jahrb. Geol. Paläontol. Monatsh.* **1980**, *6*, 321–341. (In German)

60. Dandois, P. Diagenèse et métamorphisme des domaines Calédonien et Hercynien de la vallée de la Meuse entre Charleville-Mezières et Namur (Ardennes franco-belges). *Ann. Soc. Géol. Belg.* **1981**, *90*, 299–316. (In French)

61. Duba, D.; Williams-Jones, A.E. The application of illite crystallinity, organic matter reflectance, and isotopic techniques to mineral exploration: A case study in southwestern Gaspé, Quebec. *Econ. Geol.* **1983**, *78*, 1350–1363.

62. Brauckmann, F.J. *Hochdiagenese im Muschelkalk der Massive von Bramsche und Vlotho*; Institut für Geologie, Ruhr-Universität Bochum: Bochum, Germany, 1984. (In German)

63. Bailey, S.W.; Brown, B.E. Chlorite polytypism: I. Regular and semi-random one layer structures. *Am. Mineral.* **1962**, *47*, 819–850.

64. Bailey, S.W. Classification and Structures of the Micas. In *Micas*; Bailey, S.W., Ed.; Mineralogical Society of America: Washington, DC, USA, 1984; pp. 1–12.

65. Hayes, J.B. Polytypism of chlorite in sedimentary rocks. *Clays Clay Miner.* **1970**, *18*, 285–306.

Permissions

List of Contributors

Kathryn A. Mumford
Department of Chemical and Biomolecular Engineering,
The University of Melbourne, Parkville,
VIC 3010, Australia

Brendan Pitt
Department of Environment and Geography, Macquarie
University, Sydney, NSW 2109, Australia

Ashley T. Townsend
Central Science Laboratory, University of Tasmania,
Hobart, TAS 7001, Australia

Ian Snape
Australian Antarctic Division, Kingston, TAS 7050,
Australia

Damian B. Gore
Department of Environment and Geography, Macquarie
University, Sydney, NSW 2109, Australia

Wei Chen
Department of Civil & Environmental Engineering &
Earth Sciences, University of Notre Dame,
156 Fitzpatrick Hall, Notre Dame, IN 46556, USA

Antonio Simonetti
Department of Civil & Environmental Engineering &
Earth Sciences, University of Notre Dame,
156 Fitzpatrick Hall, Notre Dame, IN 46556, USA

Quentin G. Crowley
Department of Geology, School of Natural Sciences,
Trinity College, Dublin 2, Ireland

Kyle Heron
Department of Geology, School of Natural Sciences,
Trinity College, Dublin 2, Ireland

Nancy Riggs
School of Earth Sciences and Environmental Sustainability,
Northern Arizona University, Flagstaff

Balz Kamber
Department of Geology, School of Natural Sciences,
Trinity College, Dublin 2, Ireland

David Chew
Department of Geology, School of Natural Sciences,
Trinity College, Dublin 2, Ireland

Brian McConnell
Geological Survey of Ireland, Beggars Bush, Dublin 4,
Ireland

Keith Benn
Tasiast Mauritanie Limited SA, ZRA 741, BP 5051,
Nouakchott, Mauritania, West Africa

Jorge Buzzi
Instituto Geológico y Minero de España (IGME),
Geological Survey of Spain, La Calera 1, 28760
Tres Cantos, E-28003 Madrid

Asunción Riaza
Instituto Geológico y Minero de España (IGME),
Geological Survey of Spain, La Calera 1, 28760
Tres Cantos, E-28003 Madrid

Eduardo García-Meléndez
Facultad de Ciencias Ambientales, Universidad de León,
Campus de Vegazana s/n, E-24071 León,
Spain

Sebastian Weide
Remote Sensing Data Centre, German Aerospace Research,
Deustsche Zentrum für Luft und Raumfahrt
(DLR), P.O. Box 1116, D-82234 Wessling, Germany

Martin Bachmann
Remote Sensing Data Centre, German Aerospace Research,
Deustsche Zentrum für Luft und Raumfahrt
(DLR), P.O. Box 1116, D-82234 Wessling, Germany

Ned Xoubi
Nuclear Engineering Department, King Abdulaziz
University, P.O. Box 80204, Jeddah 21589, Saudi Arabia

Christina Wanhainen
Division of Geosciences and Environmental Engineering,
Luleå University of Technology, SE 971 87 Luleå, Sweden

Wondowossen Nigatu
Gunnarn Exploration AB, Blåvagen 207 Box 149, SE 923
23 Storuman, Sweden

David Selby
Department of Earth Sciences, University of Durham,
Durham DH1 3LE, UK

Claire L. McLeod
Department of Earth and Atmospheric Sciences, University
of Houston, Houston

Roger Nordin
Exploration Department, Boliden Mineral AB, SE 936 81
Boliden, Sweden

Nils-Johan Bolin
Division of Process Technology, Boliden Mineral AB, SE 936 81 Boliden, Sweden

Myriam Kars
Laboratoire des Fluides Complexes et leurs Réservoirs, Université de Pau et des Pays de l'Adour,
UMR 5150 CNRS TOTAL, Avenue de l'Université, 64013 Pau cedex, France

Charles Aubourg
Laboratoire des Fluides Complexes et leurs Réservoirs, Université de Pau et des Pays de l'Adour,
UMR 5150 CNRS TOTAL, Avenue de l'Université, 64013 Pau cedex, France

Pierre Labaume
Géosciences Montpellier, Université de Montpellier 2, UMR 5243 CNRS, Place E. Bataillon, 34095 Montpellier cedex 5, France

Thelma S. Berquó
Physics Department, Concordia College, 901 8th St. S., Moorhead, MN 56562, USA

Thibault Cavailhes
Géosciences Montpellier, Université de Montpellier 2, UMR 5243 CNRS, Place E. Bataillon, 34095 Montpellier cedex 5, France

Nouhou Diaby
Institut de Minéralogie et Géochimie, Université de Lausanne, CH-1015 Lausanne, Switzerland
Laboratoire de Traitement des Eaux usées, Institut Fondamental d'Afrique Noire (IFAN), Université Cheikh Anta Diop, BP 206 Dakar, Senegal

Bernhard Dold
Institut de Minéralogie et Géochimie, Université de Lausanne, CH-1015 Lausanne, Switzerland
SUMIRCO (Sustainable Mining Research & Consult EIRL), Casilla 28, San Pedro de la Paz 4130000, Chile

Sema Tetiker
Department of Geological Engineering, Batman University, 72100 Batman, Turkey

Hüseyin Yalçın
Department of Geological Engineering, Cumhuriyet University, 58140 Sivas, Turkey

Ömer Bozkaya
Department of Geological Engineering, Pamukkale University, 20070 Denizli, Turkey

www.ingramcontent.com/pod-product-compliance
Lightning Source LLC
Chambersburg PA
CBHW080703200326
41458CB00013B/4944